Smart Water Management

Blockchain in Sustainability and Efficiency

Editors:

Yaser Sabzevari
Department of Water Science and Engineering
College of Agriculture
Isfahan University of Technology, Isfahan, Iran

Saeid Eslamian
Department of Water Science and Engineering, College of Agriculture
Isfahan University of Technology, Isfahan, Iran

Faezeh Eslamian
Department of Bioresource Engineering
McGill University, Montreal, Quebec, Canada

CRC Press
Taylor & Francis Group
Boca Raton London New York

CRC Press is an imprint of the
Taylor & Francis Group, an **informa** business
A SCIENCE PUBLISHERS BOOK

First edition published 2027
by CRC Press
2385 NW Executive Center Drive, Suite 320, Boca Raton FL 33431

and by CRC Press
4 Park Square, Milton Park, Abingdon, Oxon, OX14 4RN

CRC Press is an imprint of Taylor & Francis Group, LLC

Library of Congress Cataloging-in-Publication Data (applied for)

ISBN: 978-1-041-12194-7 (hbk)
ISBN: 978-1-041-16712-9 (pbk)
ISBN: 978-1-003-68602-6 (ebk)

DOI: 10.1201/9781003686026

Typeset in Times New Roman
by Prime Publishing Services

Preface

Water is a fundamental resource that sustains life and plays a crucial role in various sectors, including agriculture, manufacturing, and daily consumption. As awareness of environmental sustainability increases and demand for clean water continues to grow, we need to rethink our water management systems.

Artificial Intelligence and High Computational methods are being utilized increasingly in water administration to make strides in the proficiency, maintainability, and productivity of water supply and sanitation administrations, as well as for arrival of administration when extraordinary climate wonders happen, such as floods and droughts.

In an increasingly digital world, innovative technologies are emerging as powerful tools to address challenges related to water conservation and responsible water use. Among these technologies, blockchain is an ideal solution that can change the way water resources are managed, processed, and tracked.

This preface sets the stage for exploring the intersection of blockchain technology and water management. It highlights the significance of designing water-sensitive products, enhancing water efficiency, and raising public awareness about the impact of water consumption. Moreover, it acknowledges the need for a transparent and secure system that can effectively track and manage virtual water flows.

In the following chapters, we will examine how blockchain technology can be used to record water-sensitive products, monitor digital assets, and manage mixed water. It explores how blockchain can bring greater transparency, accountability, and efficiency to water rights trading, productivity, and resource management. Chapter 1 examines and introduces the hydrologic resilient using blockchain technology. Chapter 2, reviews the different studies in the field of blockchain technology to better the use of the technology in water resources management. Chapter 3 examines the theory and methodology of blockchain and presents how it can be used in water resources management. Chapter 4, 5 and 6 present the basic concepts in blockchain technology which include Encryption, Consensus Algorithms, and Hash Functions and indicate their methodology. Chapter 7 deals with the capability and applicability of the Consensus Algorithm in Blockchain technology. Chapter 8 is about SoK on Blockchain Evolution and Taxonomy. Chapter 9 is about the Capabilities of Blockchain and the Outlook of Blockchain technology. Chapter 10 explains Broad Applications of Blockchain Technology. Chapter 11 discusses the Patent Boom in the Blockchain Frontier. Chapter 12 is about investigation of virtual water, applications, and water footprint in the Middle

East, a case study in Iran. Chapter 13 also presents application of blockchain technology for the virtual water and water rights trade. Chapter 14 discusses Application of Blockchain Technology in Climate Change, and chapter 15 gives a conclusion to block chain technique and future works in this field. Chapters 16, 17, 18, 19, 20, 21, and 22 explain about blockchain technology for secure data trust: enhancing access control, provenance, and auditability; blockchain unveiled: evolution, challenges, and transformative applications across sectors; blockchain and IoT: advancing smart water management for sustainable solutions; blockchain for hydro systems: transforming water management and sustainability; blockchain technology for environmental conservation; blockchain and environmental sustainability: addressing global challenges and advancing SDGs; and blockchain in intelligent building systems: enhancing security, efficiency, and climate resilience, respectively.

Through this exploration, we aim to provide insights into the transformative power of blockchain in preserving our environment, optimizing operational costs, and fostering a sustainable future. By embracing blockchain and its inherent advantages, we can embark on a path towards responsible water management, where both individuals and businesses actively contribute to conserving this invaluable resource.

This book presents application of blockchain technology in water resources management that is suitable for the target community of all water users such as farmers, governmental offices, and water managers.

This book is befitting for the academic environment, especially the doctoral and senior level, and the decision-making department for managers and employees of the water sector.

Yaser Sabzevari, Saeid Eslamian, and Faezeh Eslamian
Isfahan University of Technology and McGill University

Contents

List of Contributors

Aanandha Saravanan
Department of ECE, Vel Tech Rangarajan, Dr. Sagunthala R&D Institute of Science and Technology, Chennai, India.

Amit Tyagi
National Institute of Fashion Technology.

Asghar Akbari Foroud
Electrical and Computer Engineering Faculty, Semnan University, Semnan, Iran.

B. Sathyasri
Department of Electronics and Communication Engineering, Vel Tech. Rangarajan Dr. Sagunthala R&D Institute of Science and Technology, Chennai, India.

Bharat Ramdas Pawar
Assistant Professor, Department of Electronics & Telecommunication, VPM MPCOE, Velneshwar.

Bharat Ramdas Pawar
Department of Electronics & Computer Engineering, CSMSS Chh Shahu College of Engineering, Chh Sambhaji Nagar, Maharashtra, India.

Farkhondeh Jabari
Power Systems Operation and Planning Research Department, Niroo Research Institute (NRI), Tehran, Iran.

Fatemeh Dadvand
Department of Water Science and Engineering, College of Agriculture, Isfahan University of Technology, Isfahan, Iran.

G. Sasikala
Department of Electronics and Communication Engineering, Vel Tech Rangarajan Dr. Sagunthala R&D Institute of Science and Technology, Chennai, India.

Harpreet Kaur Channi
Department of Electrical Engineering, Guru Nanak Dev Engineering College, Ludhiana, Punjab, India.

Karkuzhali S.
Department of Computer Science and Engineering, Mepco Schlenk Engineering College, Sivakasi,Tamilnadu, India

Mahendra Gooroochurn
Mechanical and Production Engineering Department, Faculty of Engineering, University of Mauritius, Moka, Mauritius.

Mohammad Karrabi
Electrical and Computer Engineering Faculty, Semnan University, Semnan, Iran.

Parminder Singh
Department of Electrical Engineering, Chandigarh University, Gharuan, Mohali, Punjab.

Paven Thuat D.O.
Hong Kong University of Science and Technology.

Prateek Pandey
Jaypee University of Engineering and Technology, Madhya Pradesh, India.

Pulkit Kumar
Electrical Engineering Department, University Institute of Engineering, Chandigarh University, Mohali, Punjab India

Saeid Eslamian
Department of Water Science and Engineering, College of Agriculture, Isfahan University of Technology, Isfahan, Iran.

Senthilkumar S.
Department of Chemistry, Ayya Nadar Janaki Ammal College, Sivakasi,Tamilnadu, India.

Tanuj Surve
University of California, Berkeley.

Tulsi Pawan Fowdur
Electrical and Electronic Engineering Department, University of Mauritius, Moka, Mauritius.

Yaser Sabzevari
Department of Water Science and Engineering, College of Agriculture, Isfahan University of Technology, Isfahan, Iran.

1 An Introduction to Hydrologic Resilient and Blockchain

Yaser Sabzevari[1*] and Saeid Eslamian[2]

Hydrologic systems are increasingly exposed to pressures from climate variability, population growth, and intensifying water demand. Ensuring resilience, defined as the capacity of water systems to withstand, adapt to, and recover from disruptions, has therefore become a central objective in modern water resources management. At the same time, emerging digital technologies such as blockchain offer new opportunities for secure, transparent, and decentralized data management within the hydrologic sector. This paper provides an introductory overview of hydrologic resilience and explores the potential role of blockchain technology in enhancing system reliability, decision-making, and stakeholder coordination. Key concepts such as system vulnerability, adaptive capacity, decentralized ledgers, smart contracts, and transparent data-sharing frameworks are discussed. By connecting principles of hydrologic science with innovative digital infrastructures, this work highlights pathways for integrating blockchain into water monitoring, allocation, and governance systems to strengthen long-term resilience.

1.1 Introduction

Based on the index of the United Nations Sustainable Development Commission, if the entire annual water consumption of any country exceeds 40% of the total annual renewable resources, the country will experience a severe water crisis (Igbatayo et al., 2022; Mohamadjani and Yazdanian, 2014). As agricultural, industrial, and domestic sectors increasingly compete for scarce water supplies, global food production demands will further heighten pressure on these resources (Schultz, 2017). While rivers function as the primary lifeline for both humans and natural ecosystems, they can paradoxically turn destructive, inflicting substantial and often irreversible harm (Nasim and Selajgeh, 2006; Gregory, 2006; Eslamian and Maleki, 2021). Flood events inflict extensive damage each year, harming agricultural lands, transportation networks, hydraulic structures, and bridges, and in some instances resulting in significant loss of human and animal life. Such disasters disrupt social systems and impose substantial economic burdens (Eslamian et al., 2021; Nazif

[1,2] Department of Water Science and Engineering, College of Agriculture, Isfahan University of Technology, Isfahan, Iran.
* Corresponding author: yasersabzevari1511@gmail.com

DOI: 10.1201/9781003686026-1

et al., 2021). Consequently, comprehensive investigations and the application of suitable analytical methods are essential to understand the underlying causes of floods, diagnose their impacts, and develop effective strategies for managing and mitigating them.

In recent years, severe flood events have caused extensive destruction across many regions of the world. Owing to climatic fluctuations and intense rainfall, floods recur frequently in numerous areas (Hirabayashi et al. 2013; Khosravi et al. 2019; Riudas et al. 2022). Rising populations, expanding cities, and increased agricultural activities further heighten flood susceptibility (Rahman et al. 2021; Zhang et al., 2018). Floods also pose a major threat to human safety, as natural disasters of this kind often lead to considerable casualties and large-scale economic losses (Dottori et al., 2016). According to the Center for Research on the Epidemiology of Natural Disasters, floods have caused more direct harm to communities than any other natural hazard over the past century (Hoque et al., 2019). For these reasons, conducting comprehensive assessments of flood hazards, vulnerabilities, and associated risks is vital to minimizing future impacts (Fernandez et al., 2016; Wang et al., 2011).

Flooding is a natural process that can occur under various climatic conditions and in any region, differing in both intensity and duration depending on rainfall patterns. One of the persistent challenges facing human societies and their environmental and economic resources is the rapid movement of water accompanied by alluvial and sediment-laden materials throughout a watershed, commonly referred to as sediment-rich floods. In irrigation and drainage terminology, a flood is defined as a strong, forceful flow that causes a significant rise in river water levels, often resulting in the inundation of adjacent lands. Some researchers also use the term to describe events that have a relatively low probability of occurrence.

A numerical approach to surge definition can moreover be as follows: the surge is the yield of an irregular occasion within the frame of a hydrological cycle, in which amid a generally nonstop time interim, the input work has activated the working framework. The input work is the sum of precipitation and its transient and spatial dispersion designs, and the working framework is the geological region or watershed, which is the figure of changing over the input work into the yield work. The output or surge work is additionally the time slant, stream rate, or volume changes amid the surge. According to the set of definitions: an increment within the stream of a waterway to the degree that it surpasses its capacity covers the encompassing lands and causes harm is called a surge. Long downpours, overwhelming downpours, fast softening of snow, and unnatural components such as the disappointment of dams are the most causes of surges.

Flood-related challenges are multifaceted and inherently complex. Flood events can destroy infrastructure, cause human casualties and financial losses, and interrupt transportation systems such as highways and railways. The reason floods are categorized among the most destructive natural hazards is linked to two major drivers. The first involves natural factors, most notably the intensification of extreme rainfall and flood events associated with climate change. The second relates to human activities, including the degradation of natural resources, inappropriate land-use changes, failure to identify and respect flood-prone zones, and unregulated development within flood channels. In many cases, the severity of flood damage is predominantly shaped by human-induced conditions.

Consequently, the most effective strategy for flood mitigation is a shift in perspective, from crisis response to proactive risk management. This requires integrated planning and redirecting financial resources away from post-disaster compensation toward preventive measures such as watershed management interventions, particularly those targeting upstream catchments where flood generation begins. Adopting this revised approach transforms flood management into an opportunity to simultaneously address issues such as drought, water scarcity, and long-term ecological stability.

Risk management enhances the adaptability, efficiency, and resilience of systems. Therefore, employing effective methods within this framework plays a crucial role in strengthening flood resilience.

The reason for surge control is to dispense with or decrease the negative impacts of surge calamities. In spite of the fact that it isn't continuously conceivable to be fruitful in controlling floods, and it isn't conceivable. The reason for surge control is to dispense with or decrease the negative effects of surge calamities. In spite of the fact that it isn't continuously possible to be effective in controlling surges, and it isn't conceivable to unravel the issue of flooding absolutely, endeavors ought to be made to diminish the destructive and dangerous results of surges. There are different strategies for surge control, and the choice of any strategy or combination of strategies depends on the nearby conditions of the extent, the significance of the regions beneath extended security, and the financial conditions of the nation.

The aim of this chapter is to introduce blockchain technology for flood resilience and management.

1.2 Flood Control

1.2.1 Non-Structural Methods

Non-construction strategies incorporate the foundation and usage of laws and controls on how to utilize surge fields and different preventive measures to diminish surge harm. Overviews and ponders appear that the reason for the increment in surge harms isn't the increment within the number of surges, but the heightened use of floodplain lands, i.e., the increment within the use of floodplain lands adjoining waterways, ought to be considered the most. In general, non-structural approaches to flood management encompass a wide range of strategies. These include the regulation of land use through zoning, the application of flood-proofing techniques to buildings and infrastructure, and the provision of flood insurance to reduce financial vulnerability. They also involve implementing rules for land division and parceling, enforcing construction standards that account for flood hazards, and organizing either temporary or permanent relocation of residents, assets, and equipment from areas prone to flood passage or inundation.

1.2.2 Flood Control Structural Methods

Structural or auxiliary flood-management strategies focus on reducing flood impacts through physical facilities and engineered constructions. These methods, which have been employed for centuries, encompass a range of interventions designed to decrease the volume and intensity of floodwaters, prevent them from reaching vulnerable areas, or redirect and manage their flow. Such measures are generally categorized into several types, including limiting floodwaters with local embankments or levees; modifying and improving river channels and their cross-sections; diverting excess flows away from critical zones; applying watershed-management practices; protecting sensitive areas through floodwalls or embankments; using storage reservoirs to temporarily retain floodwater; and constructing delay or detention dams.

1.3 Flood Resilience and Risk

Hochrainer-Stigler et al. (2020) graded community resilience against floods in the form of an empirical analysis on a global scale. This study presented and analyzed a standard framework for measuring community resilience to floods. The relevant measurement tool is modeled and adapted based on a so-called "technical risk classification" approach used in the insurance sector. The method for ranking indicators is based on a two-step procedure: (1) raw data is gathered, and (2) based on this data, experts rank indicators, also known as resilience resources. Over 118 communities in 9 countries, totaling about 1 point 25 million data points were used to test this strategy. To support the findings from a qualitative viewpoint, content analysis was combined with quantitative analysis. This study's key finding is that expert judgment matters much more than the actual raw data gathered when determining resilience scores related to subjective attributes like ability, emotion, and trust. The study's conclusions regarding the importance of data and grade characteristics can help determine how to evaluate flood resilience standards in a better, more effective, and more robust manner.

Fraser et al. (2020) looked into three urban floods to determine the primary causes of small-scale disasters. Understanding how spatial factors and remote sensing contribute to meeting local risk conditions remains a difficult challenge in our understanding of the causes of natural disasters. This is particularly true in the case of small-scale, high-impact disasters. This article takes a thorough approach to discussing flood risk analysis. Three case studies of the management of coastal flooding in urban areas in the European Union did this. This article reveals the significance of local institutions in reducing local risks as a result of higher-level political and economic changes. With regard to institutional changes, local disaster risk reduction, and the connection between flooding and austerity, it offers fresh empirical evidence.

1.4 Water Management Strategies

In common, water overseeing is conceivable through three open parts (government), private segment, and private organizations. The third strategy has presently ended up the normal of water oversight, since government oversight is incapable and the oversight of the private portion is not reasonable with the current structure of water administration, which incorporates a huge number of small agriculturists. Instep, participatory water administration emphasizes that water clients ought to be included in all stages and levels of water administration (Pourzand, 2004). Water system, inappropriate support of water frameworks, failure of the government to gather sufficient capital for water clients, and destitute execution of water frameworks (Garces-Restrepo et al., 2007).

Numerous endeavors have been made in numerous nations of the world to transfer administration rights and obligations from the government to the private segment or neighborhood organizations (Vermillion, 1997; Geiger et al., 1996). Governments' money related assets are considered to extend versatility against dangers such as surges, grow the act of hydroponics, and guarantee the long-term maintainability of water frameworks (Geijer et al., 1996; Vermillion, 1997; Mounz, 2007).

Exchanging water administration leads to accomplishments such as decreasing water deficiency issues, giving water amid plant development stages, expanding trim abdicate and abundance water supply, and as a result, ideal utilization of water assets and expanding wage and execution. In this manner, it'll diminish the working, upkeep, and work force costs, increase the soul of participation between ranchers and the government, increase the sense of obligation and possession of the agriculturists towards the utilization of water structures and their security, and superior conveyance of water (Mahboubi Ardakani, 2004).

The victory and productivity of buyer administration depend on their fruitful arrangement based on investigating and considering them. The proof has appeared that on the off chance that the field ponders are not legitimately considered, the foundation, arrangement, continuation, and stabilization of client administration in rustic and agrarian communities will come up short.

The involvement of the 1950s and 1960s has appeared that despite the fact that the government's section into the field of development of expansive water supply and conveyance buildings guaranteed the fast development of farming in creating nations, this approach, social issues, and other issues forced a part on the country community. One of the negative outcomes of this altered approach has been the burden of issues such as destitute upkeep of the water system framework, water wastage, and need of inspiration to spare water. These issues began with ignoring the need of farmers' nearness within the administration and arranging of irrigation systems, and the fundamental arrangement to fathom this issue was to return the individuals to the field of administration.

Nowadays, it is accepted that to attain advancement, nations ought to utilize the nearness of people and their interest in several stages and to solidify cooperation, there ought to be arranging from the bottom-up. Exchanging administration duties from government teaching to the individuals is presently an imperative arrangement in numerous nations. So that the conducted considers that nearly all nations have acknowledged the participatory approach in water administration to diverse degrees. Nowadays, handing over water administration to water clients has ended up as a global movement, within the handle of its advancement, the part of governments within the operation and support of structures has been decreased, and instep, neighborhood organizations of water clients have the specialist to oversee the operation and support of structures (Shahrudi and Chizari, 2007). Appropriately, community support is exceptionally imperative and vital for the effective and maintainable administration of water assets (Dungumaro and Madula, 2003).

Participatory water administration is giving over social benefits to the community. By utilizing these social rights, individuals get openings and conceivable outcomes by taking control of administration, as huge socioeconomic fields, and locks in in managerial-economic exercises such as arranging, giving, and apportioning money-related assets, drawing-in credits, and observing and assessment. And indeed political communication, to extend their skills and capabilities within the areas of arranging, organizing, budgeting, and budgetary and bookkeeping issues (Zarei Dastgerdi et al., 2018).

1.5 Information Technology and Hydro Informatics

The adjustment of complex man-made and normal frameworks, such as the hydrological arrangement and urban water structures, is basic for present-day social orders to outlive within the 21st century (Beck et al., 2010), common vitality bases and renewable vitality frameworks, extraordinary climate occasions and fiascos (Ewing and Demir, 2021), the rural supply chain, and flexible foundation are all subordinate (Alabbad et al., 2021). By measuring more characteristic frameworks and gathering more information, the integration of characteristic and manufactured frameworks is made more congruous (Muste et al., 2017) and making counterfeit frameworks utilizing this information (Omatu et al., 2022; Sermet and Demir, 2020) and computerized twin situations are both doable. This request has expanded the centrality of dependable data and information preparing innovation (Hu and Demir, 2021). Due to this, over the past 20 years, data and communication innovation approaches have been quickly advancing, and computerized change has quickened in nearly all divisions (Knell, 2021). The points of view of distinctive social orders have been created concurrently on points like generation, arranging, and execution (Savic, 2021). With progressions in hydroinformatics, the areas of hydrology and water assets have been taken after a comparable design (Demir et al., 2022). The foremost outstanding progressions are modern profound learning (DL) applications in copy mix and communication (Sermet and Demir, 2021), and broad modeling and examination interior frameworks. Pluto, virtual and increased reality for hydrological modeling and instruction (Sermet and Demir, 2022), and unused programming libraries and information benchmarks. Xiang and Demir, (2022) are specified. The water division has begun to produce, handle, and store more information as a characteristic result of this fast computerized change (Haltas et al., 2021). Be that as it may, in order for hydrological expectations to be more precise, this handle requires the unwavering quality, security, and standardization of the fundamental information. Compared to this, information security and standardization are fundamental for partners in water frameworks to communicate in a reliable and secure way (Vooged et al., 2022).

1.6 Blockchain Technology

As blockchain technology's latent is understood, conviction matters are being lectured more rapidly. Data is kept in blocks that make up a chain in a blockchain database. The uniformity of data in time is due to the fact that it cannot be changed or removed on-chain without network consensus. As a result, to track orders, payments, and accounts, Blockchain technology is a sophisticated database system that allows for the sharing of transparent information in a network (Nakamoto, 2008), and other contacts will result in an absolute ledger. Internal panels in the scheme guard against illegal transactions and ensure that everyone is observing authorized transactions fairly and accurately.

The center of inquiry about and improvement over the past ten years has been on blockchain, a rising innovation that has been received by a few businesses. Vitality companies utilizing blockchain innovation may create a commercial center for the trade of power, or solar-paneled mortgage holders may utilize a comparative commercial center to offer any overabundance of power to their neighbors.

Smart meters generate transactions and record them on the blockchain, making the process mostly automated. Blockchain services can be utilized by traditional financial systems like banks and stock exchanges to manage online payments, accounts, and stock market transactions. In addition, blockchain technology can be utilized by media and entertainment companies to handle copyrighted data. Similarly, retail businesses can utilize blockchain to monitor the flow of goods between suppliers and customers (Hu et al., 2022; Jing et al., 2020, 2021; Li et al., 2022a). Blockchain technology has the potential to extend the long-term vision of managing water resources, especially by transferring management to consumers, as water becomes a sensitive topic worldwide.

1.6.1 History of Blockchain Technology

The capacity to track famous data straightforwardly, evidently, in financial applications, as well as the capacity to store information in a disseminated record, select which information to store employing an agreement calculation, characterize preconditions with a shrewd contract, and so on. Dialog dates it back to the 1990s. Issues with routine information capacity strategies and the need of depending on third parties in financial transactions are dealt with. The initial application that kept money related exchange information in real-time was Bitcoin. Despite the fact that 'block' and 'chain' were continuously utilized independently within the unique Bitcoin white paper, the term 'blockchain' was in the long run supplanted by 'blockchain' (Nakamoto, 2008). The Ethereum white paper (Buterin, 2014) declared the second major step within the advancement of blockchain innovation as well as the primary application of the utilization of recording exchange information in a database as it were in the event that both parties meet particular prerequisites.

1.6.2 Blockchain Networks

Decentralized or distributed blockchain networks are divided into four main categories: public blockchain networks, private blockchain networks, hybrid blockchains, and consortium blockchain networks. Public blockchain networks are open to everyone and have no access restrictions. The blockchain is accessible to everyone with equal access to read, edit, and verify it. In order to exchange and mine cryptocurrencies like Bitcoin, Ethereum, and Litecoin, open blockchain systems are essentially utilized (Wang et al., 2021). A single organization is in charge of private blockchains. Who can connect the organization and what benefits individuals have been decided by this specialist. Due to getting to confinements, private blockchains are as they were mostly decentralized. A few characteristics of private and open systems are combined in half-breed blockchains. Companies can make both a shared framework and a private, license-based framework. As a result, they keep some information open whereas controlling access to other information that's put away on the blockchain. To enable accomplices within the framework to affirm whether or not private exchanges have taken out, they utilize savvy contracts. For occurrence, crossover blockchains can offer shared get to computerized money whereas keeping up the privacy of bank monetary forms. A collection of businesses oversees consortium blockchain systems. Blockchain progression and information get to rights are mutually overseen by pre-selected organizations. In businesses where numerous organizations work toward a common objective and can pick up from a shared duty, consortium blockchain systems are ordinarily favored (Arooj et al., 2022).

1.6.3 Blockchain Protocols

Different blockchain stages can be utilized to make applications, which is what the term 'blockchain protocol' includes. To fit specific businesses or applications, each blockchain convention adjusts the essentials of the blockchain. Hyperledger, Ethereum, Corda, and Majority are the foremost well-known blockchain conventions. An open-source extension with a few devices and libraries is the

Hyperledger convention. Blockchain applications can be modified in this way. With extraordinary identity-management and access-control highlights, Hyperledger is an open-source, measured system. These characteristics make it appropriate for a run of applications, counting supply chain following and observing, exchange financing, motivation plans, and monetary resource administration (Hang and Kim, 2021). An elective stage is the Ethereum convention. Since it is centralized, open blockchain applications can be built with it (Estevam et al., 2021). An open-source blockchain extension called Corda is geared toward commerce. Ethereum is the premise for Majority, an open-source blockchain convention. This application is particularly expected for use in private blockchain systems, where all hubs are claimed by fair one part, or consortium blockchain systems, where distinctive individuals each possess a parcel of the arrange (Singh et al., 2022).

1.6.4 Blockchain Architecture

As a result, the taking after are the key characteristics of blockchain innovation that recognize it: Subsidiarity in blockchain alludes to the transmission of control and decision-making specialist from a central legitimate question (for case, an isolated, organization, or gather) to a disseminated arrangement. Straightforwardness could be a key precept of decentralized blockchain systems that makes a difference if members construct less belief in one another. These systems restrict clients from applying control or control over one another since doing so corrupts organized proficiency. Permanence is another significant component. As a result, once the information is put away within the blockchain, it cannot be changed or changed. By chance, a report has a mistake, the mistake must be settled by including a modern exchange, and both communications must be made open on the arrangement. The blockchain system's agreement mechanism, which establishes rules for members to concur on recording exchanges, could be a conspicuous inclusion. As it were after the larger part of arranged clients have affirmed them, unused exchanges are permitted. On the other hand, the blockchain building comprises the taking after specialized components: (a) Spread record, a shared database that's utilized to log trades inside the blockchain organization.. (b) Sharp contracts are computer programs that run right absent and are saved inside the blockchain system when certain criteria are met. To guarantee that exchanges continue safely, they perform "if-then" checks, and (c) a security highlight known as open key encryption enables members within the blockchain organization to be distinguished as it were by their open key. For organized clients, two sets of keys are produced by this instrument. An open key is one of the keys that completely arranges employment at once. One private key having a place to each member is the other key. Information within the record is opened employing a combination of private and open keys (Figure 1.1). Advanced monetary forms like Bitcoin are the essential period of blockchain. Blockchain, in addition is known as chain code, it is the moment era of smart contracts. Moreover, as businesses recognize and execute cutting edge application spaces, the third or future period of blockchain will continue to form and create.

1.7 Blockchain Application in Flood Hydrology

The returns of packing data with the most elevated level of unwavering quality are the most central of the beginning thoughts for blockchain execution in hydrology and water assets administration, as is the case in numerous other businesses. Water administration, in all its shapes, depends intensely on the exactness of information related to water. Since water may be a multifaceted characteristic asset, it can best be achieved by maximizing the edge of cronies with different needs. Water encompasses a social component in expansion to distinctive needs for amount and quality. Building belief requires exact data, almost shared water assets. (Pahl-Vostel, 2015). A straightforward dispersed computerized design built on blockchain and keen contracts makes it conceivable to straightforwardly ensure social belief within the hydrological community and the unwavering quality of the information.

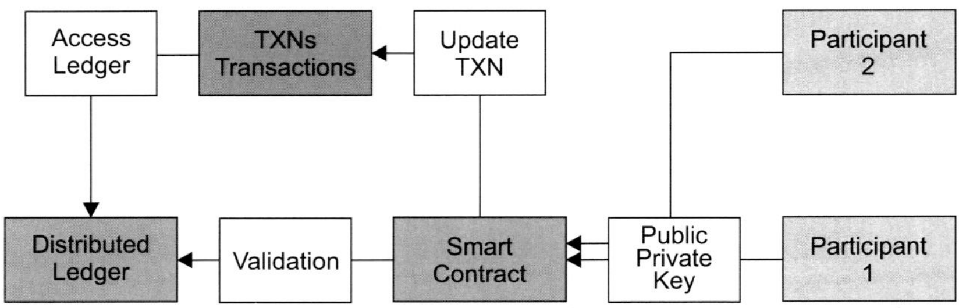

Fig. 1.1 An image of the basic architecture of the blockchain (Satilmsioglu et al., 2022).

The utilization of unchangeable recorded information can offer assistance to decrease instability in hydrological frameworks for different partners. Blockchain innovation is additionally perfect for applications related to water funds. Water markets, where water rights are exchanged, or any venture in water administration, monetary bolster to move forward water framework and surge control structures, or sanitation in creating nations are conceivable application ranges for this innovation. From the perspective of speculators or showcase members, this innovation empowers the straightforward following of all money-related exchanges. The blockchain may too give valuable arrangements when it comes to following water. Water-related traceability will be significant for agrarian water administration (Yildirim and Demir, 2022), water quality and control (Demir et al., 2009), urban water frameworks, and the water impression of a specific item.

1.8 Conclusions

Although it can be challenging to standardize and blockchain-based hydrological applications are unique in many situations, four key benefits have been identified by numerous studies, which are detailed below.

Water Trust: Because the blockchain is decentralized, none of the participants directly owns the infrastructure. Furthermore, based on smart contracts, it is guaranteed that everyone will abide by the hydrological and administrative rules. Even if they are unaware of one another, this fosters a layer of trust between parties who depend on a water resource.

Security of water data: Data in blockchain is evenly distributed among the nodes that power the system. It shields the network from common cyberattacks and power outages. Even if the data is somehow obtained, it cannot be disclosed without the user's express permission because it is stored in an encrypted format. For both natural and legal persons, it offers data security.

Immutable Water Transaction Ledgers: Data entered into a distributed ledger cannot be removed. It offers stakeholders a unified, shared database.

Water Accountability: Transaction data records are visible to everyone on the blockchain and can be checked. It is not necessary to verify data using a trusted data provider or a third-party verification system.

Blockchain technology enables the generation of an integrated data flow in flood risk management as well as brainstorming in a decentralized network of various actors.

To analyze the effectiveness of many flood risk management measures, especially non-structural measures, a long-term analysis framework is needed to analyze long-term changes in the flood system. Non-structural flood control methods are less expensive than structural methods. Also, lack of infrastructure and preventive facilities, poor and inappropriate drainage facilities, land use change, poor environmental attitude, and poor implementation of the management system are known as the most effective criteria in the event of floods. Therefore, it is necessary to involve people in flood management and resilience against it by creating secure frameworks such as blockchain.

References

1. Alabbad, Y., Mount, J., Campbell, A.M. and Demir, I. (2021). Assessment of transportation system disruption and accessibility to critical amenities during flooding: Iowa case study. *Science of The Total Environment*, *793*, 148476.
2. Arooj, A., Farooq, M.S. and Umer, T. (2022). Unfolding the blockchain era: Timeline, evolution, types, and real-world applications. *Journal of Network and Computer Applications*, *207*, 103511.
3. Beck, M.B., Jiang, F., Shi, F., Walker, R.V., Osidele, O.O., Lin, Z. and Hall, J.W. (2010, March). Re-engineering cities as forces for good in the environment. In: *Proceedings of the Institution of Civil Engineers-Engineering Sustainability* (Vol. *163*, No. 1, pp. 31–46). Thomas Telford Ltd.
4. Buterin, V. (2014). A next-generation smart contract and decentralized application platform. *White Paper*, *3*(37), 2–1.
5. Demir, I., Jiang, F., Walker, R.V., Parker, A.K. and Beck, M.B. (2009, October). Information systems and social legitimacy scientific visualization of water quality. In: *2009 IEEE International Conference on Systems, Man, and Cybernetics* (pp. 1067–1072).

6. Demir, I., Sermet, Y. and Rink, K. (2022). Next generation visualization and communication systems for earth science using immersive reality and serious gaming. *Frontiers in Earth Science, 10*, 1101538.
7. Dottori, F., Salamon, P., Bianchi, A., Alfieri, L., Hirpa, F.A. and Feyen, L. (2016). Development and evaluation of a framework for global flood hazard mapping. *Advances in Water Resources, 94*, 87–102.
8. Dungumaro, E.W. and Madulu, N.F. (2003). Public participation in integrated water resources management: the case of Tanzania. *Physics and Chemistry of the Earth, Parts a/b/c, 28*(20–27), 1009–1014.
9. Eslamian, S. and Maleki, M. (2021). Disaster resilience and computational methods for urban infrastructures. In: *Handbook of Disaster Risk Reduction for Resilience: New Frameworks for Building Resilience to Disasters* (pp. 279–298). Cham: Springer International Publishing.
10. Eslamian, S., Parvizi, S. and Behnassi, M. (2021). New Frameworks for Building Resilience in Hazard Management. In: *Handbook of Disaster Risk Reduction for Resilience: New Frameworks for Building Resilience to Disasters.* (pp. 107–130). Cham: Springer International Publishing, (pp. 107–130).
11. Estevam, G., Palma, L.M., Silva, L.R., Martina, J.E. and Vigil, M. (2021). Accurate and decentralized timestamping using smart contracts on the Ethereum blockchain. *Information Processing and Management, 58*(3), 102471.
12. Ewing, G. and Demir, I. (2021). An ethical decision-making framework with serious gaming: A smart water case study on flooding. *Journal of Hydroinformatics, 23*(3), 466–482.
13. Fernandez, P., Mourato, S. and Moreira, M. (2016). Social vulnerability assessment of flood risk using GIS-based multicriteria decision analysis. A case study of Vila Nova de Gaia (Portugal). *Geomatics, Natural Hazards, and Risk, 7*(4), 1367–1389.
14. Fraser, A., Pelling, M., Scolobig, A. and Mavrogenis, S. (2020). Relating root causes to local risk conditions: A comparative study of the institutional pathways to small-scale disasters in three urban flood contexts. *Global Environmental Change, 63*, 102102.
15. Garces-Restrepo, C., Vermillion, D. and Muñoz, G. (2007). *Irrigation Management Transfer: Worldwide Efforts and Results*. FAO.
16. Geijer, J.C., Svendsen, M. and Vermillion, D.L. (1995). Transferring irrigation management responsibility in Asia: Results of a workshop–FAO/IIMI Expert Consultation on Irrigation Management Transfer in Asia, Bangkok, and Chiang Mai, 25–29 September, 1995.
17. Gregory, K.J. (2006). The human role in changing river channels. *Geomorphology, 79*(3–4), 172–191.
18. Haltas, I., Yildirim, E., Oztas, F. and Demir, I. (2021). A comprehensive flood event specification and inventory: 1930–2020 Turkey case study. *International Journal of Disaster Risk Reduction, 56*, 102086.
19. Hang, L. and Kim, D.H. (2021). Optimal blockchain network construction methodology based on analysis of configurable components for enhancing hyperledger fabric performance. *Blockchain: Research and Applications, 2*(1), 100009.
20. Hirabayashi, Y., Mahendran, R., Koirala, S., Konoshima, L., Yamazaki, D., Watanabe, S., ... & Kanae, S. (2013). Global flood risk under climate change. *Nature Climate Change, 3*(9), 816–821.
21. Hochrainer-Stigler, S., Laurien, F., Velev, S., Keating, A. and Mechler, R. (2020). Standardized disaster and climate resilience grading: A global scale empirical analysis of community flood resilience. *Journal of Environmental Management, 276*, 111332.
22. Hoque, M.A.A., Tasfia, S., Ahmed, N. and Pradhan, B. (2019). Assessing spatial flood vulnerability at Kalapara Upazila in Bangladesh using an analytic hierarchy process. *Sensors, 19*(6), 1302.
23. Hu, A. and Demir, I. (2021). Real-time flood mapping on client-side web systems using hand model. *Hydrology, 8*(2), 65.

24. Hu, S., Huang, S. and Qin, X. (2022). Exploring blockchain-supported authentication based on online and offline business in organic agricultural supply chain. *Computers and Industrial Engineering, 173*, 108738.
25. Igbatayo, S.A., Eslamian, S., Babalola, O.O. and Makanju, A.A. (2022). Strengthening Climate Resilience and Disaster Risk Reduction: Case Study of the Sahel Adaptive Social Protection. In: *Disaster Risk Reduction for Resilience: Disaster and Social Aspects*. Cham: Springer International Publishing, pp. 221--243.
26. Jing, N., Liu, Q. and Sugumaran, V. (2021). A blockchain-based code copyright management system. *Information Processing & Management, 58*(3), 102518.
27. Khosravi, K., Shahabi, H., Pham, B. T., Adamowski, J., Shirzadi, A., Pradhan, B., ... & Prakash, I. (2019). A comparative assessment of flood susceptibility modeling using multi-criteria decision-making analysis and machine learning methods. *Journal of Hydrology, 573*, 311–323.
28. Knell, M. (2021). The digital revolution and digitalized network society. *Review of Evolutionary Political Economy, 2*(1), 9–25.
29. Li, Z., Mount, J. and Demir, I. (2022). Accounting for uncertainty in real-time flood inundation mapping using HAND model: Iowa case study. *Natural Hazards, 112*(1), 977–1004.
30. Mahboubi Ardakani, A.R. and Razavi, A. (2004). The role of participatory management in the operation and maintenance of irrigation networks and the review of Turkey's experience in handing over operation and maintenance management to users. In: *Gohran Kavir, The First Conference on Problems of Irrigation, Drainage, and Optimal Use of Agricultural Water*, June, Tehran, Iran.
31. Mohammadjani, I. and Yazdanian, N. (2014). Analysis of the water crisis in the country and the requirements of its management. *Journal of Ravand, 21*(65–66), 117–144.
32. Muste, M., Lyn, D.A., Admiraal, D., Ettema, R., Nikora, V. and Garcia, M.H. (2017). *Experimental hydraulics: Methods, Instrumentation, Data Processing, and Management* (Volume I): *Fundamentals and Methods*. CRC Press.
33. Nakamoto, S., (2008). Bitcoin: A peer-to-peer electronic cash system [online]. [Viewed 21 November 2022]. Available from: https://bitcoin.org/bitcoin.pdf.
34. Nasim, A., and Selajgeh, A. (2006). Determination of river border and bed using AUTOCAD, HEC-RAS software, case study: A section of the Karaj River (Sira-Pul Khob). In: *7th International Conference on River Engineering*, Shahid Chamran University, Ahvaz, Iran (In Farsi).
35. Nazif, S., Mohammadpour Khoie, M.M., and Eslamian, S. (2021). Urban Disaster Management and Resilience. Chapter 7, In: *Handbook of Disaster Risk Reduction for Resilience: New Frameworks for Building Resilience to Disasters* Springer, Cham, Switzerland, pp. 157--185.
36. Omatu, S., Malluhi, Q.M., Gonzalez, S.R., Bocewicz, G., Bucciarelli, E., Giulioni, G., and Iqba, F. (Eds.) (2015). Distributed computing and artificial intelligence.,In: *12th International Conference* [online]. Cham: Springer International Publishing. [Viewed 18 December, 2022]. Available from: doi: 10.1007/978-3-319-19638-1.
37. Pahl-Wostl, C. (2015). *Water Governance in the Face of Global Change* [online]. Cham: Springer International Publishing. [Viewed 21 November 2022]. Available from: doi: 10.1007/978-3- 319-21855-7.
38. Porzand, A. (2004). Recent measures to make the management of exploitation and maintenance of irrigation and drainage networks private and competitive. In: *The First Conference on Investigating the Problems of Irrigation and Drainage Networks and Optimal Use of Agricultural Water*. Tehran, Iran, June (in Persian).
39. Rahman, M., Chen, N., Elbeltagi, A., Islam, M.M., Alam, M., Pourghasemi, H.R. and Dewan, A. (2021). Application of stacking hybrid machine learning algorithms in delineating multi-type flooding in Bangladesh. *Journal of Environmental Management, 295*, 113086.

40. Ruidas, D., Chakrabortty, R., Islam, A.R.M.T., Saha, A. and Pal, S.C. (2022). A novel hybrid of meta-optimization approach for flash flood-susceptibility assessment in a monsoon-dominated watershed, Eastern India. *Environmental Earth Sciences*, *81*(5), 145.
41. Savić, D. (2022). Digital water developments and lessons learned from automation in the car and aircraft industries. *Engineering*, *9*, 35–41.
42. Schultz, B. (2017). Agricultural water management and food security in a sustainable environment. Transport of viruses and colloids in partially-saturated soil and groundwater. Seyed Majid Hassanizadeh, Gholamreza Sadeghi, Zhang Qiulan, Jack Schijven, 23. In: *International Drainage Workshop* (*IDW*) hosted by International Commission on Irrigation and Drainage (ICID).
43. Sermet, Y. and Demir, I. (2020). Virtual and augmented reality applications for environmental science education and training. In: *New Perspectives on Virtual and Augmented Reality*. Routledge, pp. 261–275.
44. Sermet, Y. and Demir, I. (2021). A semantic web framework for automated smart assistants: A case study for public health. *Big Data and Cognitive Computing*, *5*(4), 57.
45. Sermet, Y. and Demir, I. (2022). GeospatialVR: A web-based virtual reality framework for collaborative environmental simulations. *Computers & Geosciences*, *159*, 105010.
46. Shahrodi, A.A., Chizari, M., and Pezeshki Rad, Gh. (2007). Water user's cooperative effect on farmers' attitude towards the agricultural water management: A case study on Khorasan. *Journal of Agricultural Economic and Development*, *22*(2), 71–85 (In Persian).
47. Singh, A., Kumar, G., Saha, R., Conti, M., Alazab, M. and Thomas, R. (2022). A survey and taxonomy of consensus protocols for blockchains. *Journal of Systems Architecture*, *127*, 102503.
47. Satilmisoglu, T.K., Sermet, Y. and Demir, I. (2022). Blockchain Applications and Opportunities for Water Resources and Hydrology: A Systematic Review. https://eartharxiv,orog.
49. Vermillion, D.L. (1997). Impacts of irrigation management transfer: A review of the evidence. https://core.ac.uk.
50. Voogd, R., Rudberg, P.M., de Vries, J.R., Beunen, R., Espiritu, A.A., Methner, N. and Kruger, E. (2022). A systematic review on the role of trust in the water governance literature. *Water Research X*, *16*, 100147.
51. Wang, Y., Li, Z., Tang, Z. and Zeng, G. (2011). A GIS-based spatial multi-criteria approach for flood risk assessment in the Dongting Lake Region, Hunan, Central China. *Water Resources Management*, *25*, 3465–3484.
52. Wang, X., Ni, W., Zha, X., Yu, G., Liu, R.P., Georgalas, N. and Reeves, A. (2021). Capacity analysis of public blockchain. *Computer Communications*, *177*, 112–124.
53. Xiang, Z. and Demir, I. (2022). Flood Markup Language: A standards-based exchange language for flood risk communication. *Environmental Modelling and Software*, *152*, 105397.
54. Yildirim, E. and Demir, I. (2022). Agricultural flood vulnerability assessment and risk quantification in Iowa. *Science of The Total Environment*, *826*, 154165.
55. Zarei Dastgerdi, Z., Iravani, H., Shabanali Fami, H. and Mokhtari Hesari, A. (2008). Analysis of the participation of farmers in the management of water supply network of Jarghouyeh city. *Journal of Rural Development*, *12*(3), 92–73 (In Persian).
56. Zhang, Q., Gu, X., Singh, V.P., Shi, P. and Sun, P. (2018). More frequent flooding? Changes in flood frequency in the Pearl River basin, China, since 1951 and over the past 1000 years. *Hydrology and Earth System Sciences*, *22*(5), 2637–2653.

2 A Broad Review on Blockchain Technology

Saeid Eslamian[1] and Yaser Sabzevari[2*]

Blockchain technology has rapidly evolved from a foundational component of cryptocurrencies to a versatile digital infrastructure with applications across numerous sectors. Its core features, decentralization, immutability, transparency, and security, offer transformative potential for systems that rely on trustworthy data exchange and multi-stakeholder coordination. This review provides a broad overview of blockchain technology, tracing its conceptual foundations, architectural components, consensus mechanisms, and major development milestones. Key application domains, including finance, supply chain management, energy systems, healthcare, and governance, are examined to highlight the technology's expanding influence. The paper also discusses current challenges such as scalability, interoperability, regulatory uncertainty, and energy consumption. By synthesizing existing research and practical applications, this review offers a comprehensive understanding of blockchain's capabilities, limitations, and future development pathways.

2.1 Introduction

Blockchain technology has found numerous hydrological submissions in recent years. Literature searches were originally separated into 2 areas: scholarly use and non-scholarly use. A review of previous academic research includes articles, books, and scientific reports. A review of non-academic research includes white papers, web documents, public institution reports, and hackathon projects.

Aimed at the scholarly work reviewed, multiple parameters and classification variations were chosen to assess the reviewed scholarly work and multiple parameters and classification variables to monitor the present maturity of blockchain technology for hydrological uses.

The current maturity of blockchain technology for hydrological uses is as below:

[1,2] Department of Water Science and Engineering, College of Agriculture, Isfahan University of Technology, Isfahan, Iran.
* Corresponding author: yasersabzevari1511@gmail.com

DOI: 10.1201/9781003686026-2

***Magazine Form*:** Specifies where and in what format scholarly work such as research articles, congress papers, books, book chapters, articles, and expert reports are published.

***Main Emphasis*:** Classification of presented studies, focusing on sub-basin hydrology and water resources. Though the greatest studies emphasize multiple areas, the careful analysis assumed that these studies were primarily focused on a single subject area. It is categorized into management of city water, administration of water quality, water administration, water authority, administration of agrarian water, water and Sustainable Development Goals (SDGs).

***The Resolution*:** This shows the primary application case that the research or creation is intended to address.

***Blockchain Application Development Status*:** This shows how much blockchain technology is being used in research.

***Search*:** This program evaluates and demonstrates the submission points of blockchain into optimal administration of water sources besides examining the research and submission gaps. Confidential research in this area is for exploratory purposes only and does not include blockchain parameters.

This section aims to deliver a summary of research on blockchain technology in hydrology and water sources administration.

2.2 A Summary of Peer-reviewed Articles and Non-academic Content

Here it presents and reviews schematic summaries of articles and non-scientific content. The compact graphic display involves the supply of articles and non-scientific content based on publication date and grade. In addition, charts in the reviewed articles provide technical details on blockchain use cases, including applications in the subfield of hydrology, stages of development of blockchain applications, types of blockchains, and blockchain networks.

Looking at the article distribution by year, we can see an almost linear increase. This drift can be clarified by the expanding fields of use of blockchain technology and the increasing discovery and acceptance of the latest knowledge in hydrological courses. It seems that the number of articles in 2022 has not passed 2 years yet. Since the articles studied covered the period before October 15, 2022, this number might represent a more linear rise. Additionally, the extreme volatility of cryptocurrency connections may limit the application of blockchain in different regions. Aside from a sharp drop in 2020, the production of non-academic content seems to be on the rise steadily (Figure 2.1).

The articles reviewed are categorized into the following categories: "Journal Article", "Conference Paper", "Book Chapter", "Book", "Dissertation", "Working

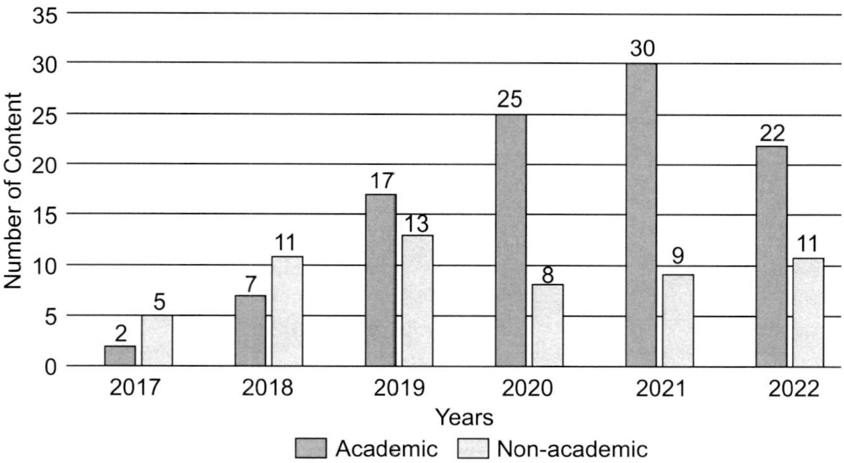

Fig. 2.1 Separation of scholarly and non-scholarly works in the year (Satilmsioglu et al., 2022).

Paper", and "Technical Report". When the results of different stages of the project were submitted at different Congresses, contributions to both conferences were reflected. Content issue, as a practical report, is considered scientific content if the publisher is a government agency, and non-scientific content is published for public information purposes by a non-governmental organization. Among the peer-reviewed papers, contributions to journals and conferences dominate, while the master's can be classified as small. The aim for this is that the incorporation among the scholarly blockchain network and the hydraulic informatics community is immature, and as an outcome, not only experienced consultants in this interdisciplinary field are encouraged to participate in the papers. It can be interpreted as difficulty in finding suitable use cases to support. Non-academic content for review falls into three types of publications: web documents, white papers, and hackathon developments (Figure 2.2). Projects that have published white papers highlight potential applications rather than offering actual-time hydrological applications. Hackathon outputs are the outcome of hackathon occurrences that deal directly with hydrological methods.

The articles reviewed are further categorized according to their main contribution to the hydrology subarea. Most of the articles focused on multiple subfields, but they were grouped under the hydrology subfield, which was considered to be the area of greatest focus (Figure 2.3).

Water Management: These articles generally cover the use of blockchain as a catalog, focusing on the resolution of managerial glitches and structuring hope between users.

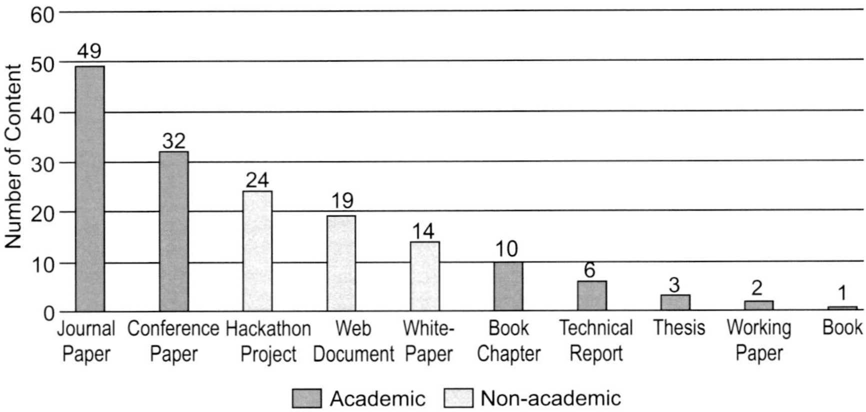

Fig. 2.2 Separation of academic and non-academic works based on the type of publication (Satilmsioglu et al., 2022).

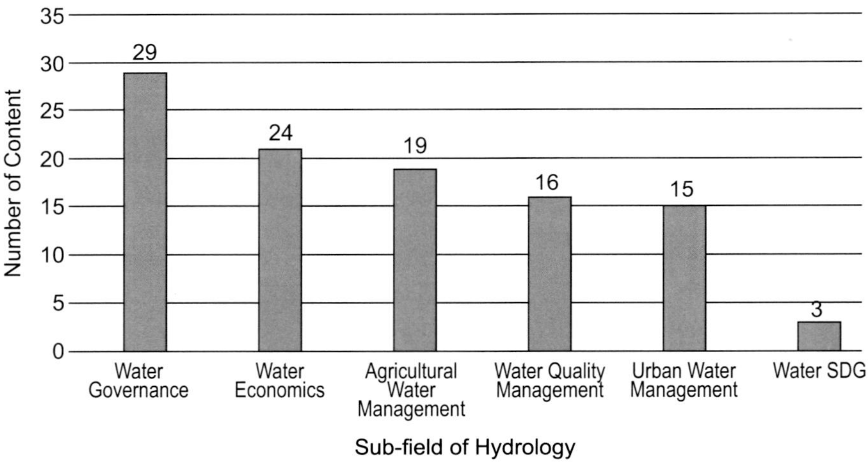

Fig. 2.3 Division of academic works based on hydrology topics (Satilmsioglu et al., 2022).

Water Quality Monitoring: These articles deal with water excellence issues such as evaluation, examination, and management of wastewater and provide clearer information on water quality standards.

***Water Management*:** These articles emphasize water market methods at different levels of water interchange, water truths, and water management use.

***Management of Water in Agriculture*:** These articles emphasize water issues connected to agriculture, such as irrigation and water quality in agriculture.

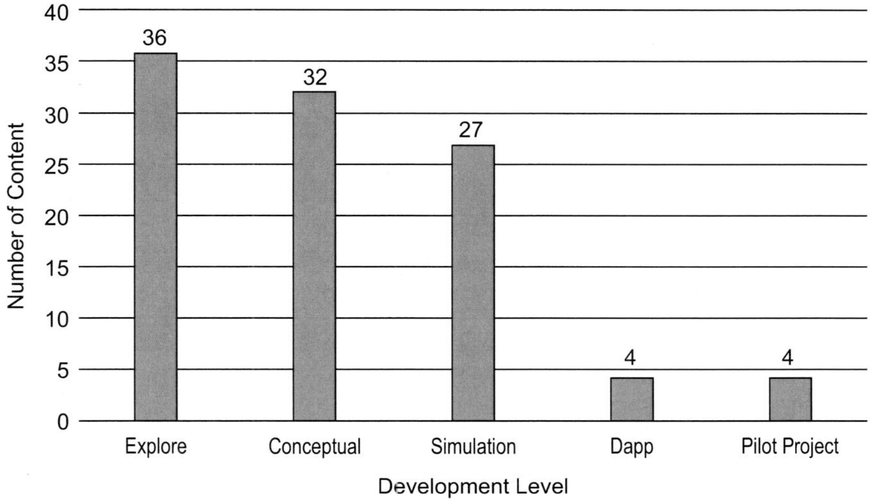

Fig. 2.4 Division of academic works based on blockchain development (Satilmsioglu et al., 2022).

Water Management: This research deals with the water market in various aspects such as water trade, water truths, and water management requests.

***Monitoring of Agricultural Water*:** This research works on problems related to water used in the agricultural sector, such as irrigation and water quality in agriculture.

***Urban Water Monitoring*:** This research works on the use of water in the urban environment, such as drinking water systems.

***Water SDGs*:** These works deal with the impact of blockchain in hydrology besides their role in realizing maintainable expansion goals.

Decentralized financial requests are the initial submissions of block-chain in numerous various issues. With more knowledge in this area, water management is: The distribution of blockchain application development status indicates that hydrology submissions are in a mature stage till date (Figure 2.4). Note that very few studies include end-product applications or pilot projects involving actual use cases. The quantity of work in the Water SDG group would increase if further work were absorbed on his probable influence on the SDGs.

The "N/A" columns in Figures 2.5 and 2.6 refer to the same articles in the "Explore" column. They do not propose architectures or systems, but merely explore potential applications of blockchain technology. Therefore, there is no dispute over the nature of blockchains or blockchain networks, as this article only evaluates probable requests. If a scheme is proposed, but the blockchain

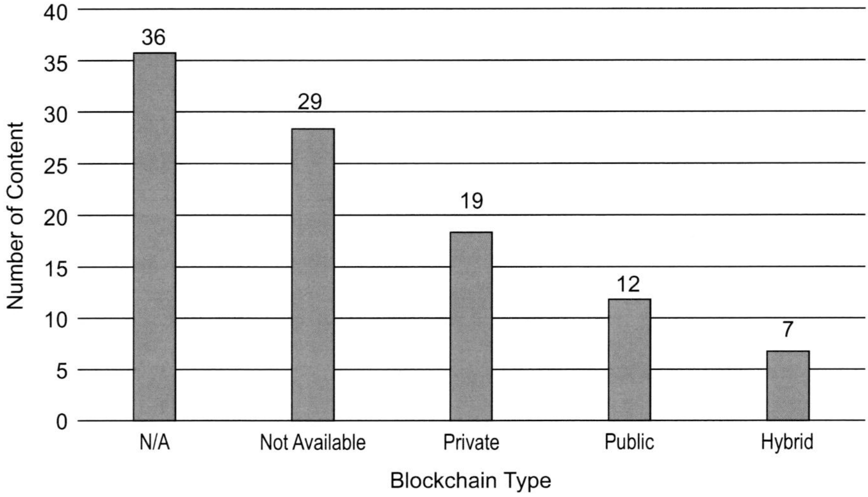

Fig. 2.5 Classification of academic works focused on blockchain type (Satilmsioglu et al., 2022).

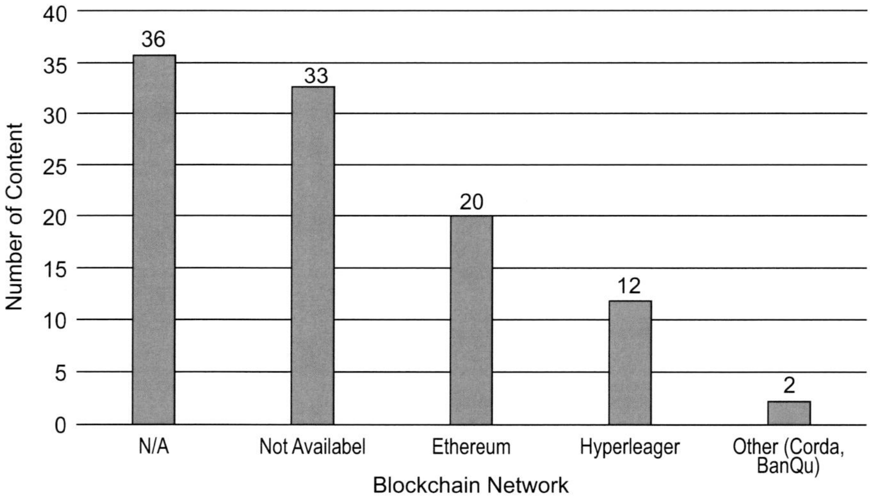

Fig. 2.6 Classification of academic works by blockchain network (Satilmsioglu et al., 2022).

kind is not specified, it will be presented as "not available". This is why private blockchains are recommended as most papers contain conceptual studies and simulations, so water-related usage data should not transparently be disclosed, even if the information is cryptographically secure. Additionally, volatility in the cryptocurrency market may raise ambiguity about fees for paid activities on public blockchains. Ethereum and Hyperledger are superior not only in the field of hydrology but also in diverse fields of application owing to their compatibility with bright contracts and personal blockchains. Other personal blockchains featured in this study were Corda and BanQu.

2.3 Assessing of Academic Works

Here is a brief list of the academic materials studied. These reviews highlight the paper's method of integrating blockchain and hydrology.

2.3.1 Water Authority

The general consent is that mutual conviction is an important principle of water policy. Assured inputs and establishing and maintaining a certain level of trust are key points. Blockchain offers an absolute catalog for securely storing water-related inputs. This provides blockchain technology a simple frame for water governance stakeholders to coordinate. Additionally, bright agreements can assist maintainable water management via formulating ecological besides financial forms. Such advances are completely possible, for example, through the progress of cryptocurrencies and the use of Smarthis bonds to create incentives for measures to conserve water resources. In 2020, Sriyono (2020) investigated the possibility of utilizing blockchain technology for better administration of water properties. The author outlined a background and structure for blockchain and water administration and proposed a blockchain-powered resolution to the water excellence issue in Puerto Rico. Dogo et al. (2019) explore the properties of integrating blockchain technology with smart water management,m and the potential of blockchain to transform water and cleanliness systems to fulfill the United Nations' SDG 6 by 2035. Punia et al. (2021) separately and simultaneously studied the spatio-temporal circulation of several drought forms in India. A blockchain-based frame is proposed to improve the recent scarcity risk administration organization and help drought-affected people get help as soon as possible. Hanging et al. (2022) investigated the capability of extensive data and distributed ledger technology in the domain of water properties, contending that blockchain can perform as a bridge connecting regionally implemented remedies and universally accessible worldwide infrastructure. I scrutinized the utilization, which is overseen by a partnership of multinational entities. Iyer and Giri (2020) evaluated water problems that could be resolved via blockchain. This becomes significant in the context of pricing and emphasizes information slides as cross-network data for reuse. Wu et al. (2022) thoroughly explored the features of blockchain and the situations and applications of blockchain in the issue of water resource conservation, for example, storing water form information, and cross-subdivisional collaboration, besides increasing citizens' contribution.

According to Ragghianti (2021), blockchain allows additional watershed evaluation and straight control of water consumption for all users of the system, while maintaining flexibility and ensuring the socio-environmental constraints imposed on watersheds increase. Xia et al. (2022) considered decentralized water data administration systems across supply, consumption, and runoff processes. Singh and Goel (2020) explored the way blockchain can be used to alert establishments of flood emergencies. Yasuno et al. (2020) worked to reexplain dam basins as astute dams by integrating upriver evaluating, dam influx forecasting, and smart contract-enabled blockchain frameworks to tune flood forecasting knowledge. The research enabled actual-time direction between investors and actual-time dissemination of disaster deterrence data.

Smart water protection systems "authentication and key arrangement could be susceptible to SPOF (single point of failure) and DDOS (distributed denial of service) occurrences. To address the issue, they proposed a blockchain-based authenticated key contract system for keen water plans. Ethereum was chosen as the blockchain network, and smart devices acted as light nodes that participated in it. Light nodes do not mine but rather download a minor portion of the blockchain system. This approach is suitable for scenarios where there are numerous keen plans. The agreement process aimed at the blockchain scheme remains Practical Byzantine Fault Tolerance (PBFT). Asgari and Nemati (2022) examined written works focusing on 3 main utilization domains: decentralized ledger technology (DLT) for intelligent water frameworks, water quality supervision, and precipitation management. Moreover, they have discussed legal, societal, managerial, and pragmatic concerns that might obstruct the acceptance of blockchain. Stankovic et al. (2020) presented a summary of potential uses of blockchain in water and cleanliness sectors across Latin America and the Caribbean.

Mughal et al. (2022) conducted a pilot project focused on blockchain-based solutions to improve data-driven choice support organizations in Pakistan. The constancy, fixity, and reliability of streaming time-series data is protected by means of a personal blockchain structure constructed on the Hyperledger Fabric stage. This typically used the Pakistan Irrigation Network Independent Distributed Management Authority to evaluate the use of Hyperledger Fabric. Nodes not only collect device information for flow forecast but also maintain persistent data storage in flow records, aggregate, and verify delivery system compliance by chain-code. The Proof-of-Authority practice lets chosen nodes make novel blocks in the irrigation network that monitor flow data, thus maintaining the overall security of the chain.

A scheme in Peru has demonstrated the feasibility of participating in blockchain hooked on the water and sanitation micro-finance business to grow the use of present micro-finance models and bring the unbanked to finance. It is intended to help you include it in your system. An experimental learning tool lets for further research through first-hand experience with limited micro-finance establishments and borrowers. The project will prioritize his BanQu private blockchain network, and a feasibility study will be conducted to demonstrate operational cost efficiency situations as a product of blockchain application.

2.3.2 *Water Economy*

The study of water economics involves analyzing the financial worth of water as a natural asset and its supplementary benefits in the context of microeconomic principles. The chief objective of this method is to ensure that water quality and quantity are maintained while adhering to economic and financial limitations and to establish effective investments, expenses, and water market strategies. Due to the fundamental requirement for reliable data exchange mechanisms and currencies, water management has emerged as a significant area of research for blockchain technology. Poberezhna (2018) explains in what way blockchain-based tools can help companies and government agencies in the water sector obtain real-time information on bazaar share, ingesting drifts, customer beak organization, and extra facilities. These were checked by me. Ikeda and Liffiton (2019) presented 2 of her use bags aimed at blockchain-based water management: 1) blockchain-improper water and hygiene subsidies, and 2) the use of blockchain gears to control water prices and ingesting. Collected rainwater information is stored directly on the shackle. Thomason et al. (2018) explored the association between block-chain and weather change economics in relation to water shortage problems in poor countries and proposed the application of block-chain-improper water interchange as a possible answer. Zhao et al. (2019) highlighted the meaning of establishing a decentralized business network with blockchain functionalities, enabling additional irrigation firms to engage in a system that distributes water securely and with transparency, consequently enhancing the overall effectiveness.

Zecchini (2019) explored possible request situations for a blockchain-enabled water praise system. Grigoras et al. (2018) introduced blockchain and keen agreements into an integrated theoretic platform for water truths interchange. Angara and Saripalle (2022) systematically reviewed the virtual water work and presented a theoretical cryptocurrency and blockchain-allowed virtual water trading scheme. Sivaramakrishnan (2020) proposed a block-chain-improper planning for a water exchange platform among agrarian investors. The Ethereum free net is encouraged to meet controlling supplies over canny agreements. Vanucci et al. (2021) spread this work and goal to assist in the examination and running of flood actions after an administrative, commercial, and fiscal perception. The research also examined a sole and vigorous blockchain-based insurance system. Zhang (2022) drafted a blockchain-based theoretical model for his source chain monetary structure for water resource companies and his transformation of a potential blockchain-based digital transformation in China; he proposed Seven Waters and presented a model to analyze the company's financial situation. A blockchain-based conceptual architecture aims to store water resource data directly on the supply chain. Miller (2021) and Ramsey et al. (2020) described and then evaluated the submission of blockchain functionality in the water truths exchange system and designed various facilities that influence blockchain functionality and commercial value. Liu and Shang (2022) proposed a mixture of block-chain methods for water truths trading.

The most issues within the continuation of this venture will be that the current water bazaar and water supply conditions are not able to coordinate.

2.3.3 Management of Water Quality

The administration of the quality of water is the complete advancement of evaluating water quality from the asset to the buyer to retain water quality, and treat the water as required. Amid this improvement, reliable data and real-time evaluation are fundamental to recognize potential defilement and prepare effectiveness.

Ortiz (2018) noted the probability of block-chain requests to advance communal consciousness and government answerability for the quality of water at altogether levels in Puerto Rico. Yan et al. (2019) presented a block-chain-based ecological assessing scheme construction that synchronizes information investigation, investor verification, and water quality information. Wan et al. (2020) introduced an AI-based effluent action plant management system besides exploring blockchain as an information storing scheme.

Quist-Aphetsi and Blankson (2019) investigated the cryptanalytic properties of Protected Hash Meaning 256, the most general hash meaning in block-chain requests, and explored a possible mixture data classification scheme that cryptographically keeps the value of water abounding to consumers. Kaur and Oza (2020) documented Ethereum's demand for observation on 20 standards (ERC20). Since a smart city perception which Ethereum-based marks necessarily adhered to in this work, Water Recycling Coins (WRC) were simulated toward making a commercial situation where everybody has equivalent access to praises founded on recycled wastewater to eliminate differences between companies of different sizes. Iyer et al. (2019) proposed a secluded blockchain net founded on Hyperledger to exploit the efficiency of industrial effluent treatment schemes. The modeling process for this study involved irregularity-finding algorithms to enforce chain rubrics for potential consequences for businesses trying to tamper with water quality devices. Berman et al. (2020) presented a blockchain system, for example, documentation and classification using independent mini-boats that can automatically move and quantify water chemistry. Crawford et al. (2021) created an R3 Corda-based DLT conspiracy in subsurface oil and discussed vaccination switch (UIC) forms to protect water quality in freshwater aquifers. Gudmundsson and Hougaard (2021) developed a model that clearly shows the effect of water quality on manufacturing incomes and provides a scheme for allocating the incomes from the finest likely contamination reduction for automated trading.

This paper presented a dispersed resolution by means of keen agreements. The DApp (decentralized application) equally classified in this piece was developed via Alharbi et al. (2021). Primarily, this training was aimed at measurement of water quality limitations in manufacturing tanks and detecting any damage using the Net of Things. Later, the private Hyperledger program was used to impose appropriate consequences on sacrilegious manufacturing facilities and ensure the correctness, reliability, and slide of violation archives. This knowledge will be intelligent to measure water quality in actual time and detect violations immediately so that appropriate penalties can be imposed. The government can approve a dispersed web request to monitor the health of registered industrial facilities' water facilities and view water damage data using easy-to-read images. Shi et al. (2019) presented an advanced IoT solution using a private Hyperledger network to ensure safe drinking water ingesting at universities. It showed that 39 universities have now applied this development in the Shancheng Region of Hangzhou, China. The development will reduce the assignment of healthcare workers and encourage a shift from routine on-site inspections to automatic remote controlling.

2.3.4 Management of Agricultural Water

The management of agricultural water for sustainability ecosystem services aims for optimal water quantity and quality for agricultural products. This optimization is conceivable through ordering the information safety connected to agrarian water organization procedures. Dong and Fu (2021), Kumar et al. (2021) investigated the conceivable application of digital resolutions based on blockchain in the agrarian sector and evaluated the procedure of transformation from old-style agriculture to a digital smart agriculture system based on blockchain. Dragolinscu et al. (2021) developed a conceptual outline and conducted imitation training to supply and display nearly corporal and biochemical water limitations of irrigation systems, such as temperature, pH, dissolved oxygen, and cloudiness. Chang et al. (2021) presented a theoretical irrigation program created by Ethereum, a scheme that collects data on agrarian sources and requests in the source shackle. Krishna et al. (2021) presented the architecture of a shrewd agrarian scheme that accumulates soil humidity and temperature on the blockchain to keep agrarian information safe.

Sakthi and DafniRose (2022) proposed a private Hyperledger net to ensure transparency and information dependability for investors in the agricultural sector, encouraging them to reduce soil and water pollution with pesticides, and fertilizers. Lin et al. (2017) deliberated possible requests of blockchain and offered an ICT-based architecture and a mixture of Ethereum network integration. Munir et al. (2019) and Ting et al. (2022) proposed a keen sprinkler scheme using uncertain logic-improper decision support process procedures and data blockchain dependability. Zeng et al. (2021) modeled a scheme to manage the coordinated use of good kernels and water sources communities. In this study, an actual kernel quality intensive system keen irrigation administration scheme developed by means of the IoT and block-chain addition. Giafreda et al. (2019), Bordell et al. (2019), and Pincheira et al. (2021) introduced Ethereum-improper agrarian water administration schemes, including isolated, mixture, and community networks, separately., These 3 studies presented a framework that promotes and awards for proper conduct in agricultural actions for definite multi-stakeholder networks. The key purpose of these works is to develop water consumption for irrigation as an extra maintainable and ecologically friendly scheme. The scheme uses Ethereum communal networks and keen pacts that allow clients to occupy energy and waterfall. ERC20 was applied for transactions and 2 cryptocurrencies which included SIST (Small Irrigation System Token) and Astral Coinage.

2.3.5 *Urban Water Administration*

The goal of Urban water management is to provide and manage water services for the city. Chief rudiments of city water management are water source, drainage, and action amenities. Responsibility of the city and the organization of water can develop more stability in the context of block-chain. Makani et al. (2022), Kim et al. (2022), and Kumar et al. (2022) explored a potential application blockchain for smart cities. Blockchain can raise water security and accountability by implementing transparent and secure measures books. In addition, water amount and quality can be monitored in the ensemble water source and request procedures with storing, broadcast, treatment, and consumption. There is community mindfulness and a layer of trust among people regarding the chief contribution of blockchain and canny urban incorporation. They extended this research and created a MATLAB toolkit in which containers act out Blockchain-enabled water supply schemes. Diverse systems harmony instruments on the blockchain are associated based on their withdrawal time and allow the operator to choose the chosen option.

Lale et al. (2020) introduced the application of block-chain knowledge by k-means machine learning algorithm to protect operator secrecy. Operators are gathered in clusters besides each cluster takes a single block-chain to supply its member information. Thakur et al. (2021) presented an incentive-improper keen water building supply and storage that syndicates block-chain knowledge by advantage of a computer. In the system, houses planned as Ethereum-based nodes net the calculation works at the edge of the network and guarantees speed water-saving stimulation structure. Rottondi and Verticale (2017) proposed public keen water measuring with blockchain building and a thoughtful platform game that promotes maintainable city water ingesting. Every crowdsensing activity consumes an incentive mechanism based on chain code and the system trust level is protected by the Hyperledger fabric.

2.3.6 Water and the SDGs

Sustainable water and sanitation are the goals of SDG-6 Everyone. Mora et al. (2021) outlined the main issues of digital sustainability currency and blockchain technologies focusing on SDG6, fresh water via disinfection. Parmentola et al. (2021) "analyzed the probability of block-chain to improve environmental sustainability. Eslamian and Eslamian (2025) investigated the probable blockchain-based information storing schemes to help achieve SDGs such as weather variation, biodiversity loss, and water shortage.

2.4 Non-academic Content Analysis

This part briefly summarizes the non-academic works assessed. Online papers tend to focus on blockchain's potential hydrological applications and highlight positive outcomes. Utilizing blockchain technology to manage water data is crucial as it enhances investor trust in the availability of quality water. The alternative matter is the benefits of digitizing traditional blockchains, easier access to currently traded water markets such as responsiveness, target market information, and reduced transaction costs.

The Office for Water Regulation, a non-ministerial UK government agency, published a report on the reliability of customer data and blockchain recommendations for UK water corporations in 2017 (Ofwat, 2017). In the numerical benefit stage, BANKEX shows the Water Coin pilot project. This project is meant to attract direct investment for providing clean water safely by eliminating intermediaries or other groups that could affect the process in Kenya. Fujitsu has launched a new system for trading water using blockchain technology that allows for safe and special plant-based water. The process of getting water from plants involves turning food waste into drinking water. Companies that want to use this plant water get it from a special factory that uses technology from Fujitsu.

Table 2.1 summarizes the areas of critical review.

Table 2.1 Blockchain research in water management.

Sr. No.	*Key Area*	*Author(s)*
1	New ways to price water and make policies using Blockchain, Artificial Intelligence, and smart ideas (like smart meters and adjusting demand) can help both the energy and water sectors.	Grigoras et al. (2018)
2	A water management system to help communities use water wisely and effectively. Blockchain technology is used to create trust between people in a community and for business resources.	Bordel et al. (2019)

Contd...

Table 2.1 *Contd...*

Sr. No.	*Key Area*	*Author(s)*
3	A combined system that lets you control and watch your home network in real-time. It keeps your data safe using blockchain technology to help manage water usage better.	Phahontu et al. (2020)
4	Create a smart water system and smart water meters that provide up-to-date information on water usage. A plan to protect user privacy is created using Blockchain and Machine Learning (k-means). This plan divides users into groups, and each group has its own private blockchain to store the information of its members.	Lalle et al. (2020)
5	IoT is used to check how good the water is in factory tanks, and if there are any problems, a fine is given. Blockchain is used to keep violation records clear, trustworthy, and accurate.	Alharbi et al. (2021)
12	Blockchain helps to achieve supply chain goals such as keeping prices low, reducing risks, and ensuring quality.	Kshetri (2018)
13	Proposed a collaborative record-keeping system that prioritizes the confidentiality of trade partners while facilitating clearer decision-making.	Wu et al. (2017)

2.5 Conclusions

This section looks at the evaluation of studies that propose using blockchain technology to solve water-related issues. We reviewed 104 scientific journals and 37 non-scholarly studies about hydrology and water sources from 2017 to October 15, 2022. During this time, we noticed that the number of publications about using blockchain in hydrology has been steadily increasing. In 2020, this trend continued to rise. The study assesses the probability of the current areas of application increasing as the application of blockchain improves expertise and realization of potential novel hydrological programs.

References

1. Alharbi, N., Althagafi, A., Alshomrani, O., Almotiry, A. and Alhazmi, S. (2021, July). A blockchain based secure IoT solution for water quality management. In: *2021 International Congress of Advanced Technology and Engineering* (*ICOTEN*) (pp. 1–8), IEEE.
2. Angara, J.S. and Saripalle, R.S. (2022). Towards a virtual water currency for industrial products using blockchain technology. *Water Policy*, *24*(6), 923–941.

3. Asgari, M. and Nemati, M. (2022). Application of distributed ledger platforms in smart water systems: A literature review. *Frontiers in Water*, *4*, 848686.
4. Berman, I., Zereik, E., Kapitonov, A., Bonsignorio, F., Khassanov, A., Oripova, A., ... & Bulatov, V. (2020). Trustable environmental monitoring by means of sensors networks on swarming autonomous marine vessels and distributed ledger technology. *Frontiers in Robotics and AI*, *7*, 70.
5. Bordel, Borja, et al. (2019). A blockchain-based water control system for the automatic management of irrigation communities. In: *2019 IEEE International Conference on Consumer Electronics* (*ICCE*), IEEE.
6. Bordel, B., Martin, D., Alcarria, R. and Robles, T. (2019, January). A blockchain-based water control system for the automatic management of irrigation communities. In: 2019 *IEEE International Conference on Consumer Electronics* (*ICCE*) (pp. 1–2), IEEE.
7. Chang, Yunyan, Jian Xu and Kayhan Zrar Ghafoor. (2021). An IOT and Blockchain Approach for the Smart Water Management System in Agriculture. *Scalable Computing: Practice and Experience 22.2*, (2021), 105-116.
8. Crawford, J., Folsom, A., Vo, V., Tante, A. D., John, P. Y. and Lei, C. (2021, May). California oilfield underground aquifer injection monitoring by blockchain technology. In: *2021 4th IEEE International Conference on Industrial Cyber-Physical Systems* (*ICPS*) (pp. 283–288), IEEE.
9. Dogo, E.M., Salami, A.F., Nwulu, N.I. and Aigbavboa, C.O. (2019). Blockchain and internet of things-based technologies for intelligent water management system. *Artificial Intelligence in IoT*, 129–150.
10. Dong, N. and Fu, J. (2021, April). Development path of smart agriculture based on blockchain. In: *2021 IEEE Asia-Pacific Conference on Image Processing, Electronics, and Computers* (*IPEC*) (pp. 208–211), IEEE.
11. Eslamian, S. and Eslamian, F. (2025). *Blockchain Technology in Water and Environmental Systems*. CRC Press, Taylor and Francis, USA, 282 p.13.
12. Grigoras, G., Bizon, N., Enescu, F.M., Guede, J.M.L., Salado, G.F., Brennan, R. and Alalm, M.G. (2018, June). ICT based smart management solution to realize water and energy savings through energy efficiency measures in water distribution systems. In: *2018 10th International Conference on Electronics, Computers, and Artificial Intelligence* (*ECAI*) (pp. 1–4), IEEE.
13. Gudmundsson, J. and Hougaard, J.L. (2021). *River Pollution Abatement: Decentralized Solutions and Smart Contracts* (No. 2021/07). IFRO Working Paper.
14. Hangan, A., Chiru, C.G., Arsene, D., Czako, Z., Lisman, D.F., Mocanu, M., ... & Sebestyen, G. (2022). Advanced techniques for monitoring and management of urban water infrastructures: An overview. *Water*, *14*(14), 2174.
15. Ikeda, J. and Liffiton, K. (2019). *Fintech for the Water Sector: Advancing Financial Inclusion for More Equitable Access to Water*, World Bank.
16. Iyer, L.S. and Giri, S.V. (2020). Harnessing technology for mitigating water woes in the city of Bengaluru. In: *Journal of Physics: Conference Series* (Vol. 1427, No. 1, p. 012004). IOP Publishing.
17. Iyer, S., Thakur, S., Dixit, M., Katkam, R., Agrawal, A. and Kazi, F. (2019, July). Blockchain and anomaly detection based monitoring system for enforcing wastewater reuse. In: *2019 10th International Conference on Computing, Communication, and Networking Technologies* (*ICCCNT*) (pp. 1–7), IEEE.
18. Kaur, E. and Oza, A. (2020). Blockchain-based multi-organization taxonomy for smart cities. *SN Applied Sciences*, *2*(3), 440.

19. Kim, S., Zhang, A., Liao, R., Zheng, W., Hu, Z. and Sun, Z. (2022). Sampling blockchain-enabled smart city applications among South Korea, the United States and China. *Journal of Smart Cities and Society*, *1*(1), 53–70.
20. Krishna, K.M., Borole, Y.D., Rout, S., Negi, P., Deivakani, M. and Dilip, R. (2021, September). Inclusion of cloud, blockchain and IoT based technologies in agriculture sector. In: *2021 9th International Conference on Cyber and IT Service Management* (*CITSM*) (pp. 1–8). IEEE.
21. Lalle, Y., Fourati, L.C., Fourati, M. and Barraca, J.P. (2020, July). A privacy-protection scheme for smart water grid based on blockchain and machine learning. In: *2020 12th International Symposium on Communication Systems, Networks, and Digital Signal Processing* (*CSNDSP*) (pp. 1–6), IEEE.
22. Lin, Y.P., Petway, J.R., Anthony, J., Mukhtar, H., Liao, S.W., Chou, C.F. and Ho, Y.F. (2017). Blockchain: The evolutionary next step for ICT e-agriculture. *Environments*, *4*(3), 50.
23. Lin, Y.P., Petway, J.R., Anthony, J., Mukhtar, H., Liao, S.W., Chou, C.F. and Ho, Y.F. (2017). Blockchain: The evolutionary next step for ICT e-agriculture. *Environments*, *4*(3), 50.
24. Makani, S., Pittala, R., Alsayed, E., Aloqaily, M. and Jararweh, Y. (2022). A survey of blockchain applications in sustainable and smart cities. *Cluster Computing*, *25*(6), 3915–3936.
25. Miller, B. (2021). *Application of Blockchain Capabilities to the Management of Water Rights and Water Markets*. Information Systems Water and Catchment Group Department of Environment Land Water and Planning Victoria, Australia: Victoria, Australia.
26. Mora, H., Mendoza-Tello, J.C., Varela-Guzmán, E.G. and Szymanski, J. (2021). Blockchain technologies to address smart city and society challenges. *Computers in Human Behavior*, *122*, 106854.
27. Mughal, M.H., Shaikh, Z.A., Ali, K., Ali, S. and Hassan, S. (2022). IPFS and Blockchain based Reliability and availability improvement for integrated Rivers' streamflow data. *IEEE Access*, *10*, 61101–61123.
28. Parmentola, A., Petrillo, A., Tutore, I. and De Felice, F. (2022). Is blockchain able to enhance environmental sustainability? A systematic review and research agenda from the perspective of Sustainable Development Goals (SDGs). *Business Strategy and the Environment*, *31*(1), 194–217.
29. Pérez Ortiz, Y. (2018). *How Blockchain Technology Could Improve the Quality of Drinking Water in Puerto Rico*. Available at SSRN 3266166.
30. Pincheira, M., Vecchio, M., Giaffreda, R. and Kanhere, S.S. (2021). Cost-effective IoT devices as trustworthy data sources for a blockchain-based water management system in precision agriculture. *Computers and Electronics in Agriculture*, *180*, 105889.
31. Poberezhna, A. (2018). Addressing water sustainability with blockchain technology and green finance. In: *Transforming Climate Finance and Green Investment*. Elsevier.
32. Quist-Aphetsi, K. and Blankson, H. (2019, May). A hybrid data logging system using cryptographic hash blocks based on SHA-256 and MD5 for water treatment plant and distribution line. In: *2019 International Conference on Cyber Security and Internet of Things* (*ISCIOT)*) (pp. 15–18). IEEE.
33. Ragghianti, V.T. (2021). Tecnologia Blockchain: instrumento de gestão dos recursos hídricos em Santa Catarina. ResearchGate.
34. Ramsey, E., Pesantez, J., Fasaee, M.A.K., DiCarlo, M., Monroe, J. and Berglund, E.Z. (2020). A smart water grid for micro-trading rainwater: Hydraulic feasibility analysis. *Water*, *12*(11), 3075.
35. Rottondi, C. and Verticale, G. (2017). A privacy-friendly gaming framework in smart electricity and water grids. *IEEE Access*, *5*, 14221–14233.
36. Sakthi, U. and DafniRose, J. (2022). Blockchain-enabled smart agricultural knowledge discovery system using edge computing. *Procedia Computer Science*, *202*, 73–82.

37. Satilmisoglu, T.K., Sermet, Y. and Demir, I. (2022). Blockchain Applications and Opportunities for Water Resources and Hydrology: A Systematic Review. https://eartharxiv,orog.

39. Shi, Z., Liang, J., Pan, J. and Chen, J. (2019). How IoT and blockchain protect direct-drinking water in schools. *IEEE Internet of Things Magazine*, *2*(4), 2–4.

39. Singh, M. and Goel, S. (2020). Development of 5G enabled IoT framework for flood disaster monitoring using Blockchain Technology. *Solid State Technology*, *63*(5), 3283-3292.

40. Sivaramakrishnan, V. (2020). *Peer to Peer Energy and Water Trading in the Wheatbelt: A Sustainable Move towards Achieving Energy and Water Independence for Farm Communities* (Doctoral dissertation, Murdoch University).

41. Sriyono, E. (2020). Digitizing water management: Toward the innovative use of blockchain technologies to address sustainability. *Cogent Engineering*, *7*(1), 1769366.

42. Stankovich, M., Hasanbeigi, A. and Neftenov, N. (2020). *Use of 4IR Technologies in Water and Sanitation in Latin America and the Caribbean*. IDB.

43. Thakur, T., Mehra, A., Hassija, V., Chamola, V., Srinivas, R., Gupta, K.K. and Singh, A.P. (2021). Smart water conservation through a machine learning and blockchain-enabled decentralized edge computing network. *Applied Soft Computing*, *106*, 107274.

44. Thomason, J., Ahmad, M., Bronder, P., Hoyt, E., Pocock, S., Bouteloupe, J., ... & Shrier, D. (2018). Blockchain—powering and empowering the poor in developing countries. In: *Transforming Climate Finance and Green Investment with Blockchains* Academic Press, pp. 137–152.

45. Ting, L., Khan, M., Sharma, A. and Ansari, M.D. (2022). A secure framework for IoT-based smart climate agriculture system: Toward blockchain and edge computing. *Journal of Intelligent Systems*, *31*(1), 221–236.

46. Vannucci, E., Pagano, A.J. and Romagnoli, F. (2021). Climate change management: a resilience strategy for flood risk using Blockchain tools. *Decisions in Economics and Finance*, *44*, 177–190.

47. Wan, K., Guo, Z., Wang, J., Zeng, W., Gao, X., Shen, Y. and Yu, K. (2020, August). Deep learning-based management for wastewater treatment plants under blockchain environment. In: *2020 IEEE/CIC International Conference on Communications in China* (ICCC Workshops) (pp. 106–110). IEEE.

48. Wu, G., Li, E. and Wang, M. (2022, July). Application and prospect analysis of blockchain technology in water resources protection. In: *2022 International Conference on Blockchain Technology and Information Security* (*ICBCTIS*) (pp. 182–187). IEEE.

49. Xia, W., Chen, X. and Song, C. (2022). A framework of blockchain technology in intelligent water management. *Frontiers in Environmental Science*, *10*, 909606.

50. Yan, J., Zhang, F., Ma, J., An, X., Li, Y. and Huang, Y. (2019, October). Environmental monitoring system based on blockchain. In: *Proceedings of the 4th International Conference on Crowd Science and Engineering* (pp. 40–43).

51. Yasuno, T., Ishii, A., Amakata, M. and Fujii, J. (2020). Smart dam: Upstream sensing, hydro-blockchain, and flood feature extractions for dam inflow prediction. In: *Advances in Information and Communication: Proceedings of the 2020 Future of Information and Communication Conference* (*FICC*) (Volume 1, pp. 139–158). Springer International Publishing.

52. Zecchini, M., Chatzigiannakis, I. and Vitaletti, A. (2019). *Data Collection, Storage, and Processing for Water Monitoring Based on IoT and Blockchain Technologies* (Doctoral dissertation, University of Rome).

53. Zeng, H., Dhiman, G., Sharma, A., Sharma, A. and Tselykh, A. (2023). An IoT and Blockchain-based approach for the smart water management system in agriculture. *Expert Systems*, *40*(4), e12892.

54. Zhang, R. (2022). Research on financial development of water resources enterprises based on blockchain technology. *Mobile Information Systems*, *2022*(1), 3289301.
55. Zhao, G., Liu, S., Lopez, C., Lu, H., Elgueta, S., Chen, H. and Boshkoska, B.M. (2019). Blockchain technology in agri-food value chain management: A synthesis of applications, challenges, and future research directions. *Computers in Industry*, *109*, 83–99.

3 Theory and Methodology of Blockchain

Yaser Sabzevari[1*] and Saeid Eslamian[2]

Blockchain technology has emerged as a foundational framework for secure, transparent, and decentralized data management. This paper presents a comprehensive examination of the underlying theory and methodology of blockchain systems. Fundamental concepts such as cryptographic hashing, distributed ledger structures, peer-to-peer networking, and consensus algorithms are reviewed to establish the theoretical basis of blockchain functionality. Methodological approaches, including protocol design, smart contract development, performance evaluation, and security analysis, are discussed to demonstrate how blockchain systems are built, validated, and optimized. The study also highlights methodological challenges related to scalability, interoperability, governance models, and system verification. By integrating theoretical principles with practical design methodologies, this work provides a holistic understanding of how blockchain systems operate and how they can be effectively engineered for diverse applications.

3.1 Blockchain Theory

Blockchain technology is being noticed more and more for its capacity to create a reliable and shared space for different activities. Its importance is especially clear in the supply chain industry, where it is very important to know the exact location, check where products come from, and keep trustworthy delivery records. Also, using blockchain in the water industry is becoming more popular because it can help improve transparency and efficiency in how water is managed.

3.2 Blockchain Structure Feature (Christian et al., 2018)

- Offers benefits to all participants involved in the supply chain, including manufacturers, brokers, transporters, processors, wholesalers, retailers, and consumers.
- Enables various advantages for each participant, such as increased value, improved access to insurance and finance, and enhanced capabilities.

[1,2] Department of Water Science and Engineering, College of Agriculture, Isfahan University of Technology, Isfahan, Iran.
* Corresponding author: yasersabzevari1511@gmail.com

DOI: 10.1201/9781003686026-3

- Utilizes blockchain technology, a decentralized ledger system, to facilitate the recording and storage of transactions. Each transaction is stored in a cryptographically signed block structure that is replicated across the network.
- Ensures data integrity and trust through mathematical validation mechanisms.
- Allows every participant to contribute to consensus building by acting as a node in the network. Consensus is achieved through the execution of complex algorithms.
- Stores essential information and detailed transaction data within each block.

3.3 General Rules

- Product manufacturers are required to register on the blockchain platform, providing specific details such as location information and status updates on product components.
- Each product is assigned a unique product ID and includes information about its embedded water, which represents the approximate amount of water used in its production.
- During the product assembly process, manufacturers consider factors like quality and price when selecting components. However, they also assess the "embedded water" parameter to make informed decisions regarding the assembly.

  Blockchain technology safely keeps track of all transactions in the network, so there's no need for a middleman to verify them. This improves productivity and trust in the supply chain system.
- Each owner can see information from other owners, which can help make the process of matching data easier and cheaper. This creates openness and trust. This also lowers the chance of fake products. This lets everyone involved approve of the clever agreement, which is an important part of blockchain. It makes everyone involved agree to rules and steps that can help make sure the agreement is followed and checked.

You can store very private information in a separate secret record, along with the public record. The private ledger is only shared with trusted partners. The partners follow an agreement called a Memorandum of Understanding (MoU). This feature is not required and is shown in the diagram (Figure 3.1). It shows a picture of how a blockchain transaction works. In making a product, many parts are used, and each part comes from a different place or area. Also, these parts might include other smaller parts, creating a chain that can reach back to the use of basic materials. Please remember that the simplified image does not show the details of how products move along the supply chain. To build trust and be clear, all competitors should share their water usage information. This information is very important for building trust between the people involved. Once information is put on the blockchain, it can't be changed or deleted. So, the people in charge of parts need to be careful when sharing their data. It's going to be hard to change the information later because the blockchain doesn't allow changes. This, in turn, helps the participants take responsibility for themselves.

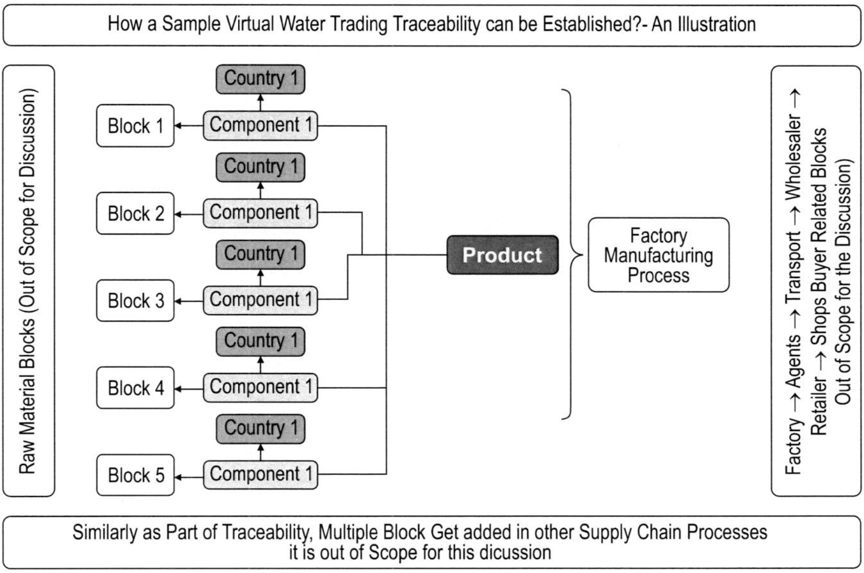

Fig. 3.1 Diagram of virtual transaction block of water-blockchain (Jayasri et al., 2022).

A usual block has the company ID, location, water used to make a part, and a link to other blocks with different parts (since a product is made of different parts), if there are any, as shown in Figure 3.2.

3.4 Key Points of the Process

- Transactions on a blockchain are confirmed and secured using a method where many people agree, making it safe with encryption.
- Transactions are organized into blocks, and miners check them to make sure they are correct. Miners act like outside helpers.
- When a new transaction happens, it is shared with computers all around the world. These computers work on complicated math problems and check if transactions are approved, then they add them as blocks to the chain.
- The first block in the chain is called the Genesis block, and each block after it is connected to the one before it using a special key.
- Each block has a special code called a hash, and if anything in that block is changed, it will become invalid and affect the whole chain, making it very secure against tampering.
- Hashing and proof of work together keep the blockchain safe. Proof of work makes it harder to create new blocks, which slows down the process. This action stops hackers from easily controlling the whole system.
- Consensus mechanisms, like Proof of Work (PoW) and Proof of Stake (PoS), are used to help all the computers in the network agree with each other. This helps make the network more secure.

- The blockchain uses strong security and agreement systems, and it works on a shared network where everyone involved has their own copy of the whole blockchain. This copying process makes security better. Each computer that joins the network is called a node. When a new block is added, every computer checks it completely before it is accepted. This process of working together to make an agreement is called consensus. As shown in Figure 3.3, all the parts of the network agree to add each new block. If a changed block is added, the nodes will not accept it because of the rules they follow. This makes sure that only correct and unchanged blocks are added to the blockchain.

Fig. 3.2 Virtual water data structure of blockchain (Jayasri et al., 2022).

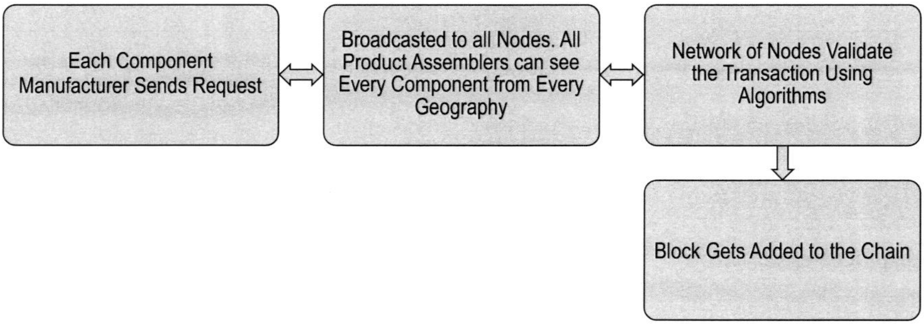

Fig. 3.3 Water-blockchain virtual transaction process (Jayasri et al., 2022).

The utilization of virtual water in the production or importation of goods emphasizes the significance of considering the temporal and spatial value of water. It plays a crucial role in shaping water policies at both governmental and organizational levels, providing a framework for defining guidelines and strategies concerning water management.

3.5 Methodology

As digital technology and blockchain have improved, using computers has become essential for making systems and processes better. New technologies like blockchain are quickly taking the place of old systems and helping society move forward faster. A blockchain, which is also called a distributed ledger, is a digital record system that keeps changing and connects different parts using common rules. Each block in the chain has a special code from the last transaction, the time it happened, and details about the exchange. The way distributed ledgers are built makes sure that information can't be changed. This is why blockchain can be used in many different industries and fields (Table 3.1).

Instead of thinking of blockchain as something that disrupts existing systems, the main challenge is to see it as a comprehensive solution that can provide various business opportunities. The main benefit of distributed ledger technology is that it can help create new action plans on various platforms and make resources digital (Swan, 2015). Blockchain is a new technology that helps companies and organizations change how they work and do things in a better way. In farming water services, blockchain technology could help make support, rules, and legal tasks easier. This includes things like billing customers and managing water resources online.

Table 3.1 Blockchain applications.

Class	*Examples*
Currency	Bitcoin, Ethereum, Litecoin, Monero, Dash, ICON, XRP, Hypercash
General	Escrow transactions, bonded contracts, third-party arbitration, multiparty signature transactions
Financial transactions	Stock, private equity, crowdfunding, bonds, mutual funds, derivatives, annuities, pensions
Public records	Land and property titles, vehicle registrations, business licenses, marriage certificates, death certificates
Identification	Driver's licenses, identity cards, passports, voter registrations
Private records	IOUS, loans, contracts, bets, signatures, wills, trusts, escrows
Attestation	Proof of insurance, proof of ownership, notarized documents
Physical asset keys	Home, hotel rooms, rental cars, automobile access
Intangible assets	Patents, trademarks, copyrights, reservations, domain names

Source: Adapted from Swan (2015, p. 10).

Table 3.2 Summary of registered documents of proof of existence.

Sample Document Digest	*Timestamp*
320c891df43f3505acdfa3bd4db02fbce691fdda6 8e10bfa65c42902cbfc7cc1	2020-02-11 09:09:54
45ff928c5364b1556bf2480bc64cf74bc88e6ec140 fe5648d368fdeae3091f8b	2020-01-29 01:05:00
071f1ef9de294ebab5858b06f4ee1b54fd8ec748653 962cce289298518d6c517	2019-11-30 03:55:30
c8ef33b52bd36c975ec91a7e29307e46b4 cdf22777d80e3ca2240d5386d4b580	2019-09-03 14:22:35

Source: Calculated using SHA-256 Hash.

Instead of needing separate plans for each organization to put measures in place and store data, blockchains create a shared space where organizations can work together with their partners. This often creates confidential conditions between members in many manual procedures and procedures that require compromise between members. The procedures related to money, it is about income and expenses. It is the creation of a circular economic activity with complex revenue models that will ultimately be economically and biologically flexible.

A present situation of a dispersed record submission is the criticality of discounting that works on diverse ledgers to provide the required levels of safety, immutability, and consistency. This implementation will reduce the costs of discounted work vitality for workplace re-exchange. In the same way, blockchain does more than just set rules and give approvals for water exchange. By keeping track of these items in a shared record, information about other parties involved can become easy to handle, a process that often costs millions of dollars in this industry.

3.6 Applications of Blockchain for the Water Industry

Recent studies have looked into how blockchain technology can help solve problems with distribution, reduce flood risks, and keep track of water management systems in new ways. Blockchain can be a useful tool for keeping records. It helps water organizations keep track of rules and agreements better than the methods they use now (Poberezhna, 2018). All data, like usage and value records, can be safely kept in a permanent ledger that cannot be changed. By using blockchain and smart contracts, water companies and government offices can quickly access information about different parts of the industry, their plans, how they manage operations, customer billing, and other viewpoints. Using a blockchain system for keeping records can greatly lower the costs compared to old-fashioned ways of record-keeping.

Also, blockchain technology can help lower the risk of data breaches. Setting up a shared record system on a network of computers, where each one keeps a copy of the records, allows regulators to keep track of and control the status of water permits, and access to assets step-by-step. This method improves security and makes sure that everyone involved in managing water resources gets the same information. It moves away from the old way of central administration. The safe and decentralized record system in blockchain technology ensures security and helps people share information easily in the private water resource management community.

This argument shows that new technology, like distributed ledgers, can help improve the way we manage water exchanges. This is important in a sensitive area like water management. This is achieved by making information clear to reviewers and controllers by sharing accounts and activities on the blockchain. To help with water management, communities need a digital representation and a way to keep track of water resources. Blockchain technology can be used for recording information, ensuring rules are followed, and checking records. A shared record system in the water industry can help keep track of the tools needed by government rules. Writing down activities and their results permanently on a blockchain helps keep a clear record for managers and makes it easier to connect different systems. This change would save a lot of time and effort (and money) for financial institutions by making the administrative work easier. At the same time, it would improve quality, accuracy, and reliability.

Blockchain water management provides reliable, fast, and clear information, allowing partners to be involved in real-time instead of just receiving results after the fact. The new plan for some parts of the distributed ledger can create great but lightweight systems that have central features. These include easy adjustments to different systems, strong protection of data for high security, and flexible copying in different places.

Finally, bonds can serve a purpose in the blockchain dedicated to water management. Improving the use of distributed ledger technology in the water industry can be done by creating a shared platform for various offices involved in approving requirements, issuing bonds, following government rules, and protecting customer interests. This includes handling green bonds and allowing the exchange of bonds. Keep the green loan separate. Water management can use blockchain to gather money and handle investments in areas related to water.

It is thought that these actions can be seen as a smart way to manage water. Smart water management means using technology to quickly and automatically gather information about water resources, the environment, weather, and climate. This helps us use water more effectively (Dogo et al., 2019). Many countries have set up smart water management systems to help with early warning for floods and droughts.

Using smart management for water resources can be very helpful and has a lot of possibilities (Dogo et al., 2019). This involves making sure we use the right amount and quality of water for irrigation, keeping an eye on the safety of natural resources, and handling risks connected to water-related disasters, like floods and droughts. This information can later help solve problems related to managing water resources effectively. Another use is to keep track of water infrastructure in real-time through safety checks and projects to improve dams. This includes finding different ways to make the best use of water resources.

3.7 Blockchain Technology in Disaster Management

Blockchain technology offers a secure, transparent, and reliable method for storing and sharing information without the need for a central governing body. It is used in digital money, helping with disaster aid, online government services, health care, environmental studies, and managing products (Sheela and Priya, 2021). The main goals of a disaster management system using blockchain technology are to: (1) check the accuracy of information from various sources, (2) combine data from different sources to give clear and trustworthy information, (3) keep the personal information of victims safe and secure, and (4) create a cost-effective system for managing disasters that also protects people's identities.

Information systems have gathered information about natural hazards with some level of automation. The information gathered from various technologies at different times during the disaster is scattered, and there is no system that brings together data from suppliers and local authorities. Since the systems work separately, we can't ensure that the data is accurate or reliable. Because of this, teams managing disasters are having a hard time making a clear way to give out victim records and financial transactions.

3.8 Digitization of Water Resources Management

Recent reports, like the World Bank's *Climate Change, Water, and the Economy* show that not having enough clean water, worsened by environmental changes, impacts a few regions but affects a large part of the economy. It impacts the economy (GDP), increases movement of people, and raises the chance of conflict.

Water is very important for life and for the food system. About 70% of the world's water is used for farming, 20% for industries, and 10% for everyday household use. People's ways of earning a living use 15% of the world's water, and this might grow by 20% in under 15 years. To help improve the economy, we need to look at how safe and accessible water is, as it is closely connected to our nutrition, and how we use these water resources. In the same way, water funding needs interest in plans, technology, organizations, and connections. These complicated tasks can be handled with technologies like blockchain, artificial intelligence, or big data (Table 3.3).

There are a few problems in the water management area based on local indicators. These problems include floods, droughts, lack of water, old and broken buildings, water quality issues, wasted water, project delays, and poor management of water use. The most important task is to take accurate measurements and understand the factors that affect water. This helps to manage current water supplies, improve water quality, reduce pollution, and strengthen the overall water system. The water sector, different from other Sustainable Development Goals (SDGs), needs careful planning, improvement, and teamwork (Iman, 2019).

Table 3.3 Sustainability criteria in water management.

Criteria	*Indicators*	*Technology*
Ecological sustainability	• Annual recharge rate • Pollution rate	Al, big data
Social sustainability	• Water-related disease • Access to water	Al, big data
Economic sustainability	• Cost price of water • Efficient use of water • Highest economic return	Blockchain, IoT

Contd...

Table 3.3 *Contd...*

Criteria	*Indicators*	*Technology*
Technical sustainability	• Durability of the water system • Promcting efficient use of water system	Blockchain, IoT, AI
Political sustainability	• Integrating water policies • Institutional capacity • Open and democratic planning	Blockchain

Despite the critical role of water worldwide, it tends to be overlooked in favor of other important rescurces. Low income and benefits, unclear issues, outdated laws, partial ownership, complicated rules, connected business interests, and political reasons can all make water management less effective. A World Bank study from 2016 found that people were interested in this sector in developing countries. They invested \$1.9 billion in 2016, which was about half of the usual amount of \$4 billion over five years. The report says that the water sector made only \$669 million in benefits for private companies, which is a big decrease from the average of 88% over the last five years.

Food security relies heavily on water availability, and since the demand for food will double in the next 20 years because of population growth, it is essential to finance limited resources like water for farming. This is especially important for areas that depend on water or have nutritional needs. There are a few action plans that can help improve water security and sustainable development: (1) using boreholes to help with food security and water availability, (2) creating a plan to lower water risks over time, (3) working together with different groups to achieve common goals, and (4) raising money by taking manageable risks. Suggested overall plan can be seen in Figure 3.4.The main money and environmental problems with rare assets are getting smaller. This important problem needs a new and well-thought-out approach. This approach should be used in both government and private businesses, and then used again in government. Because problems in water management can be unexpected, we need a new and better plan that uses green funding, new technology, and other important parts of managing water. Bringing together different community groups and private organizations is key to reaching sustainable development goals. This includes improving industries, making cities better, using resources wisely, and creating a supportive environment for these goals.

A private shared ledger will be used to improve current and future records. It will make the audit process better and easier, and help provide clear and consistent information when needed, especially about natural impacts and support lists related to corporate social responsibility (CSR). CSR, ESG, and detailed assessment tools (besides Thompson Reuters, MSCI, Trucost, or S&P Dow Jones) give managers the information they need to change how they act.

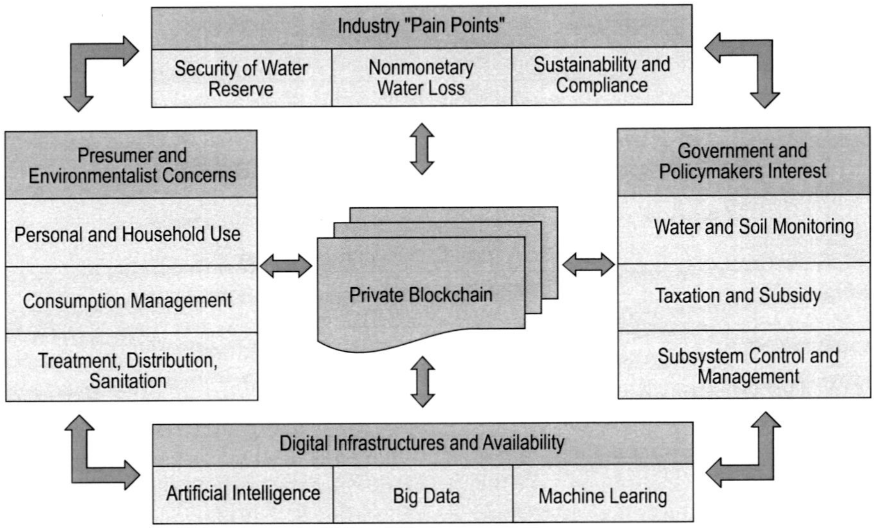

Fig. 3.4 Proposed framework for water management blockchain (Jayasri et al., 2022).

Using new technology like a shared record system lets us check information and data in a way that everyone can see. This brings clear openness and trust to the market. The ability to win, gather, and trade transactions is done, checked, and saved in a shared record. The water management blockchain helps protect against false information or lies. The ability of water to move and change is important for keeping water systems able to provide income during very dry times. It's also necessary for coping with slow changes in the environment.

Distributed ledgers help manage water by providing financial rewards for governments, boosting income for farmers, supporting sustainable practices, and allowing extra water to benefit the environment. To get results quickly, everyone needs to work together, and both public and private organizations should have a plan. The idea of blockchain could be a great way to think creatively and work together to address the important problem of not having enough water.

In summary, we can confidently say that the distributed ledger can be used for managing water use as well as for water recycling. For example, we could set up direct water trading platforms where people can exchange treated wastewater and rainwater. This could help create a bigger market for buyers. People can get wastewater to clean, reuse, recycle, or throw away at any stage of water production. So, blockchain can really change how we handle water resources. It can help us use clean and fresh water better, manage payments for water use, and track water usage over time.

3.9 Hydro-Blockchain for Flood Forecasting

To make flood predictions more accurate, we need to use different information like rainfall, water levels, and man-made water sources. Different people and groups are part of this event, and each one handles the data in their own way using their own systems. Data isn't always shared and can come with costs due to rules about using private and sensitive information. To fix this problem, we need to create a system that uses data safely and fairly.

In this section, we look at a system called "Hydro-blockchain for Flood Forecasting" (see Figure 3.5). This system uses blockchain technology to collect and use large amounts of data about water conditions over time and space. This system helps manage, share, and exchange data safely using blockchain technology.

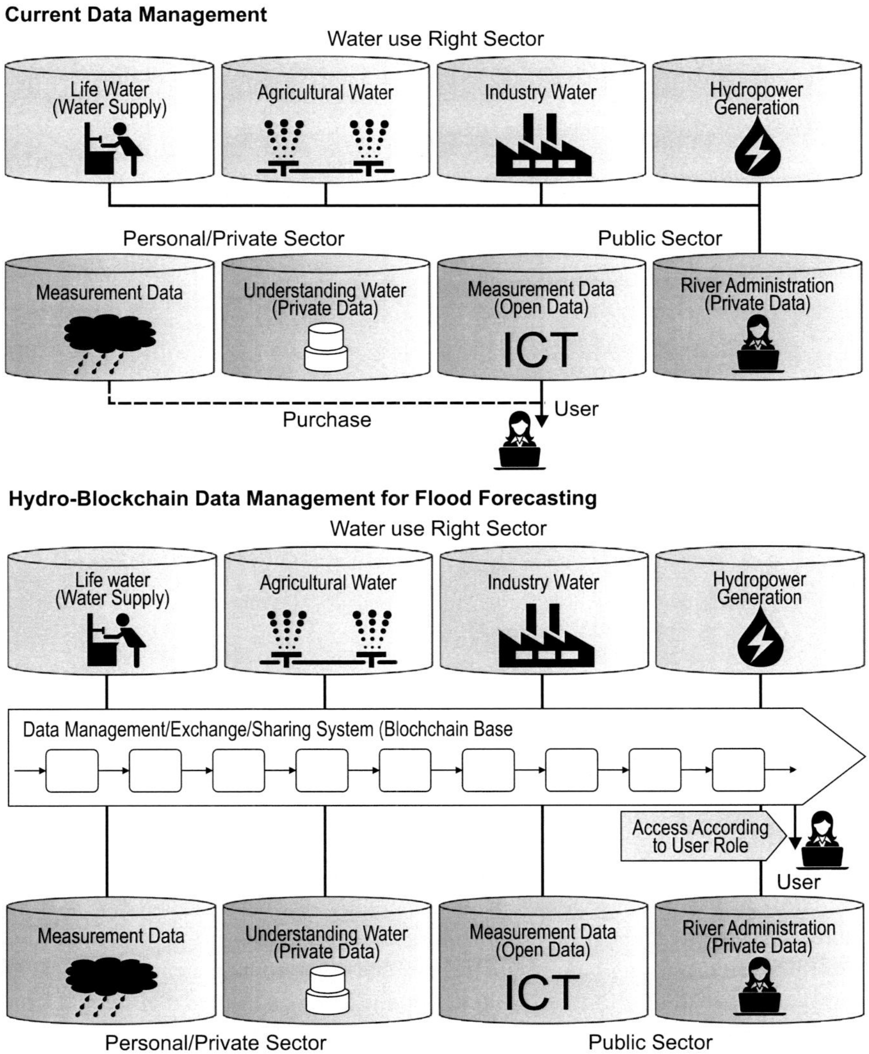

Fig. 3.5 Hydroblockchain data management schematic (Yasuno et al., 2020).

Figure 3.6 shows a simple picture of how data from water level sensors is saved to a database and used with the hydro-blockchain. The sensor ID labels the data, and it gets saved in the database using blockchain every 10 minutes. The person using the information has permission to use it based on their job. Because data is shared using secure transactions, it can be used easily and fairly while keeping privacy and data protection in mind. This system helps keep important data about floods from getting lost or mixed up, and it also makes predictions about how much water will flow into dams more accurately. Also, by using smart contracts, sharing information like dam water flow forecasts will help a) quickly and smartly coordinate between different groups with different interests, and b) ensure that disaster prevention information is shared easily and on time. He does it. The main takeaway is that this system is effective in multiple drainage zones, not solely in one.

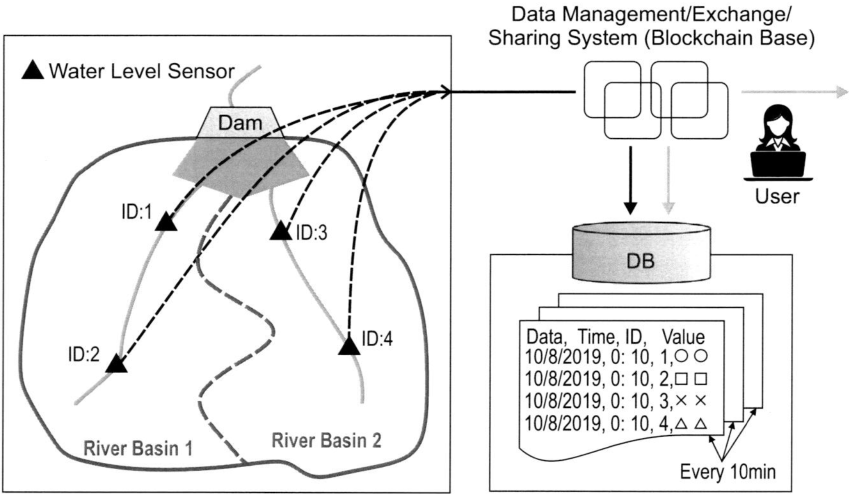

Fig. 3. 6 Schematic of the data recording process from water level sensors to the database (DB) and using the data through the hydro block chain (Yasuno et al., 2020).

3.10 Blockchain-based Technology for Flood Risk Management

Blockchain is a permanent digital record that keeps track of transactions (Poonia et al.,2021). Even though blockchain technology has been around for a while, studies on how it can be used for managing floods are just starting (Eslamian and Eslamian, 2025). Using blockchain technology for flood management helps keep and organize a lot of information about flood victims. It ensures their privacy and keeps costs low. This technology lets people work on their own without needing a central authority or a middleman. This method makes the system work better by lowering transaction costs. Blockchain is helpful in disaster management because

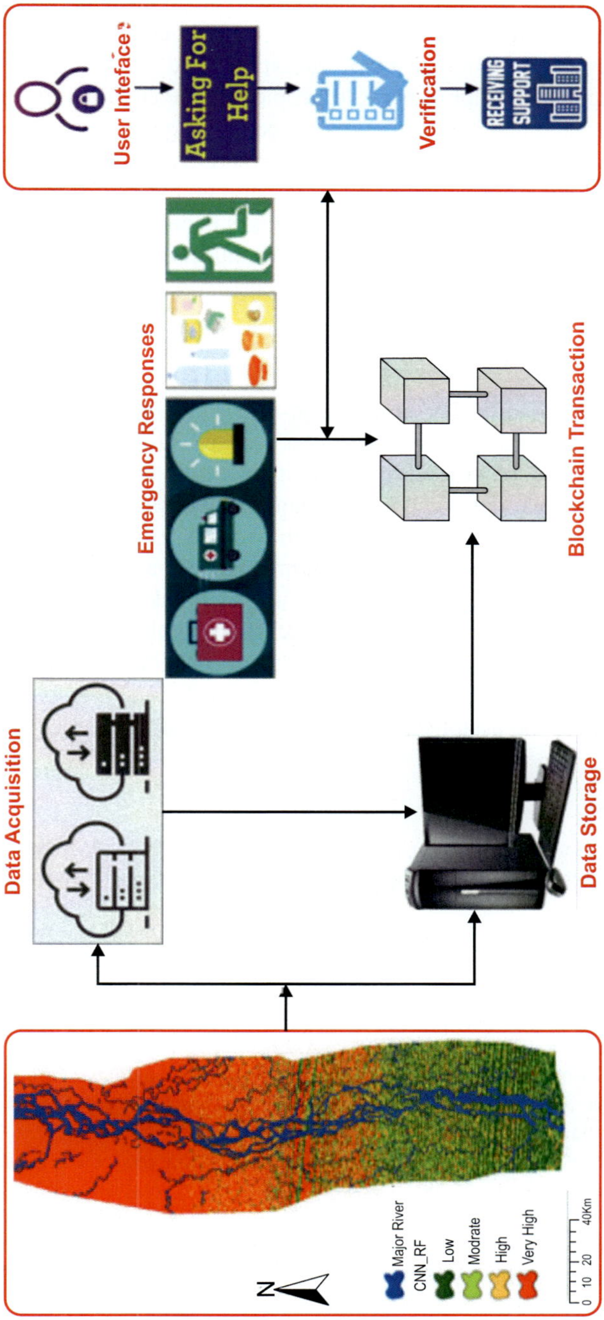

Fig. 3. 7 An overview of blockchain applications for flood risk management.

it makes emergency evacuations clear and open to the public. (Dogo et al., 2019; Engin and Treleaven, 2019). This technology has caught the interest of people from many different industries, including healthcare, smart cities, etc.

Many natural and human-made disasters, like the ongoing COVID-19 pandemic, often cause a lot of harm to people and the environment globally. It's essential to develop technical measures aimed at mitigating the effects of natural disasters. Blockchain technology offers numerous advantages for simplifying data access, as it allows information to be stored immutably and shared in real-time. For disasters and emergencies, blockchain makes it easy to collect data and check how well emergency actions are working (Vannucci et al., 2021; Nguyen et al., 2021; Gupta et al., 2021).

3.11 Conclusion

Currently, water rights play a crucial role in the utilization of water resources. Managing water rights trading wisely is crucial for protecting the environment and supporting the economy. Even though it's not fully used yet, blockchain is an important new technology trend today. Using blockchain makes it easier to trade water rights for farming, helping manage them better. Blockchain is an emerging technology that is expected to play a significant role across various industries. We consider the issues surrounding the use of blockchain in business to be significantly important for academic study and community initiatives. We propose a few interesting areas for research. For instance, understanding how blockchain enhances company and market operations, safeguards and protects data, manages digital currencies, and establishes guidelines for the online marketplace.

References

1. Dogo, E.M., Salami, A.F., Nwulu, N.I. and Aigbavboa, C.O. (2019). Blockchain and internet of things-based technologies for intelligent water management system. *Artificial Intelligence in IoT*, 129–150.
2. Engin, Z. and Treleaven, P. (2019). Algorithmic government: Automating public services and supporting civil servants in using data science technologies. *The Computer Journal*, *62*(3), 448–460.
3. Eslamian, S., and F., Eslamian (2025). *Blockchain Technology in Water and Environmental Systems*. CRC Press, Taylor and Francis, USA, 282.
4. Gupta, V. and Consensys, L.L.C. (2017). *Building the Hyperconnected Future on Blockchains*. World Government Summit. http://internetofagreements. com/files/WorldGovernmentSummit-Dubai2017. pdf.
5. Iman, N. (2019). Revisiting sustainability with a business management lens. *Sustainability: The Journal of Record*, *12*(1), 34–39.
6. Nguyen, G.N., Le Viet, N.H., Elhoseny, M., Shankar, K., Gupta, B.B. and Abd El-Latif, A.A. (2021). Secure blockchain enabled Cyber–physical systems in healthcare using deep belief network with ResNet model. *Journal of Parallel and Distributed Computing*, *153*, 150–160.

7. Poonia, V., Goyal, M.K., Gupta, B.B., Gupta, A.K., Jha, S. and Das, J. (2021). Drought occurrence in different river basins of India and blockchain technology based framework for disaster management. *Journal of Cleaner Production*, *312*, 127737.
8. Sheela, K. and Priya, C. (2021). *Enhancement of Iot Security Solutions Blockchain Technology. Blockchain Applications for Secure IoT Frameworks: Technologies Shaping the Future*. India: Bentham Science Publishers.
9. Vannucci, E., Pagano, A.J. and Romagnoli, F. (2021). Climate change management: A resilience strategy for flood risk using Blockchain tools. *Decisions in Economics and Finance*, *44*, 177–190.

4 | Encryption

Saeid Eslamian[1] and Yaser Sabzevari[2*]

Encryption plays a central role in modern information security, providing mechanisms to protect data confidentiality, integrity, and authenticity in digital communication and storage systems. This chapter presents a comprehensive overview of encryption concepts, principles, and techniques. It begins with the theoretical foundations of cryptography, including the distinction between symmetric and asymmetric encryption, key management, and essential mathematical underpinnings. Major encryption algorithms, such as AES, RSA, and ECC, are introduced, along with their strengths, limitations, and typical applications. The chapter also explores practical considerations such as secure key distribution, cryptographic protocols, end-to-end encryption, and the challenges posed by emerging technologies like quantum computing. By integrating theoretical knowledge with practical examples, this chapter aims to equip readers with a clear understanding of how encryption safeguards digital systems and why it remains indispensable in cybersecurity, blockchain networks, communication systems, and data protection frameworks.

4.1 Introduction

"Cryptography" is a knowledge that investigates and recognizes the principles and methods of transferring or storing information in a secure manner (even if the path of information transfer and communication channels or the location of information storage are insecure) (Panwar et al., 2023).

Cryptography utilizes mathematical techniques to secure data and protect it from unauthorized access. Encryption is a way of changing a message or information so that only someone with a password can understand it. Only someone who knows the key and the method can understand the important information from the encrypted data. If someone doesn't know either the key or the method, they can't access the information (Chai et al., 2022; Eslamian and Eslamian, 2025). Cryptography is built on ideas from information theory, number theory, and statistics. Today, it is mainly studied and applied in the field of telecommunications. Cryptography is a very popular technology used all around the world today, and it has some interesting technical aspects (Xiao et al., 2021; Chai et al., 2022). Cryptography has been around for a long time and started in ancient Greece and

[1,2] Department of Water Science and Engineering, College of Agriculture, Isfahan University of Technology, Isfahan, Iran.
* Corresponding author: yasersabzevari1511@gmail.com

DOI: 10.1201/9781003686026-4

the Roman Empire. Long ago, people sent messages using a code where they moved letters around. Only someone with a special key could read the messages. Today, encryption is used in many tech applications, like personal messaging, keeping company data safe, and sending money. This chapter starts by discussing the history of cryptography, then explains how cryptographic algorithms work, and lastly looks at the different types of cryptography.

4.2 Encryption

Encryption serves as a method to protect a message or file by employing a password for security. By using this method, only specific people can see the message or file. Encryption starts by mixing up the input data. Then the receiver uses a code to unscramble the data. This process is called Decrypt (Li et al., 2023).

Keys play a crucial role in the field of cryptography. Nowadays, the creation of keys is accomplished through computer software that functions similarly to random number generators. In contemporary times, keys are produced with the help of computer programs that operate like random number generators (Wang et al., 2022). For example, some of today's sophisticated algorithms use the user's mouse movement to generate a random phrase as an encryption key.

In other words, in cryptography, the existence of information or sending a message is not hidden at all, but the storage of information or the sender of the message is known, but only the intended people can retrieve the original information. On the contrary, in secrecy, the essence of the existence of information or the sender of a secret message is kept hidden, and no one other than the sender and the receiver knows about the sender of the message (Gan et al., 2022).

In encryption, the content of a text is changed letter by letter and in some cases bit by bit, and the goal is to change the content of the text, not to change its linguistic structure. On the other hand, encoding is a transformation that replaces a word with another word or symbol and changes the linguistic structure of the text (Su & Wang, 2022).

In ancient Greece, a tool called Scytale was used for easy encryption of messages using displacement encryption. Shift encryption is done by changing the letters of a message and moving them around. Messages sent using encryption could not be read or interpreted even if discovered by other people and opened. Of course, such a style of encryption could be deciphered with a little effort. However, this method is known as the first encryption method in history.

Julius Caesar invented a special method for encryption, which is also known as Caesar's cipher. In this method, which is considered one of the simplest encryption methods, each letter of the alphabet is moved to the right or left side of the list of letters by a specified number. For example, with the Caesar cipher, the word GEEK is written as JHHN by shifting the English letters four places to the right of the list. The recipient of the encrypted message could easily interpret the message only by knowing the Caesar code, i.e., the number required to move the position of the letter. Other people would have had a challenge understanding the message without having the code.

Another traditional encryption method is known as Polybius Square. This square specified a position for each letter in certain rows and columns. As a result, instead of writing letters, numbers were used in the messages. Decoding was possible by accessing the main square that showed the position of each letter.

4.3 Enigma Car

During World War II, the Nazi army used professional encryption machines called Enigma. These machines created many challenges for the Allied forces and the possibility of decoding them was almost zero. It took years for Allied scientists led by Alan Turing to decipher the advanced Enigma machines. Historians say that deciphering Enigma machines played a vital role in the victory of the Allies during World War II.

4.4 History of Modern Cryptography

Modern cryptography methods have many complexities. It seems a bit difficult to explain them in ordinary language, but we will try to explain in the best possible way some of the conventional concepts in the world of cryptography. Horst Fistel and his colleagues at IBM were among the first to develop modern cryptography in messaging-focused electronic communications. They introduced the concept of pattern ciphers, which operate on fixed-length groups of bits (called blocks or patterns).

The United States was a pioneer in standardizing encryption methods. In 1976, they first adopted the Data Encryption Standard, or DES, as a format cipher. The American National Standards Organization adopted DES as the Federal Data Processing Standard that year. In the following years and decades, other countries gradually adopted the DES standard.

The original DES standard was not efficient enough for security professionals. Security concerns and slow performance have prompted researchers to develop modern and more efficient methods. Several frame ciphers were developed in the 1980s and 1990s, the most important of which are RC5, Blowfish, IDEA, NewDES, SAFER, CAST5, and FEAL. Different types of encryption algorithms with different goals and uses have different speed and performance (Ponuma et al., 2019).

Entering the new century, the need to implement and use more advanced encryption methods was felt more than ever. DES had failed in various instances of decryption, and its security as a comprehensive standard could no longer be guaranteed. Vincent Ridgeman and John Daman were cryptographers who developed a new method with higher security. They used a combination of their names to name the new method, and finally, Rijndael was born. After some time, the United States government approved this method as a more advanced and newer encryption standard, and the name AES or Advanced Encryption Standard was chosen for it. During the approval of the new method, 15 other algorithms were present until finally Rijndael was chosen as the standard method in 2001 (Ping et al., 2022)

4.5 Encryption Algorithm

Cryptographic algorithm is any algorithm or mathematical function that is used in cryptographic protocols due to having properties required in cryptography. The term cryptographic algorithm is a comprehensive concept and it is not necessary that every algorithm of this category is directly used to encrypt information, but only the existence of cryptographic applications is considered. In the past, organizations and companies that needed encryption or other encryption services used to design a unique encryption algorithm. Over time, it became clear that sometimes there are big security weaknesses in these algorithms, which makes it easy to break the password. For this reason, today encryption based on keeping the encryption algorithm hidden is obsolete, and in new encryption methods, it is assumed that the complete information of the encryption algorithm has been published and what is hidden is only the password key.

Therefore, all security obtained from standard encryption algorithms and protocols relies on the security and concealment of the encryption key, and the full details of these algorithms and protocols are published to the public (Li et al., 2023). According to the definition above, the functions and methods used in cryptography can be grouped into these main categories:

- Functions without keys
  - Scrambling functions*
  - One-way transformations*
- Key-based functions
  - Symmetric key algorithms
  - Pattern cipher algorithms
  - Sequential cipher algorithms
  - Message acknowledgment functions*
- Asymmetric key algorithms
  - Algorithms based on the analysis of integers
  - Algorithms based on discrete logarithm
  - Algorithms based on elliptic curves
- Digital signature algorithms*

There are many encryption algorithms, but only a few of them have been standardized.

4.6 The Difference between Cryptographic Algorithms

As mentioned, there are many cryptographic algorithms in the world today, each of which is used for specific tasks. The difference between cryptographic algorithms can be seen in simple language in two main characteristics. They differ in the degree of protection of the data and the original message, as well as the speed and efficiency in encryption and decryption.

To compare the speed of different cryptographic algorithms, the benchmark tool provided in TrueCrypt can be used. Benchmarks show that AES is still the fastest in cryptography. However, other algorithms have advantages and disadvantages compared to the standard method in terms of power or speed, and the choice among them is highly dependent on the user's needs and efficiency. For example, a user can use faster algorithms for traditional encryption of their data or encrypt data twice with the same algorithm.

4.7 Encryption Protocol

A cryptographic protocol is a set of rules and maths that explain how to use cryptographic methods together to offer a particular security service in a certain application (John & Kumar, 2023).

A cryptographic protocol usually specifies that:

- In what format should the available information be placed
- What method should be implemented to convert information into mathematical elements
- Which of the encryption algorithms and with which parameters should be used
- How to apply mathematical relationships to numerical data
- What information should be exchanged between the sending and receiving parties
- What communication mechanism is needed to transfer information?

For example, we can refer to the Diffie-Hellman* key exchange protocol for creating and exchanging a common encryption key between two parties.

4.8 Types of Encryption

Every encryption technique can be divided into two fundamental groups. The primary distinction among algorithms stems from this category, while additional differences pertain to their construction method (John & Kumar, 2023).

- Symmetric key algorithms use the same key for both locking (encrypting) and unlocking (decrypting) information.
- Asymmetric key algorithms use one key to encrypt information and a different key to decrypt it. This kind of encryption is also called Public-Key.

4.9 Encryption Algorithm with Symmetric Key

Concrete cases are the most excellent devices to get cryptographic calculations. The illustration of informing with mail is regularly utilized to clarify calculations, which we used in this segment. The source puts his mystery message inside a box and closes it with an uncommon bolt that he has the key to. The collector incorporates a comparative test of the same key that he already received from the sender in any way. Utilizing the key, he opens the box and peruses the message. Symmetrical encryption calculations are isolated into two categories: consecutive encryption and square encryption. In consecutive encryption, all bits of the message are scrambled at once. In square or design encryption, a piece containing a certain number of bits (usually 64 bits) is scrambled. There are different sorts of symmetric encryption calculations, among the foremost celebrated ones, the taking after can be specified:

- Twofish
- Serpent
- AES
- Blowfish
- CAST5
- RC4
- TDES
- IDEA

4.10 Advantages of Symmetric Encryption Algorithms

Symmetric encryption is very popular because it is easy to use and works quickly. Using the same key to lock and unlock information makes it easy and quick (John & Kumar, 2023). So, we can say that symmetric cryptography is:

- Quicker than asymmetric encryption.
- Needs less computer power.
- Keeps going fast.

Because of these benefits, it's suggested to use symmetric encryption for protecting very large amounts of data.

4.11 Common Expressions in Cryptography

Each technology comes with its own unique terminology, and comprehending these terms and their applications is essential for understanding the key concepts of that technology. Next, we will look at important words related to encryption. A collection of random bits is utilized to obscure and connect data, ensuring its confidentiality and accessibility. Keys are special and made entirely by chance, so it's hard to guess them. Longer keys are tougher to break. In symmetric encryption, keys are usually 128 bits long, while in asymmetric encryption, they are typically 2048 bits long.

Cipher: A method used to change data into a secret code and back again. An algorithm is a series of stages and procedures that happen one after the other. There are two kinds of algorithms: Block and Stream.

Algorithm: It is the order and steps that are performed in the encryption process. Today, there are a variety of cryptographic algorithms that are classified depending on the level of security and the subject and field of activity.

Decryption: A process in which a text with code or unintelligible data is converted into meaningful data using a decryption key.

Cryptanalysis: The science of studying codes and encryption systems. In criminal applications, by using this science, it is possible to access the main data without having the keys after discovering the weak points of the encryption system. In the old days, a method called frequency analysis was used, which found the repetition of some letters in the messages, and somehow extracted the original message. Of course, the mentioned method is not very efficient in dealing with today's modern algorithms.

4.12 The Working Method of the Most Famous Symmetric Algorithms

As mentioned, all kinds of symmetric algorithms are available to users in the world of technology. DES, which is one of the oldest standards, is placed in the symmetric category, and its principles formed the basis for the development of later standards. DES uses a 56-bit (or seven-byte) key for encryption, and its structure is developed based on the Feistel structure (the famous cryptographic achievement of Horst Forstel).

In the DES algorithm, data including plain text with a length of 64 bits is encrypted in one block. First, the data is divided into two 32-bit parts, and then the encryption process is performed on each of them separately. The said process includes 16 steps in which various mathematical operations are performed on the data. Finally, 64-bit encrypted data is obtained as the output of the algorithm. One of the main weaknesses of DES, which led to its obsolescence, is the shortness of the encryption key. Many security experts managed to crack the DES cipher, which eventually led to its obsolescence in 2005.

An improved version of DES called 3DES was introduced. This algorithm became famous in the 1990s and somehow optimized some DES structures. As the name of the 3DES algorithm suggests, in its encryption processes, the DES algorithm is executed three times in each data block. As a result, it will be more difficult to break the password. After some time, the 3DES algorithm was used as a popular standard in payment systems, financial standards, and technologies. The same algorithm was used in many encryption protocols such as TLS, SSH, IPsec, and OpenVPN.

All encryption algorithms face a penetration crisis after some time. 3DES finally faced the same challenge and security experts found the vulnerability known as Sweet32 and found the possibility to break into it. As a result, other experts looked for optimal ways to encrypt, and in 2019, the 3DES algorithm was also obsolete. According to the new announcement of the standards organizations, the applications that are developed from 2023, do not have the right to use 3DES. The standard TLS 1.3 protocol also no longer uses 3DES.

DES and 3DES were deprecated after the ciphers were cracked by security experts.

As we said earlier, the advanced AES standard has been introduced as an optimal replacement for outdated cryptographic algorithms. AES encryption consists of a set of pattern ciphers that have cipher keys of different lengths as well as blocks of different sizes.

The AES algorithm works based on the principles of permutation and permutation. First, the data is divided into multiple blocks in the form of plain text. Then, using the encryption key, the encryption process is applied to the blocks. The encryption process itself has several sub-processes that perform various operations on the data. Depending on the length of the encryption key, 10, 12, or 14 steps of sub-processes are applied to the data.

One of the advantages of the advanced AES algorithm is its speed, security, and flexibility. The new algorithm has a much faster speed than DES, and the difference in the length of the keys is considered its best security advantage. Do not forget that the longer the keys, the more difficult it will be to decrypt. AES is known as the most widely used cryptographic algorithm today and is widely used in the following cases:

- Security of wireless communications
- Processor security and file encryption
- SSL and TLS website security protocols
- Wi-Fi security
- Encryption of mobile applications
- VPN tools

Many security and government organizations around the world, including the US National Security Agency (NSA), use AES encryption.

4.13 Encryption Algorithm with Asymmetric Key

Within the illustration of topsy-turvy encryption, the collector and sender of the message have two diverse locks with interesting keys. Really, unlike the symmetrical strategy, there's no news of a bolt with numerous comparative keys. Within the taking after, we depict a generally straightforward clarification of the calculation usage strategy. Of course, due to the complex nature of cryptography, this straightforward case moreover appears a bit complicated.

The sender to begin with inquires the collector to send him his private key. At that point the sender puts the message in a box and locks it with the uncommon bolt of the collector. After sending a message by normal mail and in a bolted box, as it were, the beneficiary will be able to open the message, since he has the uncommon key. Indeed, the sender himself will not be able to open the box after locking it. As a result, in this sort of informing, sometime recently sending the message, you must have the recipient's extraordinary bolt.

4.14 Advantage of Asymmetric Encryption Algorithms

A critical benefit of irregular encryption is that neither the source nor the receiver is required to share their private keys. Actually, they keep the private keys with them and only the locks are moved. Don't forget that moving the shared key in symmetric encryption would increase the possibility of spying and information theft. Imagine that the recipient and the sender use a regular mail service to exchange shared keys, and someone copies the shared key in the meantime. He can decode all subsequent messages. Another important point is that if, for example, the receiver in asymmetric encryption reduces his private key, only his messages can be spied on. In fact, the sender's messages cannot be read by other people, because each one is sent with a special lock for the same receiver.

As mentioned, in asymmetric encryption, different keys are used in encryption and decryption. The recipient of the message produces two types of community and private keys (Public and Private). He then sends the public key to the senders of the message. They encrypt the messages sent using the public key. After receiving the message, only the original receiver can decrypt it, and he does it with his private key.

The key benefit of symmetrical encryption is that no sensitive security key or message is moved to use the algorithm. The isolated keys used to decrypt the messages are unique to the receiver and he does not need to share the key with the senders. Public keys are easily sent to others and do not need to be secure. On the other hand, the private key can be easily stored in a safe place on the personal computer.

4.15 The Working Method of Famous Asymmetric Encryption Algorithms

Among the popular and popular asymmetric encryption methods, we can mention RSA and ECC, each of which we will examine below.

The RSA algorithm was the result of the collaboration of three experts named Ron Rivest, Adi Shami, and Leonard Adleman, who used the initials of their surnames for the name of the algorithm and introduced it in 1977. Today, RSA is known as the most widely used asymmetric encryption algorithm. The main strength of the aforementioned algorithm can be seen in the so-called Prime Factorization method, which is used as its basis. In this method, two very large random numbers are used, which are multiplied together to make a large random number. To crack RSA ciphers, you need to identify the two numbers that make up the final and super key.

Asymmetric algorithms also add authentication capabilities to the process.

Looking at the whole problem, we come to the conclusion that it will be impossible to solve it and find two main numbers. Of course, in 2010, researchers were able to break the RSA-768 cipher with 1,500 years of processing time (distributed among hundreds of computers). Don't forget that today RSA is used with 2048-bit keys and it is actually impossible to break them.

One of the advantages of the RSA algorithm is its excellent scalability. The mentioned algorithm can be used with keys of various lengths from 768 and 1024 to 2048 and 4096. As a result, when smaller keys are broken, the encryption process can be made more complicated by significantly increasing the key length.

The RSA algorithm uses a simple mathematical operation. As a result, its implementation in asymmetric methods will be simple. This simplicity and fast implementation have made RSA more than all algorithms used in asymmetric cryptography today. The areas of use of RSA include SSL and TLS internet licenses, passwords and email encryption.

Another famous algorithm in the asymmetric category is called ECC, which was developed in 1985. Neal Koblitz and Victor S. Miller created the algorithm that suggested using elliptic curves for security in technology. Their idea started being used in 2004, about 20 years later, and was called ECC or Elliptic Curve Cryptography. In the ECC encryption process, an elliptic curve shows how many points satisfy a certain math equation ($y^2 = x^3 + ax + b$). Like RSA, the ECC algorithm is based on the idea that certain processes cannot be reversed. In simple terms, it's easy to solve a math problem to find the answer. But going back to figure out the original numbers used can be much harder.

In the ECC algorithm, a number that stands for a point on a curve is multiplied by another number, which gives a different point on the curve. To fix the issue, you need to locate the new spot on the line. The math ideas behind ECC are made so that even if you have the starting point, you can't figure out the new point.

The advantages of ECC compared to RSA include much higher security, which provides an equal level of security by using much shorter keys. As a result, by using long keys in ECC, it becomes almost impossible to break passwords. The advantage of shorter ECC keys can be seen in increasing the speed and efficiency of encryption processes. Shorter keys require less processing power and network pressure. In situations where ECC is used in SSL/TLS protocols on websites, we will see faster loading. The applications of ECC include application encryption, digital signatures and similar applications. Of course, currently, a large number of servers do not support ECC and we have to wait for the development of appropriate and coordinated applications.

4.16 Hybrid Cryptography: Symmetric Cryptography Alongside Asymmetric

Hybrid or combined cryptography takes the best features of both symmetric and asymmetric methods to work as efficiently as possible. As can be seen in the examples above, symmetric and asymmetric methods have their own disadvantages along with many advantages. For example, symmetric encryption, despite its high speed in encrypting large data, does not cover the unique authentication and validation part, which is especially critical in Internet security. On the other hand, asymmetric encryption, by assigning special keys to each user, provides authentication and high security; but in heavy and bulky applications, it provides very low speed and performance.

In many applications, especially in web security, we need high-speed encryption of voluminous data. In addition, authentication must be done in all processes to ensure the security of each user. In such cases, hybrid or combination encryption is used.

4.17 Comparison of Symmetric Key and Asymmetric Key Cryptography

Basically, symmetric key encryption and asymmetric key have two different natures and have different applications; Therefore, it will not be accurate to compare these two types of encryption regardless of the intended application and system. But if the criterion of comparison is, specifically, the volume and the required computing time, it should be said that, taking into account the equivalent security scale, symmetric cryptographic algorithms are much faster than asymmetric cryptographic algorithms.

4.18 Securing Web Communications with Encryption

For many years, the Secure Sockets Layer (SSL) has been protecting the information exchanged between a user's browser and websites. This means that if someone tries to break into the conversation, they won't be able to read the information. SSL works using a system that is both simple and appealing, called an asymmetric algorithm.

First, the browser asks the server for a safe webpage using a secure address starting with https. Then the server sends its public key to the browser along with its license. The browser makes sure the server's license is valid and that it relates to the original server. The browser uses the public key it got. This key is used to lock a secret key that keeps information safe. Next, the key made from the encrypted web address and other information is sent to the server. The server first unlocks the secret key using its own private key. In the end, this key is used to unlock the URL and other information. In the last step, the browser gets the HTML documents it asked for, along with other data that is kept secure using a special key. The browser shows the data by unlocking it with the same key.

Cryptography has become a vital part of the technology world today. Almost all processes that involve data transfer require encryption to ensure the identity and security of users in cyberspace. As you can see, the techniques for breaking cryptographic algorithms have also evolved over the years. For this reason, no algorithm can be considered safe forever. On the other hand, with the possibility of the supply and use of quantum computers becoming more serious, the current encryption methods are at risk more than ever. Maybe in the future with the presence of quantum computers, quantum encryption algorithms will be developed.

4.19 Kershoff's Six Principles

August Kershoff gained his fame from his linguistic research and the books he wrote on this subject and the Volapük language. In 1883, he published two articles entitled "Military Cryptography". In these two articles, there were six basic principles, the second of which is still used by advanced cryptographers as one of the cryptographic rules:

- The encryption system should be unbreakable not only in theory but also in practice.
- The encryption system should not have any hidden or confidential information. Rather, the only thing that is secret is the password key.
- The password must be selectable in such a way that: 1) it can be easily changed, and 2) it can be remembered and there is no need to memorize the password.
- Cipher texts must be transmissible through telegraph lines.
- Encrypting devices or encrypted documents must be transportable by one ton.
- The encryption system should be easy to set up.

The key should be the only confidential information in a cryptographic system for several main reasons:

1. Keeping a 512-bit string secret is (typically) much simpler and more practical than keeping an algorithm or implementation method secret.
2. If the encryption key is disclosed, a new and secure cryptographic system can be achieved again only by adopting a new key. If the security of the cryptographic system depends on the algorithm and implementation method, by disclosing this information, it is difficult to reproduce a new and very secure cryptographic system. It will be difficult.
3. When the encryption key is the only confidential information of a cryptographic system, a common system (with different keys) can be used to communicate with different receivers/senders, if it were otherwise, the same algorithm would be used to communicate with each sender/receiver. And a new implementation method was needed.
4. When the algorithm and the implementation method are available to everyone, the problems and security holes of the algorithm will be investigated and fixed by security researchers before being exploited by an attacker, and therefore cryptographic systems that are based on clear algorithms and they are usually more reliable.

4.20 Advanced Encryption

With the emergence of computers and the increase of their computing power, cryptography entered the field of computer science and this phenomenon caused three important changes in cryptography issues:

1. The existence of high computing power made it possible to create more complex and efficient encryption methods.

2. Cryptographic methods, which were mainly used to encrypt messages, found new and many applications.
3. Until then, encryption was mainly done on textual information using alphabets; but the arrival of the computer made encryption on all types of information based on bits.

4.21 Applications of Cryptography

Before computers entered the field of encryption, the use of encryption was almost limited to encrypting the message and hiding its content. But in advanced encryption, various services are provided, including the following:

- **Maintaining confidentiality or content security*:** Sending or storing information in such a way that only authorized people can know its content, which is the main and primary service of hiding the contents of the message.
- **Maintaining data integrity or content health*:** It means ensuring the accuracy of information and not changing its original content during sending. Changing the initial content of the information may be accidental (due to problems in the transmission path) or intentional.
- **Authentication or authentication of content:** It means to recognize and ensure the identity of the sender of information and the impossibility of falsifying the identity of people.
- **Non-repudiation*:** It means that the sender of the information cannot in the future deny that it was sent or deny its content.

The above four items are considered the main uses of cryptography, and other purposes and uses of encryption can be achieved by combining the above four items.

These applications are comprehensive concepts and can be implemented and used for various applications. For example, content authenticity service is important in both commercial transactions and is used in military and political issues. Different encryption protocols are used to provide each encryption service, depending on the type of application.

4.22 Decryption

Decryption refers to all actions based on mathematical and scientific principles, the purpose of which is to destroy the security of encryption and finally unlock the code and access the main information. In cryptography, an attempt is made to find possible security weaknesses in the encryption system by examining the details of the encryption algorithm or encryption protocol used and using any available external information, and in this way, the encryption key or information content is somehow obtained. The encrypted data is to be extracted.

Cracking is sometimes done in order to break the security of a cryptographic system and as sabotage and an anti-security activity, and sometimes it is done in order to evaluate a cryptographic protocol or algorithm and to discover its possible weaknesses and vulnerabilities. For this reason, cryptanalysis is not inherently a hostile activity; but they usually consider the part of vulnerability assessment and discovery as a part of necessary operations when designing new algorithms and protocols, and as a result, decryption brings to mind more destructive and anti-security activities. According to this article, the term cryptographic attacks is used to refer to such activities.

Cracking basically refers to the mathematical examination of the algorithm (or protocol) and the discovery of its possible weaknesses; But in many cases, instead of the mathematical principles and basis, in many cases, the malicious activity examines a specific implementation of that algorithm (or protocol) in a specific application and tries to break the password and find the password by using various facilities. This class of malicious actions is called lateral attacks.

4.23 Substitution Codes

In substitution cryptography, each letter or group of letters is switched with a different letter or group of letters to change the message. One of the earliest ciphers is the Caesar cipher, which is believed to have been created by Julius Caesar. In this method, the letter a changes to *d*, *b* changes to *c*, *e* changes to *f*, and it continues like this until *z*, which changes to *a*.

4.24 Multiphase Encryption

Today, multiple encryption methods cannot provide sufficient security. The multiphase encryption method greatly increases the complexity of encryption algorithms. This idea is different from existing network encryption methods and information security on wireless networks. As in encryption protocol, increasing the complexity of data encryption methods increases the security of data transmission on wireless channels. The large number of encrypted data encryptions increases the complexity of data encryption, which also makes its decryption very complicated.

4.25 Function

In fuzzy encryption methods, each encryption cycle uses a different key. In this encryption method, decoding is done in the reverse direction. In multiphase encryption, many processes are repeated which increases the complexity of encryption/decryption and also improves data security.

Cryptographic algorithms and key sizes are chosen to ensure encryption, personal identity verification (PIV), and applications. Multiphase cryptography may reduce the key management problem in existing technology (PIV); Because it uses cryptographic algorithms with a fixed key size instead of a large number of variable-length keys.

4.26 Advantages

Multiphase data encryption is a two-way method for data and information security and plays an important role in modern encryption. Multiphase encryption performs the advanced complexity of data encryption with various operations of fuzzy encryption methods. The advantage of multiple ciphers is that they provide better security; because even if some components of the ciphertext are broken or some of the secret keys are known, the confidentiality of the original data can still be maintained with other cryptography.

The implementation of multiphase encryption is a positive step towards defining a standard for network security. However, it increases confidential data communications over unsecured wireless networks.

4.27 Redundancy

The first rule is that all secret messages need to have some extra information. In other words, you don't have to send the real data as it is. An example can help to understand the reason for this need. Suppose a company called TCP accepts purchase orders with 60,000 products through the electronic mail system. The programmers of the TCP company think that they write efficient and effective programs, the product order messages consist of 16 bytes of the customer name, followed by three bytes of the data field (including one byte for the product number and two bytes for the product number), and the last three bytes are encrypted by a very long key and this key is known only by the client and the TCP company.

4.28 Newness of Messages

The second basic principle in cryptography is that it must be checked to make sure that every received message is new, or in other words sent to a target. This check is required to prevent old messages from being sent again by an active intruder. If such checks are not done, our fired employee. It is able to create a secret branch of the phone line to repeatedly send authentic messages that have already been sent, even if it does not know what the content is.

Such computing can be anticipated by placing a time stamp in the messages so that the messages are valid for, for example, ten seconds. They have a time stamp and will be recognized as old messages and deleted.

4.29 Hardware Encryption

Encryption algorithms can be implemented both in hardware (for higher speed) and in software (for more flexibility). Substitution and permutation methods can be implemented with a simple electronic circuit. P-box is a tool that is used to swap the bits of an eight-bit input. With the internal wiring and programming of this p-box, it is able to perform any bit permutation virtually at a speed close to the speed of light, because no computing is required and there is only signal propagation delay.

This design follows the principle of Kershoff, that is: the attacker is aware of the general method of swapping bits, what he does not know is which bit is written to which bit, that is the key of the code.

4.30 Vocabulary

In the following list, according to the direct connection of Cryptography, some terms that exist in the discussion of network and computer security are mentioned.

Steganography: In cryptography, the act of hiding a file, message, photo, or video inside another file, message, photo, or video is called Steganography. The advantage of obfuscation over cryptography is that, unlike encryption, obfuscation does not attract attention. Encrypted messages that are obvious attract attention and arouse interest, no matter how indecipherable they are; Therefore, while the focus of cryptography or cryptography is only on protecting the message, Steganography is on hiding the existence of the encrypted message. Encryption involves hiding information within computer files. In digital encryption, electronic communications may involve covert encryption within a transport layer such as a document file, photo, video, program, or protocol. Media files are an ideal option for covert transmission due to their high volume.

Decryption: It is the reverse of Encryption and in Cryptography it is called decoding, extracting, and revealing hidden information.

Plain text: The initial text that is not encrypted or the final decrypted text is called plain text.

Cipher: An algorithm used to lock or break a lock as a series of defined steps. Cipher is used so that less time is spent on encryption. The method of encryption is the cipher method in such a way that the information of the plain text is converted into a password or code.

Decryption: Methods of decrypting cipher texts without knowing their key are said.

Disruptor: Literally, it means disturbing or disrupting, and in cryptography, it means someone who has a copy of an encrypted message and intends to decrypt it. The meaning of breaking the password is to decrypt that text, which itself is of two types. Active intruder who can exchange and change the information on the line and passive intruder who can only have the information on the line and does not have the ability to change them.

Protocol: It is a method or a contract between two or more people to exchange information.

Points of Intrusion: Points where an intruder can gain access to valuable information.

Internal Access Point: Systems that are located in the room or the internal network and no security (Local-Security) has been set on them and there is a possibility of attacking them.

External Access Point: Equipment that connects us to an external network such as the Internet or applications that work through the Internet and there is a possibility of attacking them.

Attack: Anything that bypasses the system's security mechanism and causes damage is called an attack.

The following types of attacks can be mentioned:

1. DoS
2. DDoS
3. Spoofing (MAC Spoofing, IP Spoofing و Web Spoofing)
4. Man-in-the-Middle
5. Password Guessing

Key: Information that can be used to convert cipher text into plain text or vice versa, in simple words, a text encrypted by a key with a suitable algorithm is converted into simple text.

4.31 Conclusion

In general, security can be defined as the protection of what is valuable to us. This protection has three basic features, confidentiality, accuracy and availability. To establish and maintain these three features, the use of cryptography and secure digital signatures is vital and unavoidable. Therefore, wherever there is a need for information and communication security, we have to use encryption and digital signatures. Today, all over the world, a variety of symmetric and asymmetric encryption methods are used. But since according to Shannon's law, in order to maintain complete confidentiality, the size of the key space must be greater than or equal to the size of the message space, it is impossible to create complete confidentiality in practice, and the cryptographic methods we mentioned, in practice, create computational confidentiality for us. They do. Computational secrecy means that these codes cannot be broken in an acceptable time using today's classical computers.

References

1. Chai, X., Fu, J., Gan, Z., Lu, Y. and Zhang, Y. (2022). An image encryption scheme based on multi-objective optimization and block compressed sensing. *Nonlinear Dynamics*, *108*(3), 2671–2704.
2. Eslamian, S. and Eslamian, F. (2025). *Blockchain Technology in Water and Environmental Systems*. CRC Press, Taylor and Francis, USA, p. 282.
3. Gan, Z., Song, S., Zhou, L., Han, D., Fu, J. and Chai, X. (2022). Exploiting compressed sensing and polynomial-based progressive secret image sharing for visually secure image selection

encryption with authentication. *Journal of King Saud University-Computer and Information Sciences*, *34*(10), 9252–9272.

4. John, S. and Kumar, S.N. (2023). 2D Lorentz Chaotic Model Coupled with Logistic Chaotic Model for Medical Image Encryption: Towards Ensuring Security for Teleradiology. *Procedia Computer Science*, *218*, 918–926.
5. Li, T., Wang, Z., Zou, L., Chen, B. and Yu, L. (2023). A dynamic encryption–decryption scheme for replay attack detection in cyber–physical systems. *Automatica*, *151*, 110926.
6. Panwar, K., Kukreja, S., Singh, A. and Singh, K.K. (2023). Towards deep learning for efficient image encryption. *Procedia Computer Science*, *218*, 644–650.
7. Ping, P., Yang, X., Zhang, X., Mao, Y. and Khalid, H. (2022). Generating visually secure encrypted images by partial block pairing-substitution and semi-tensor product compressed sensing. *Digital Signal Processing*, *120*, 103263.
8. Ponuma, R., Amutha, R., Aparna, S. and Gopal, G. (2019). Visually meaningful image encryption using data hiding and chaotic compressive sensing. *Multimedia Tools and Applications*, *78*, 25707–25729.
9. Su, Y. and Wang, X. (2022). A robust visual image encryption scheme based on controlled quantum walks. *Physica A: Statistical Mechanics and Its Applications*, *587*, 126529.
10. Wang, X., Liu, C. and Jiang, D. (2022). An efficient double-image encryption and hiding algorithm using a newly designed chaotic system and parallel compressive sensing. *Information Sciences*, *610*, 300–325.
11. Xiao, D., Zhao, M. and Wang, M. (2021). Low-cost and secure multi-image encryption scheme based on P-tensor product compressive sensing. *Optics and Laser Technology*, *140*, 107077.

5 | Consensus Algorithms

Yaser Sabzevari[1*] and Saeid Eslamian[2]

Consensus algorithms form the core of distributed systems, enabling multiple nodes to agree on a single, consistent state even in the presence of failures, delays, or malicious actors. This chapter provides an in-depth examination of the principles, mechanisms, and practical applications of consensus in decentralized networks. Foundational models such as crash-fault tolerance (CFT) and Byzantine-fault tolerance (BFT) are introduced to explain system reliability under different failure assumptions. Major consensus algorithms, including Proof of Work (PoW), Proof of Stake (PoS), Practical Byzantine Fault Tolerance (PBFT), Raft, and Paxos are analyzed with respect to their design logic, operational characteristics, security assumptions, and performance trade-offs. The chapter also explores emerging consensus innovations aimed at improving scalability, energy efficiency, and decentralization. By connecting theoretical foundations with real-world implementations, this work provides readers with a comprehensive understanding of how consensus algorithms maintain trust, integrity, and robustness within blockchain platforms, distributed databases, and large-scale multiagent systems.

5.1 Introduction

Consensus algorithm is a mechanism that allows a user to coordinate with other users and the blockchain of a cryptocurrency (Mingxiao et al., 2017). The agreement method needs to make sure that everyone in the network uses the same resource. If some users leave, it should not harm the system overall. In simple terms, the consensus algorithm helps the system keep working well even when there are problems (Ongaro & Ousterhout, 2014).

A consensus algorithm is actually a process that allows users to agree on a single data value across distributed processes or systems. Using consensus algorithms, users can agree on a specific piece of information. This is reaching an agreement on a part of the information that has always made blockchain protocols reliable and reliable. The most attractive feature of blockchain networks is their decentralization. In centralized systems, all network decisions are made by only one transaction party. The party who made the decisions can change the data as they like, but this is not the case in decentralized systems. In decentralized systems,

[1,2] Department of Water Science and Engineering, College of Agriculture, Isfahan University of Technology, Isfahan, Iran.

* Corresponding author: yasersabzevari1511@gmail.com

DOI: 10.1201/9781003686026-5

data is recorded in a database that is the blockchain network. Of course, everyone in the blockchain network must have the same version of the database so that contradictory information is not provided. Public key encryption also prevents network users from spending each other's coins. However, there should be a fact-checking authority that network users can refer to (Bhardwaj & Datta, 2020).

This is why Satoshi Nakamoto, the creator of Bitcoin, proposed the "Proof of Work" consensus algorithm. This algorithm forces users who want to own blocks in the network to provide a contribution and actually stake a part of their digital assets. A stick is a type of value that a validator provides. This value, recorded by the network, guarantees the honest behavior of that validator in the blockchain network (Eslamian and Eslamian, 2025). If the validator cheats, they lose their stake. The stake may be his computing power, digital currency, or reputation

The reason this is done in a staking blockchain network is that solving math problems in the blockchain network will be beneficial for validators. Also, this process makes the mechanism of decentralized networks transparent, which is another main feature of blockchain networks. Consensus algorithms help identify the following (Bamakan et al., 2020):

- Determining whether users can perform a distributed transaction on the database.
- Determining a node to manage some distributed tasks in the network.

5.2 The Difference Between a Centralized System and a Decentralized System

In a centralized system, a single center manages the entire system. In most cases, the admin center can make any changes it wants and there are no complicated steps to make the changes. Centralized systems such as banks and governments are institutions that are managed and changed by a central authority.

But in a decentralized system, the story is completely different. Suppose all users are making changes on a distributed system. How to maintain the unity of the system so that everyone does not go their own way and everyone agrees on the same version of the system? This is where consensus algorithms come into play. The challenge of lack of trust between nodes (network users) who have no knowledge of each other creates an opportunity for blockchain. An opportunity where trust is based on all users and not just on a central authority (Chaudhry & Yousaf, 2018).

5.3 Types of Consensus Algorithms

Some of these algorithms are the following:

- Burn Proof
- Capacity Proof
- Elapsed Time Proof
- Activity Proof

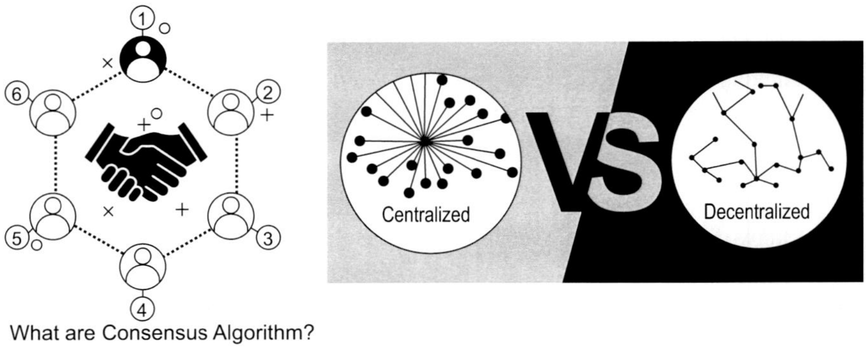

Fig. 5.1 View of the consensus algorithm.

- Authority Proof
- Delegated Proof of Stake
- Byzantine Fault Tolerance
- Proof of Importance
- Direct Acyclic Graphs

All blockchains are built based on consensus algorithms, and in fact, these algorithms are the main part of any blockchain. These are the algorithms that distinguish different blockchain protocols. In fact, if there are no consensus algorithms, there will be no blockchain, because without these algorithms, the blockchain becomes an inflexible database that no longer fulfills the capabilities we expect from a blockchain (Bach et al., 2018).

5.4 Proof of Work Algorithm

The Proof of Work (PoW) system is the original method used for reaching agreement in blockchains. The Bitcoin network was the first one to use the PoW system. In the PoWk system, people called miners work to find a special answer by mixing up the data they want to add. A hash is a mix of letters and numbers made when you put data through a special process called a hash function. Even a tiny change in the detailed information of the blockchain network will result in different data. In the PoW system, the rules explain what needs to happen for blocks to be seen as valid. The issue with PoW systems is that they require a lot of computing power to work (Hao et al., 2018).

The PoW algorithm relies more on accurate data and information and is not suitable in terms of transaction speed, and the low transaction speed in this algorithm is considered one of its weaknesses. If the problem is too complex, it will take a long time to generate a block, thus delaying transactions. In fact, it was this scalability issue that led to the creation of many other consensus algorithms that we use today.

5.5 Proof of Stake Algorithm

The Proof-of-Stake (PoS) method was suggested as a different way to do things than the PoW method. To use the PoS method for mining, you don't need super strong equipment like you do for the PoW method. You just need a working computer and a good internet connection. In the PoS algorithm, a minimum amount of money is set for staking. These funds are stored in a wallet and can't be taken out until the staking period is over.

Validators typically agree on which transactions should be part of the next block. In fact, validators randomly bet on which block will be selected, and eventually, the protocol selects a block. If the desired block of a validator is selected, a percentage of the transaction fee based on that validator's stake will be paid to him. The higher the amount of staked funds of the credit meter, the more profit it will earn. If you try to cheat by offering fake trades, you will lose some or all of your stake. In addition, the two big advantages of the PoS algorithm are that it reduces the risk of attack by 51% and its transaction speed is faster than the PoW consensus algorithm (Ferdous et al., 2020).

5.5.1 Delegated Proof of Stake (DPoS) Consensus Algorithm

The ceded or granted proof of stake consensus algorithm is a traditional proof of stake algorithm. This system is highly reliable and brings a new level of flexibility. If you are looking for a fast, efficient, and decentralized consensus algorithm, then the DPoS algorithm is the best solution. By using this algorithm, the problem of stakeholders is solved in a completely democratic way. Using the DPoS consensus algorithm, each network component can become an "agent". In this case, the nodes are called "agents" instead of miners or validators. This system can perform transactions in less than one second by determining block generation.

Using the DPoS algorithm, witnesses can produce blocks of information. A vote may also be taken to select the best witnesses. Of course, voting happens only when the system is completely sure of its decentralization. In addition, the interest of all witnesses is paid immediately after the block is produced. This rate is determined in advance through the voting system (Ferdous et al., 2020).

5.6 Leased Proof of Stake Algorithm (LPoS)

Another kind of PoS system is called the leased proof-of-stake system (LPoS). This new agreement method was created by the Waves blockchain platform. Like other blockchain platforms, Vives says it can offer better features while using less energy. The original PoS system had limits on how much you could stake. In the original system, people with only a few coins might not be able to join the staking process. To keep the network working well, only a few people with more money would be allowed to stake. This process lets the system build a central community on a decentralized platform, which is not good at all. With the LPoS algorithm, store owners can finally have a chance to earn rewards by staking. In this system, people can lend their coins to the network and earn money from it (Xiong et al., 2022).

5.7 Proof of Elapsed Time (PoET) Consensus Algorithm

The Proof of Elapsed Time Algorithm, or PoET, is one of the most effective ways for computers to agree on things. This algorithm is mainly used in private blockchain networks where you have to get approval to join the network. These groups need to make decisions about who can mine or how to vote. PoET algorithms use a special method to make everything clear in the network so that rules are followed correctly. These consensus methods help keep logins safe because they require miners to complete verification steps before they can get in (Huang et al., 2019).

The PoET Algorithm picks winners in a fair way. By using this method, everyone will be treated the same and have the same chances. Everyone in the network has to wait for a while, but how long they wait is different for each person. A shareholder who has waited a good amount of time is added to the list to help create a new block (Huang et al., 2019). To explain these situations, the algorithm needs to think about two things:

First, did the winner choose a random number? Someone might pick a random number quickly and win first.

Did that person really wait for that exact amount of time?.

PoET needs a special feature on the CPU called Intel Software Guard Extension. This Software Guard Extension helps run special codes on the network. PoET uses this system to make sure the winner is chosen fairly.

5.8 Proof of Authenticity (PoA) Consensus Algorithm

The last consensus algorithm we discuss in this article is the PoA Algorithm or PoA. A Validation algorithm is used in centralized systems. This means that authentication is done by verified accounts (chosen by system administrators) across the network. The centralized nature of the authentication consensus algorithm is mainly used in private networks due to its nature (Alsunaidi & Alhaidari, 2019).

5.9 Proof-of-Weight Consensus Algorithm (POWeight)

Proof of Weight (PoWeight) algorithm is a blockchain consensus mechanism that gives users "weight" based on the amount of digital currency they have! A weighted approach means that you use a combination of information from several independent sources to provide enough evidence to satisfy an information need.

In a network that uses PoWeight, users with more weight are very important for reaching an agreement. Each user gets a "weight" based on a chosen value that shows how much they help the network. To stop double-spending attacks and other dishonest activities on the blockchain, most of the vote (more than two-thirds) needs to come from honest users. (Alsunaidi & Alhaidari, 2019).

5.10 Proof of Token Burning (PoB)

Users send tokens to a token-burning address that cannot be returned and are rewarded depending on the amount of tokens. In this mechanism, similar to the certificate of deposit or PoSA mechanism, users stake or deposit tokens; but unlike PoS, the deposited tokens are not returned to users. Tokens in this mechanism are sent to a public address that is randomly generated without a private key and cannot be returned. Tokens sent to this address are inaccessible and will be discarded. Also, like a proof of stock, no processing hardware is purchased. Also, the 51% attack in this algorithm has no economic benefits; because in the PoB algorithm, the more tokens the nodes burn, the more likely they are to win to create a new block (Li et al., 2019).

5.11 Practical Byzantine Fault Resilient Consensus Algorithm (PBFT)

Practical Byzantine Fault Tolerance (PBFT)) is a way for computers to agree on things, and it was created in the late 1990s. PBFT is made to work in systems where things don't happen at the same time. This algorithm is used to help everyone agree in a network that is spread out, like blockchain.

The PBFT method helps different parts of a network agree on things, even if some parts fail or send wrong information. The PBFT system helps keep the network safe from mistakes by making decisions together. The goal of making decisions together is to lower the chances of bad or dishonest people affecting the outcome (Li et al., 2019).

5.12 SBFT Consensus Algorithm

SBFT is a new version of the BFT algorithm that addresses the challenges of scalability and decentralization. Unlike many previous BFT systems that only work well when the network is centralized, SBFT is optimal for a fully decentralized network. SBFT provides a smart contract execution environment based on Ethereum Virtual Machine (EVM) code (De Angelis, 2018).

5.13 DBFT Consensus Algorithm

DBFT is an improved version of PBFT and is suitable for public blockchain systems because it has high comparability. The DBFT algorithm is a consensus algorithm used in the NEO blockchain. The DBFT algorithm provides the possibility to reach a consensus by voting for a representative. In the NEO ecosystem, these representatives are referred to as bookkeepers. Because in DBFT, one representative is given representation and consensus is made among a small number of representatives, this algorithm works more efficiently compared to PBFT (De Angelis, 2018).

5.14 DAG (Directed Acyclic Graph) Algorithm

In token-based cryptography, each sphere in the structure represents a transaction. There is no concept of blocks here, and like blockchain, mining is not required for network continuity. So instead of aggregating transactions into blocks, each transaction builds on another. Of course, there is a small POW operation in DAG that when a node sends a transaction, it must confirm the previous transaction in order to register it. This ensures the stability of the network and validates previous transactions (Li et al., 2019).

5.15 Proof of Significance (POI) Consensus Algorithm

Proof of Importance (POI) is a blockchain consensus algorithm first introduced by NEM. PoS is a mechanism for determining which network stakeholders (nodes) are eligible to add blocks to the blockchain.

The block confirmation process in NEM is called Harvest. The person who does the harvesting is called the harvester. In exchange for mining a block, nodes are able to collect transaction fees on that block. Accounts with a higher importance score are more likely to choose a block (Li et al., 2019).

5.16 Proof of Capacity (POC) Consensus Algorithm

Proof of Capacity (POC) is a relatively new mining algorithm in which miners use the free space of the hard drive and plot it to participate in the network. Each plot is a way to solve the mine problem and the plotting process is complex and time-consuming. As a result, miners calculate and store solutions to problems before starting the mining process, considering the time to create each block, which is approximately 4 minutes. In each round of consensus, the solution that reaches the result first is the winner, and that miner writes the block and gets the reward (Pahlajani et al., 2019).

As mentioned, there are different types of consensus algorithms. But the two types that are more used in blockchain are:

- Proof of Stake (PoS) algorithm
- Proof of Work (PoW) algorithm

Currently, most digital currencies such as Bitcoin and Ethereum use PoW and because this method increases energy consumption by increasing the difficulty of the network, these networks intend to move to other algorithms such as PoS. Accordingly, Ethereum prepares its network on PoS. In the following, we will examine each of these algorithms, their advantages and disadvantages.

5.17 What is the Proof of Work Algorithm?

In the PoW algorithm, the process of adding blocks to the blockchain is called mining or extraction, and the person who does this is called a miner or extractor (Pahlajani et al., 2019). The computer used for mining is also called a node.

Every block in the blockchain has a special code called a hash. A hash is a group of random numbers. If the details in the block change, the block's ID will also change. Each block's hash includes a piece of the hash from the block before it. To add a block to the blockchain, miners must find its hash. To do this, they need to solve math problems and puzzles from the blockchain to find the right hash.

Hashing operations and finding the correct answer to the puzzle consume a lot of electricity. The incentive for miners to consume this energy and hardware resources is the reward they receive from the network for confirming transactions. The reward is the native currency of the network. For example, in the Bitcoin blockchain, miners perform expensive and difficult mining operations in the hope of receiving a network reward in the form of Bitcoins.

It has been said that in Bitcoin's PoW algorithm, every four years, or in other words, after every 210,000 blocks are mined, the mining reward is halved. This is called halving. Currently, the mining reward is 6.25 bitcoins.

Some of the famous cryptocurrencies that use this algorithm are Bitcoin, Litecoin, and Dogecoin.

5.18 Advantages of the Proof-of-Work Algorithm

- Lack of concentration
- Reward mechanism
- No need for a license
- Good and very high experience and history.

5.19 Disadvantages of Proof of Work

- Energy loss
- 51% attack possibility
- Low speed of adding blocks to the blockchain
- Low scalability
- Formation of mining pools.

5.20 The Proof of Stake Algorithm

The most important criticism of the PoW algorithm is its energy consumption. According to a report from the BBC website, Bitcoin consumes about seven gigawatts of electricity, equivalent to 0.21% of the world's electricity. This power is equal to the power that seven Dungeness nuclear power plants produce simultaneously (Pahlajani et al., 2019).

In a report published by Coin-telegraph, the amount of electricity consumed by the Bitcoin network is estimated to be equivalent to 10 million American households. The waste of energy and environmental problems related to the PoW algorithm provided the basis for the construction of another algorithm called the PoS algorithm.

In the PoS process, people do not perform mining operations. There is no miner, and instead, the validator performs the process of adding blocks to the blockchain. In PoS, there is no equation to solve and there is no need for expensive and time-consuming mining equipment. Because of this, it is much cheaper than PoW. In the process of PoS similar to PoW, network members approve information and issue permission to be added to the blockchain. In the PoW algorithm, people bear a lot of expenses, including equipment and electricity consumption. They do this in hopes of receiving a reward. They are only rewarded when they confirm the correct information. In case of fraud, the result will be detrimental to the miners.

But in PoS, there is no mention of mining costs. In PoS, vandals and fraudsters are fined or so-called Slash. That is, if a person tries to confirm fraud or a false block, all his assets will be zero (Pahlajani et al., 2019).

5.21 The Process of Participating in Proof of Stake

In the PoS process, network members must first procure the network's native currency. In the next step, assign it to the network (Stake) through a contract; or in other words, Stake. In PoS, there is no mining reward and members receive rewards based on the amount of currency they have staked.

5.22 Benefits of the Proof of Stake Algorithm

- Faster transactions
- Receive rewards and interest using the staking process
- Very low probability of 51% attacks
- Environmentally friendly and low electricity consumption
- High scalability.

5.23 Disadvantages of Proof of Stake

- Compared to other methods, it is less tested and has low experience.
- Increasing the power of the rich and major shareholders
- The need for currency for staking and the price is relatively high.

5.24 Which Consensus Algorithm is More Popular?

Today, various algorithms are used in the blockchain network to achieve consensus and coordination among members. Although the PoW algorithm has been very successful among other algorithms, there is still a lot of opportunity for new methods and solutions. Proof of Capital Algorithm, Proof of Elapsed Time Algorithm, Proof of Work Delay Algorithm, and Proof of Reputation Algorithm are just some of the blockchain algorithms that have recently been used in the field of cryptocurrencies.

Because consensus algorithms are the only reason for the success of blockchains and their lack of a central authority, they are very important. Among all consensus algorithms, PoW is still the main proposed option, and a more secure and reliable alternative has not yet been provided for it. That being said, tremendous research and development has been done in the field of PoS replacement, and it is likely that more networks will use this algorithm in the coming years.

5.25 Consensus Algorithm and Blockchain

In blockchain, users' assets are stored in a database, and it is essential that all users have a single copy of this database. Otherwise, very soon users will face many conflicts, which is against the main purpose of blockchain (Bamakan et al., 2020).

The encrypted public key ensures that users cannot spend each other's coins. But the problem of double-spending still remains.

5.26 The Most Important Consensus Algorithms in Blockchain

- Byzantine fault tolerance (PBFT)
- Proof of Work (PoW)
- Proof of Stake (PoS) algorithm
- Proxy Proof of Stake (DPoS)
- Proof of Burn (PoB)
- Proof of Capacity (PoC)
- Proof of Storage.

5.27 The Nature of the Consensus Algorithm

In the mechanism of blockchain algorithms, we first of all need users who want to add blocks to the blockchain so that new transactions can be registered. We call these validator nodes. It is necessary for each validator to add value to the network that prevents him from cheating and encourages honest activity in the network. If any validator commits a dishonest act in the network, he loses his stake in the network. This stock can be computer processing power, cryptocurrency, or even his reputation (Beiranvand et al., 2026).

But why should a validator risk his assets? Here comes the question of reward. A reward, usually of a currency on the same blockchain, is given to the user to encourage him to stay on the network. This reward is either composed of fees paid by other users or new coins that are hidden in the new block and awarded to the user (or both).

The last thing we need in this direction is transparency. We need to make sure that no one cheats on the network. In a system with a consensus algorithm, cheating on the network is very costly and unprofitable for anyone, while honest activity is profitable and low-cost.

5.28 The Efficiency of Consensus Algorithm in Blockchain

The community algorithm in blockchain allows the network to agree on a common version of data. One of the most important features of blockchain networks and other distributed offices; it is decentralized. In fact, the decentralization of these networks means that all information and data are not stored and processed on the main server, but are stored and maintained on several different computers. The servers responsible for maintaining this data are called nodes. On the other hand, the main reason for using distributed ledgers like blockchain is to increase the level of security and ensure that data is not destroyed or damaged. In fact, if the information on one of the nodes is damaged and destroyed for various reasons, there are countless other nodes that have stored this information on themselves. But when using blockchain or other types of distributed ledgers, there is a very critical issue; in fact, when the data is to be continuously updated, this update must be done on all nodes or servers, and to do this, a mechanism or algorithm called "consensus mechanism" is used. Basically, in the field of blockchain networks and cryptocurrencies, the consensus algorithm is very important. So that this mechanism plays the most important role in the security of distributed systems and blockchain networks (Bamakan et al., 2020).

Consensus algorithms are ways for everyone in a blockchain network to agree on the current state of the shared record. So, consensus algorithms help make sure that the blockchain network is reliable and that unknown users can trust each other in a system where the computers are spread out.

The consensus protocol makes sure that each new block added to the blockchain is the only correct version and has been agreed upon by all nodes.

The blockchain consensus mechanism aims for a few key things: every node has the same rights, they work together, they agree on things, and all nodes must be involved in the process. So, the consensus algorithm works to make sure everyone in the network agrees on the same decision.

5.29 Consensus Algorithm Applications in Blockchain

The consensus algorithm has various applications in the blockchain. Its most important use, as mentioned before, is the coordination and acceptance of all network members on a change or update in the network. But this blockchain feature also has other features:

- Confirmation of transactions
- Selecting the node to represent the network in the distribution office
- Ensuring the uniformity of information in the network.

Agreement on each of the mentioned functions is done through voting among network members. Each node or node has the same voting rights as others. If a node or node confirms the eligibility of a transaction to be registered in the blockchain, the transaction will be done, and if it rejects it, the transaction will be canceled.

5.30 The Consensus Process Has Three Important Uses in the Blockchain Network

- Integration of information on the system of network members
- Network management by selecting some members as leaders
- Check and decide on the validity of a transaction and store it on the distributed ledger.

For example, in a company, if employees have different opinions on a subject, the company manager will be the final decision-maker. With the help of a consensus algorithm, reaching an agreement between nodes is done through a voting system. In this way, if a user sends a transaction (which includes any type of data) to the network, the information sent is checked by all nodes. If according to the previous information, the sent transaction is correct, the node sends a confirmation to the network.

A set of transactions form a block. If more than 51% of the nodes approve the addition of this block to the blockchain, the nodes will add the new block to their system and the transactions inside it will be successful and final.

The most well-known and first use of blockchain is Bitcoin. In the Bitcoin network, every block has details about transactions (who sent the money, who got it, and how much was sent), a special code called the block hash, and the code for the block before it. In the Bitcoin network, computers check transactions and make new blocks using a process where everyone agrees.

5.31 Conclusion

In this chapter, we discussed the algorithm for recording data in the blockchain, that is, the consensus algorithm. We got acquainted with the most important consensus algorithms, i.e., PoS and PoW,, and the advantages and disadvantages of each were examined. We found that the PoW algorithm used in the blockchain network is expensive due to the required hardware and high energy consumption, but on the other hand, the PoS consensus algorithm imposes a lower cost on members. But there are other algorithms, some of which are a combination of the PoS and PoW algorithms, and others are completely different.

References

1. Alsunaidi, S.J. and Alhaidari, F.A. (2019, April). A survey of consensus algorithms for blockchain technology. In: *2019 International Conference on Computer and Information Sciences (ICCIS)* (pp. 1–6), IEEE.
2. Bach, L.M., Mihaljevic, B. and Zagar, M. (2018, May). Comparative analysis of blockchain consensus algorithms. In: *2018 41st International Convention on Information and Communication Technology, Electronics, and Microelectronics (MIPRO)* (pp. 1545–1550), IEEE.
3. Bamakan, S.M.H., Motavali, A. and Bondarti, A.B. (2020). A survey of blockchain consensus algorithms performance evaluation criteria. *Expert Systems with Applications*, *154*, 113385.
4. Beiranvand, F.S., Sabzevari, Y. and Eslamian, S. (2026). Underground dam: Theory and application. In: *Hydrosystem Restoration Handbook*. Elsevier, (pp. 445–461).

5. Bhardwaj, R. and Datta, D. (2020). Consensus algorithm. In: *Decentralised Internet of Things: A Blockchain Perspective* (Springer Nature Link). pp. 91–107.
6. Chaudhry, N. and Yousaf, M.M. (2018, December). Consensus algorithms in blockchain: Comparative analysis, challenges, and opportunities. In: *2018 12th International Conference on Open Source Systems and Technologies* (*ICOSST*) (pp. 54–63). IEEE.
7. De Angelis, S. (2018). Assessing security and performances of consensus algorithms for permissioned blockchains. *arXiv preprint arXiv:1805.03490.*
8. Ferdous, M.S., Chowdhury, M.J.M., Hoque, M.A. and Colman, A. (2020). Blockchain consensus algorithms: A survey. *arXiv preprint arXiv:2001.07091.*
9. Eslamian, S., and F., Eslamian. (2025). *Blockchain Technology in Water and Environmental Systems*. CRC Press, Taylor and Francis, USA, p. 282.
10. Hao, Y., Li, Y., Dong, X., Fang, L. and Chen, P. (2018). Performance analysis of consensus algorithm in private blockchain. In: *2018 IEEE Intelligent Vehicles Symposium* (*IV*) (pp. 280–285), IEEE.
11. Huang, D., Ma, X. and Zhang, S. (2019). Performance analysis of the raft consensus algorithm for private blockchains. *IEEE Transactions on Systems, Man, and Cybernetics: Systems*, *50*(1), 172–181.
12. Li, Y., Wang, Z., Fan, J., Zheng, Y., Luo, Y., Deng, C. and Ding, J. (2019, October). An extensible consensus algorithm based on PBFT. In: *2019 International Conference on Cyber-enabled Distributed Computing and Knowledge Discovery* (*CyberC*) (pp. 17–23), IEEE.
13. Mingxiao, D., Xiaofeng, M., Zhe, Z., Xiangwei, W. and Qijun, C. (2017, October). A review on consensus algorithm of blockchain. In: *2017 IEEE International Conference on Systems, Man, and Cybernetics* (*SMC*) (pp. 2567–2572), IEEE.
14. Ongaro, D. and Ousterhout, J. (2014). In search of an understandable consensus algorithm. In: *2014* (*USENIX*) *Annual Technical Conference* (*USENIX-ATC*) (Vol. 14, pp. 305–319).
15. Pahlajani, S., Kshirsagar, A. and Pachghare, V. (2019, April). Survey on private blockchain consensus algorithms. In: *2019 1st International Conference on Innovations in Information and Communication Technology* (*ICIICT*) (pp. 1–6), IEEE.
16. Xiong, H., Chen, M., Wu, C., Zhao, Y. and Yi, W. (2022). Research on progress of blockchain consensus algorithm: A review on recent progress of blockchain consensus algorithms. *Future Internet*, *14*(2), 47.

6 | Hash Functions

Yaser Sabzevari[1*] and Saeid Eslamian[2]

Hash functions are fundamental components of modern cryptography and digital security, providing a mechanism for transforming data of arbitrary size into fixed-length outputs that uniquely represent the original input. This chapter presents a comprehensive overview of hash function theory, design principles, and practical applications. Core properties, such as determinism, pre-image resistance, second pre-image resistance, and collision resistance, are explained to highlight the security requirements that underpin strong cryptographic hashing. The chapter examines widely used hash algorithms, including SHA-2, SHA-3, and legacy functions like MD5 and SHA-1, discussing their structures, strengths, and known vulnerabilities. Practical applications in blockchain systems, digital signatures, data integrity verification, password storage, and distributed protocols are also explored. Additionally, emerging challenges and future directions, such as post-quantum hashing and advanced hash-based cryptography, are considered. This chapter equips readers with a thorough understanding of how hash functions support secure, reliable, and efficient digital systems.

6.1 Introduction

Hash function, abstract function or mathematical function is said to convert a large amount of data (probably an unknown amount of data) into a natural number. The natural number resulting from the shuffling function is used as the index of an array. The values obtained from this function are usually called the hash value or just the hash or the hash. These functions have wide applications in computer science, including maintaining data integrity (Sateesan et al., 2023; Mishra et al., 2021; Sabzevari et al., 2025).

Shuffle functions are mostly used to speed up table lookups, data compression, insertion, or deletion for dynamic sets whose number of elements changes over time, such as searching for something in a database, detecting duplicate records in a large volume of data, or similar stretches. In DNA sequence and many similar applications (Hu et al., 2022; Eslamian et al., 2025; Karthik and Bala, 2022). In this article, we talk about the structured data called the hash table and the hashing method, which, with proper design, performs all these actions in the worst case and also in the average case with O (1) cost.

[1,2] Department of Water Science and Engineering, College of Agriculture, Isfahan University of Technology, Isfahan, Iran.

* Corresponding author: yasersabzevari1511@gmail.com

DOI: 10.1201/9781003686026-6

Assume the keys are integers ranging from 1 to 100, with approximately 100 entries present. A good way to organize data is to make an array called S that has 100 slots. Then, put the record with key i in slot Si. The recovery happens right away, and the keys are not matched. If there are around 100 records and the keys are 9-digit ID numbers, the same approach can be used. However, it saves a lot of memory because you would need an array with a billion spots to keep just 100 records. Instead, we can make a list with 100 spots, numbered from 0 to 99, and match each key to a number between 0 and 99. A hashing function is a process that changes each key into a number called an index. When we use the hashing function with a particular key, it is called a "hash key". In the case of identification numbers, a possible hashing function is:

$$h(\text{key}) = \text{key}\ \%100 \qquad ...(1)$$

(% means the remainder of dividing the key by 100). This function simply returns the last two digits of the key. If a hash key is hashed to i, we store its key and record it in Si. This strategy does not keep the keys in order, meaning this function can only be used if there is never a need to efficiently retrieve the data in order. If this is needed, one of the previously discussed methods should be used. This method is effective as long as no two keys point to the same spot (Figure 6.1). But this usually doesn't happen when there are a lot of keys. For instance, if there are 100 keys, the number of cases in which no two keys are hashed to the same index, compared to the number of cases in which each key is hashed to each of the 100 indices, is:

$$\frac{100!}{100^{100}} \approx 9.3 \times (10^{-43}) \qquad ...(2)$$

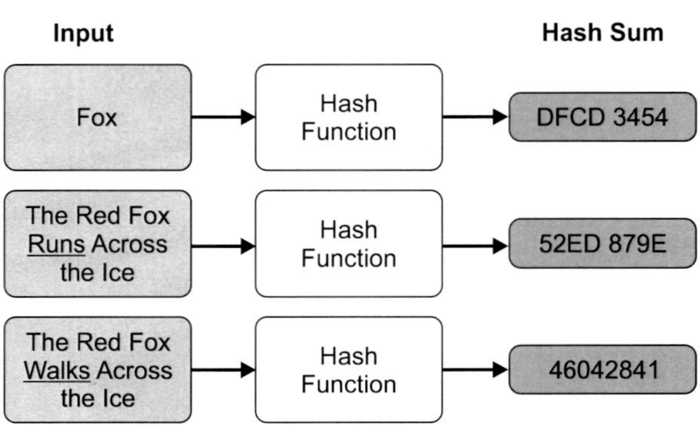

Fig. 6.1 A list of hash functions.

6.2 Direct Addressing Table

In direct addressing, it is assumed that the global set U is relatively small and k, where m is not too large, to display the dynamics of these elements, an array called the direct addressing table is used, in which only one element with key i can be placed in the i-th directory. And other folders are empty or null. But there is still the problem that this method is not responsive for large m and to solve it, the hashing table is used (Orea-Flores et al., 2022).

6.3 Show the Addition of a New Key

We are fairly confident that a minimum of two keys will converge in the same location.An event like that is called a crash (Wang et al., 2021). There are different ways to settle disputes. One of the best methods is to use open interleaving, also called open addressing. In open hashing, we make a storage space (called a bucket) for each possible hash value. We then place all items that have the same hash value into their matching bucket. Open hashing is often done using linked lists. For example, if we take the last two digits of a number and use them, we make a list of pointers that go from 0 to 99. All the keys that match i are stored in a linked list that starts at [Bucketi]. This is shown in Figure 6.2.

There is another method to solve the collision called the chain method, which will be explained later.

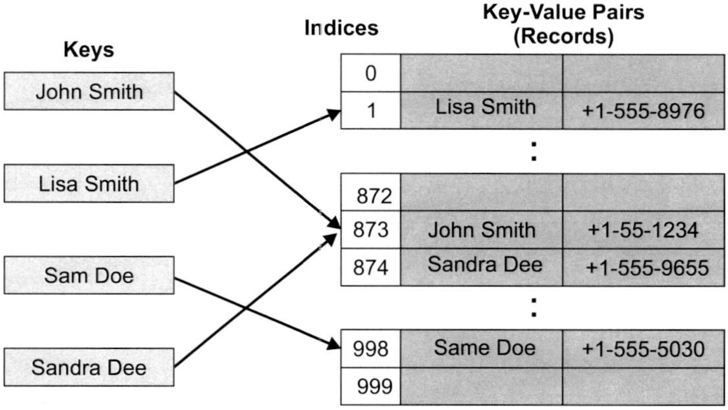

Fig. 6.2 Representation of open hashing and hashing table.

The quantity of buckets does not have to match the quantity of keys. For example, if we only look at the last two digits, we need 100 buckets. However, we can keep 1000, 200, 100, or any amount of keys in those buckets. The more keys we have, the more chance there is for them to overlap. If there are more keys than buckets, we will definitely have a collision. Since the bucket only keeps one pointer, it doesn't use up much space. So, it's usually a good idea to have a number of buckets that match the number of keys. When looking for a key, we need to check the list of items (linked list) where the key is stored, one by one. If all the keys are hashed into a bucket, the search reduces to an ordinal search. How likely is it to happen? If we have 100 keys and 100 buckets, the probability that the keys will all end up in the same bucket is:

$$100 \times \left(\frac{1}{100}\right)^{100} = 10^{-198} \qquad ...(3)$$

So, it's very unlikely that all the keys will fit into the same bucket. What about the probability that 70, 80, 90, or any number of keys end up in a bucket? Our main goal is to know how likely it is that shuffling, on average, leads to a better search than binary search. We show that if it is large enough, this is also almost a certainty; but first, we show how much better things get with hashing. Obviously, the best that can happen is that the keys are evenly distributed in the buckets. That is, if there are n keys and m buckets, each bucket contains n/m keys. In fact, each bucket contains exactly n/m keys when n is a multiple of m. If not, we will have a fairly uniform distribution. The following theorems show what happens when the keys are uniformly distributed. For simplicity, we express them for an exactly uniform distribution (that is, for the case where n is a multiple of m).

6.4 A Chain Method for Collision Resolution

In the chain method, we make the elements that are written to a directory of the hashing table by the hashing function into a one-way linear list and place its starting address in the same directory, i.e., $T[i]$. In the general case, the elements with different keys *k*, whose value is equal, are placed in the linked list, and the empty contents of the table are also empty lists. In this method, the insertion is performed in time, but the search and deletion operation is accompanied by an element search operation, which in the worst case has a cost proportional to the length of the list.

Scrambling functions

A good scrambling function is a function that is simple and monotonous, in the following we describe different simple methods to design a scrambling function (Hu et al., 2020).

6.5 Multiplication Method

In this method, first multiply the key *k* by a fixed number *A* (and get its decimal part. That's mean:

$$h(k) = \lfloor m(kA)\,\mathrm{mod}(1)) \rfloor$$

$$(kA)\,\mathrm{mod}(1) \leftrightarrow kA - \lfloor kA \rfloor \qquad ...(4)$$

Suppose that the computer word is *w* bits and *k* fits in one word. We choose A to be equal to $\frac{s}{2^w}$, where *s* is an integer in the range $0 < s < 2^w$. First, we multiply *k* by the *w*-bit integer $s = A.2w$, and the product is a 2*w*-bit number $r_1 2^w + r_0$, where high-value words and low-value words are the product of the multiply. The p-bit number of the desired hashing function is the most significant p-bit.

6.6 Double Interleaving

If two records are transferred to the same place in the file in the hashing functions, it is called collision (Rashidi, 2021). The ideal way to deal with collisions is to find a transformation algorithm that avoids collisions altogether.

One of the ways to prevent collisions is to use the linear verification method.

The linear verification method, having an auxiliary matching function called $g : U -> 0, 1, \ldots m - 1$, uses the following function $h(k, i) = (g((k) + i) \bmod (m)$ for hashing: for $i = 0, 1, \ldots m - 1$, for a key k, we check first $T[g(k)]$, then$T[g(k) + 1]$ we go to and then to $T[g(k) + 2]$. And we continue this work in a circular manner until we reach. In $T[g(k) + m - 1]$ this case, only m distinct check sequences are generated. Double hashing is a verification method in which the functions in the hash are as follows:

$$h(k, i) = (h(k) + ig(k)) \bmod(m)$$

$$g(k) = 1 + ((k) \bmod(11))\ h(\mathrm{k}) = (k) \bmod(13)\ g(k) \bmod(m)\ T(h(k))$$

Double shuffling works better than linear and degree-2 checking, because $m^2(\theta)$ compare $m(\theta)$ produces more than the number of checks. Therefore, it seems that the efficiency of this scrambling is closer to the efficiency of monophonic scrambling.

6.7 Global Hashing

When using hashing functions, if the function we are using is understood by the user, the user can enter his data in such a way that the number of occurrences is high. To prevent this from happening, we use global hashing.

Assume a family of hashing functions called H.

We say that H is a family of global hashing functions if for both keys $k1$ and $k2$ that are members of the set of keys we want, the number of hashing functions like h that are members of H is at $h(k1) = h(k2)$ most equal $\frac{|H|}{m}$.

When the user enters a set of data, we randomly select one of the hashing functions and store the entries in memory using that function. In this case, the user can no longer guess what function is selected and as a result cannot determine the input values in such a way that the number of collisions is high.

6.8 Complete Hashing

Sometimes in hashing we deal with static keys, which are already known. In this case we call a hashing method complete if, in the worst case, the number of memory accesses for a search operation. The main idea of this type of hashing is to use two hashing levels and a global hashing function for each level.

6.9 Dynamic Shuffling

In the methods we have discussed so far for shuffling, it is assumed that the number of elements is at most equal to the size of the table, the open method, or almost equal to it, the chain method, while in many cases the number of elements is not predetermined and cannot be a good upper limit. For the size of the hashing table, it predicted that it would not waste memory. To solve this problem, the dynamic hashing method is used, in which the size of the table is variable according to the need; we have to spend money to change the size of the table, but it has been shown that even with the necessary cost to change the size of the tables, the total cost remains the same .

6.10 Hashing

In the hashing process, inputs with different values are converted into an output with a fixed value through a mathematical formula. In other words, by hashing a value that can be an image, letter, or numbers with an unknown length, we convert it into an output with a specific and definite length. This process is also called the scrambling function (Figure 6.3).

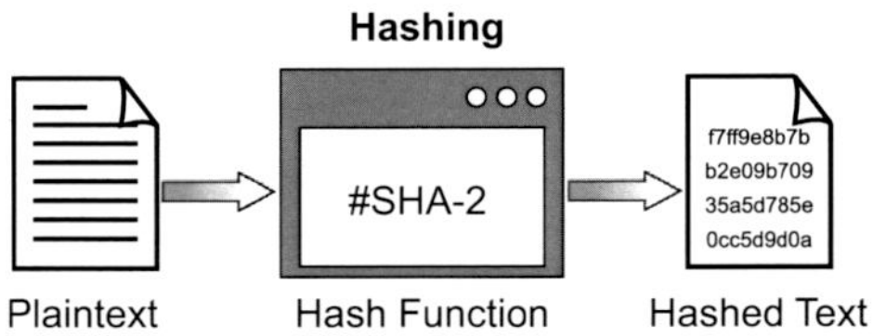

Fig. 6.3 An example of the hash function.

The hashing process is very useful in the basics of information security, and its most important application is when a person intends to send a message in a completely private manner through the creation of a formula, only to one recipient. The output of hashing is called Hash or hash value and its mathematical function is known as Hash function.

Hash Function

Hash functions are mathematical algorithms that receive data with variable size and length and convert them into the same and definite output. Hash functions are the main pillar of the hashing process.

In other words, it can be said that the Hash function receives a variable-length code string and converts it into a fixed-length output.

Main features of the hash function

- Each hash function must include special features that we will introduce below:
- The length of the hash function output or hash value is always constant.
- As long as the input does not change, some output values of the Hash function are definite and fixed.
- The hash value is usually much smaller than the input. For this reason, the hash function is also called a compressor.
- The operation of the hash function is different from cryptography.
- Hash functions are designed one way. In other words, in hashing, it is possible to get the output from the input, and its opposite operation, that is, getting the input from the output, is almost impossible.

6.11 Hash Function Applications

Hash function in blockchain has 3 main uses:

1. Save Password
One of the main applications of these algorithms is to store passwords on the web. The hash function does not store passwords transparently and stores them as a hash formula. When attacking the system, hackers can only obtain the hash output, which according to the explanation we gave a little earlier regarding the one-way nature of the hashing operation; It is impossible to get input via Hash value. As a result, saving passwords in this way is highly secure.

2. Maintaining Security
Hash functions play an important and fundamental role in maintaining the security of blockchain networks and the cryptocurrency space. The encryption process in blockchains is done through these algorithms. The data on each block is converted into a string of code through the hashing process and joined to the main chain.

3. Checking the Accuracy of the Data
Another application of hash algorithms is to produce Check-Sum on files. Checksum gives users assurance of data integrity and helps them quickly notice any changes to files. However this system does not guarantee the authenticity of the file, and hackers can generate a new Hash by changing the file and gaining access to the file data. Using Checksum is useful when the user is confident about the authenticity of the file.

6.12 How the Hash Function Works

As we said, Hash functions produce outputs of different sizes according to their construction style; But the output size of a particular type of Hash function is always the same. For example, the SHA-256 algorithm, which is the hash function of Bitcoin and the most widely used hash in the world. It is only capable of producing 256-bit output. Whatever the length and value of the input, the hash value will be 256 bits.

In the following example, we put the word Ircc with 2 cases of uppercase and lowercase in the Hash function of Bitcoin, which is SHA-256:

Ircc:
65592d00da56347062e6dc84e8ff978ba021267b288bbe3c7f2cb5c04d797bd4

Ircc:
478734e921cc9b571d7ab5fc226dcb2ce59fcab83fb73258ab8082382d86c6ee

As you can see, the output hash code changes by changing the lowercase and uppercase of just one letter, but both forms produce the same 256-bit output value. So far, we have mentioned many times that no matter what the input value is, a Hash function has a constant output value.

Note: Output with a fixed length increases security. If someone tries to decrypt and attack the network, he will not be able to recognize the input with this method.

Another popular function is SHA-1, which stands for Secure Hash Algorithms. This expression contains a set of cryptographic hash functions, which are SHA-0, SHA-1, SHA-2, and SHA-3.

The SHA-256 function, along with SHA-512 and other variants, are part of the SHA-2 family. Currently, only SHA-2 and SHA-3 groups are considered secure.

Now if we put the above example in the SHA-1 function. The following function values will be obtained:

Ircc: a6f9af856be33de008c9f5ffb942e2c4e2d64

Ircc: 3ec8e2b221402e786dfa6f0884a0e7f7be11c

This function, like all functions, produces constant values from inputs with variable length, and its Hash value changes if the input code changes.

What is a cryptographic hash function?

The cryptographic hash function is the same Hash function whose features we have described so far. The difference is that this algorithm uses the encryption process to double the security.

We said earlier that getting the input from the hash value is very difficult and almost impossible. This has made the hashing process safe and reliable. However, developers have tried to multiply this security through encryption in the encrypted Hash function.

These functions are a special type of Hash algorithms that have their own characteristics; These features have caused them to be used for identity recognition, maintaining greater security, creating digital signatures, and verifying the authenticity of data in various digital systems.

6.13 Features of Cryptographic Hash Function

The cryptographic hash function is one of the main foundations of a secure blockchain (Karthik & Bala, 2022). Below we review its features:

Certainty:
A hash function returns a fixed Hash value for a given input. It doesn't matter how many times this input is placed in the function. The output will always be constant. This feature makes tracking data easier. If a different hash value is provided each time for a fixed input. Users will face a lot of hash codes that are difficult and time-consuming to collect and track.

High speed in calculation:
The hash function should quickly convert the input into a hash value. If this calculation is not fast enough, the system is inefficient.

One-sidedness:
The cryptographic hash function is a one-way function. That is, in contrast to the input value, the output is easily obtained. However, the opposite operation of this process cannot be said to be completely impossible, but it is difficult and impractical.

Imagine that we put the numbers between 5 and 10 in this function and get their fixed outputs. In this short list and according to the certainty of the Hash value, it is possible to discover the input value of each output by replacing the inputs in the function.

However this operation is a very difficult and time-consuming process when we are faced with a long list of input values, and it can be said to be impossible. In this situation, comprehensive search, which is also called unwise or raw search, is used.

In this way, try all the outputs one by one and randomly with the input data to get the final and correct result. In this search, the chance of finding the right answer is very low and, as we said, it is a far from wise method.

Output changes with input changes:
The slightest change in the input value of the cryptographic hash function will completely change the output Hash value. For example, when you enter your Instagram password, the system will grant you access to the user account based on the previous hash value it has stored for this entry. If the slightest change is made in the password you enter, the previous Hash output will no longer work and the system will not allow access to your profile and will warn you that the password is incorrect.

Different hash values for each entry:
Each input in the hashing process will have its own Hash value, and it is not possible for the hash of two different input values to be similar or equal.

It can be said that two different input values will never have the same Hash output under any circumstances. If input A has a hash value of H3, it is not possible for input B to produce the same output.

Consistency of the output amount:
During the hashing process, for each input value, even with very large sizes, the length of the hash value will always be constant and the same. If we put the expressions "Hello, have a nice hot summer day" and "Hello" in a 256-bit hash function, the output Hash of both will be the same, and the longer phrase will not have a longer Hash. This method shortens the calculation time and also makes it easier for users to collect and track the output data.

Using the hash function in digital currency mining:
In the process of mining digital currencies, for example, Bitcoin, one of the steps that miners must go through is the use of the hash function. The use of hashing by many miners during the process of extracting new blocks is one of the main reasons for blockchain security.

A miner has to check multiple inputs to create the hash block it is mining. They can only finalize the validation of their new block when they find a Hash value with a specified number of leading zeros. The value of these zeros determines the difficulty of the mining process, and their number depends on the hash value and is determined by blockchain networks.

Hash rate:
If you have read about the use of Hash functions in the process of mining digital currencies; Undoubtedly, you have come across the term hash rate. Hash rate or Hash rate indicates the amount of network processing power in Bitcoin mining and is a measure to measure the performance of a mining device. It can be said that the hash rate shows the number of hashing operations that are running in the context of network mining processes.

If the hash rate of the network increases due to the increase in digital currency mining operations. The system automatically adjusts the mining difficulty so that the average required to mine each block in the Bitcoin network remains the same as 10 minutes. In another case, if the hash rate decreases because several miners stop mining at the same time; Again, the network adjusts the mining difficulty in such a way that the average mining time does not change.

Hash rate calculation:
Hash per second is the most common scale for measuring hash rate, which shows how many hashes are done per second. A number of other hash rate measurement scales that have actually only been converted to the unit are:

- 1 KH/s or thousand Hash per second
- 1 MH/s or one million Hash per second
- 1 GH/s or one billion Hash per second
- 1 TH/s or one trillion Hash per second
- 1 PH/s or one quadrillion Hash per second
- 1 EH/s or one quintillion Hash per second

The above digits are named Kilohash, Megahash, Gigahesh, Terahesh, Petahesh, and Exahash, respectively.

6.14 Commonly Used Hash Functions

In this section, the most commonly used Hash functions are presented (Tutueva et al., 2020).

MD5

MD hash function, which is short for Message Digest and means message summary. It includes MD2, MD4, MD5, and MD6 hash functions. In 1991, Ronald Rivest designed the MD5 function as a replacement for its previous version, MD4. MD5 produces a 128-bit hash and was designed as a cryptographic hash that was used for a long time, but was gradually abandoned from cryptographic processes due to its many flaws. Nowadays MD5 hash function is used for checksum.

SHA

We mentioned earlier that SHA is a secure hash algorithm. This hash function was designed by the American National Institute of Standards and Technology (NIST) and other security organizations. SHA-0, the first member of the SHA family, was introduced in 1993, and a little later, due to significant problems, a newer version was presented to fix the shortcomings of the previous version. The name of the new function was SHA-1.

SHA-1 was a 160-bit hash function similar to the MD5 function and was designed to be part of the digital signature algorithm. In 2005, security problems were also discovered in this hash function, which was likely to be penetrated by hackers.

So, the developers introduced the new and safe SHA-2 algorithm which consisted of 6 hash functions with different Hash values of 224, 256, 384m, and 512 bits. The names of these functions are also chosen according to the hash value produced by them; SHA-224, SHA-256 SHA-384, SHA-512, SHA-512/224, and SHA-512/256. The main difference between these algorithms is the length of their hash value.

We mentioned above that the SHA-256 algorithm is used in the Bitcoin blockchain and is the most widely used Hash function in the world. Also, this algorithm is used in TLS, SSL, PGP, SSH, multipurpose e-mail development, Internet/secure and IPsec, or Internet Protocol security.

SHA-3, the latest member of the secure hash algorithm family, was released in 2015 by NIST. In 2012, this function was selected as the best Hash function during a competition related to new algorithmic design. SHA-3 is actually a subset of a wider family called Keccak.

RIPEMD

RIPEMD is short for "RACE Integrity Primitives Evaluation Message Digest". This function, like the SHA algorithm, is a set of Hash functions that a Belgian research group named COSIC introduced in 1996. In the design of this function, the principles used in the design of the MD4 function and the SHA-1 functional system have been used. The functions of this set have different values of 128, 160, 256, and 320 bits and have high security.

Whirlpool

This algorithm is another encryption Hash function that was designed in 200 by Vincent Rayman and Paolo Bartou. Whirlpool is built on a modified version of the Advanced Encryption Standard (AES). This function receives any message with a length of less than 2256 bytes and converts it into a 64-byte hash value.

The Whirlpool hash function has been revised twice since 2000. In the first revision in 2001, the S-box was modified to make it easier to implement this algorithm on hardware. In 2003 and the second revision, a security problem was discovered in it and was quickly fixed by the developers of the function. The initial and original algorithm is called Whirlpool-0, the first revision is Whirlpool-T, and the final version is Whirlpool.

Blake

The Blake hash function is among the top 5 algorithms selected by the US National Institute of Standards and Technology (NIST). This algorithm, like SHA-2, consists of 2 different functions. One uses 32-bit words and the calculated hash length is at most 256 bits. The other uses 64-bit words with a hash length of 512 bits.

Blake2 was released in 2012. The purpose of designing this algorithm was to introduce an alternative to the widely used but vulnerable MD5 and SHA-1 algorithms in applications that required high performance. This function itself is divided into two subgroups Blake2b and Blake2s. Blake2 is said to be faster than MD5, SHA-1, SHA-2, and SHA-3, and is on par with SHA-3, the latest version of the secure hash algorithm, in terms of securityhash function; a member of the big world of data and cryptography.

Normal hash functions are used in processes such as searching in databases, analyzing files and heavy and voluminous data, managing data and information, and many other fields. Encrypted hash functions are also widely used in the field of information security, including message authentication, digital fingerprints, payment verification, and other information security issues. In the Bitcoin system, cryptographic hash functions play a role in the infrastructure, the process of extracting and creating addresses, public and private keys.

Also, these functions help us when we are faced with a large amount of information. In this case, a large file with a large amount of information can be given as input to a hash function, and then its output, which has a much smaller volume, can be easily used.

In general, it can be said that if there were no Hash functions, there would be no news of security in the web platform or the world of digital currencies today. The peace of mind we have when using the Internet and the world of digital currencies is due to the presence of these useful and vital functions that maintain the security of data and messages on these platforms.

The hash function works to maintain the integrity of the data in the blockchain network and is considered one of the main foundations of this network.

According to the studies conducted by experts in this field, the SHA-256 function is the most widely used in the network and is considered one of the best hash functions.

6.15 The Fastest Hash Function

For long network messages and data, the MD5 function is the fastest at 400 MB/s (Chen et al., 2022).

Cryptocurrencies are heavily dependent on hashing for their survival and security. Because most of the operations performed in their blockchain are based on hashing information in order to maintain their security. In the following, we discuss the role of hashing in digital currencies in the important section.

6.16 Authentication and Digital Signature

A digital signature serves as a means to verify the authenticity and reliability of digital data.

In simple terms, a digital signature is like a special code added to a message or document. This code shows that the message has not been changed while it was sent from the sender to the receiver. The idea of keeping communication safe with secret methods has been around for a long time. However, digital signatures became possible in the 1970s when public key cryptography was invented.

Public key cryptography isn't good for signing long documents. To save time, most digital signature methods use a short code of the message called a hash instead of encrypting the whole message.

Hashing is an important part of the digital signature system. In digital money, a digital signature system has three steps: creating a hash, signing it, and verification.

Hashing the data

The first step in the digital signature is data hashing. This process begins by giving information to a hashing algorithm and receiving the hashed value. As mentioned earlier, input messages may differ in size, but ultimately all hashed values by a particular algorithm have the same and fixed length.

Hashing the data is not necessary to create a digital signature, as the private key can be used to sign the information. But in the context of digital currencies, data is always hashed because dealing with fixed-length data is very effective in facilitating the process of blockchain operation.

Signing

After hashing the information into a code, the person sending the message needs to sign it. This is where public key encryption is used. There are different kinds of digital signature methods, and each one works in its own way. In simple words, the sender will use their private key to create a special version of the message. The person who receives the message can check if it's real by using the sender's public key.

In simple terms, if the private key isn't used to create the signature, the person receiving it won't be able to use the matching public key to check if the information is real. The person sending the message makes both the public and private keys, but only the public key is given to the person receiving the message or some other people.

It's important to remember that digital signatures are linked to the information in each message. Unlike handwritten signatures that look the same no matter what the message is about, each digital signature is unique for every message.

Validation

Let's go through the whole validation process step by step using an example. Imagine that Mary wrote a message to Muhammad, created a unique code from it (called a hash), and then mixed that code with her secret key to make a digital signature. This signature will serve as a one-of-a-kind digital mark for that message. Therefore, keeping the private key secret and secure is essential for anyone. If someone else has access to Maryam's private key, they can easily pretend to be Maryam and generate digital signatures that lead to fraud and access to her private information. In the context of digital currencies, this means that anyone with access to Maryam's private key can withdraw coins from her account without her permission.

6.17 Blockchain Immutability

In blockchain, hashes are used to represent the current state of the network, so that the input data represents everything that has happened in the blockchain. In fact, every transaction that has taken place up to that moment is combined with the new data that is added. This means that the output hash is based on all previous transactions that happened on the blockchain (Eslamian and Eslamian, 2025).

The first hash for the first block or genesis block is calculated according to the transactions recorded in that block. For each new block that is created from now on, in addition to the transactions of the desired block, the hash of the previous block is also given as input to the hash function and the hash value of that block is calculated. In this way, a chain of blocks is created.

The new hash block is created by referring to the hash block before it. This hashing system is actually a guarantee that proves to us that no manipulation of transactions is possible. Because if only a small part of the transaction information is changed, as a result, the hash block in which the transaction is recorded will also change, and in the same way, the hash blocks after it will also change.

To change a block in the blockchain, the hash of all previous blocks must also be changed because each block has the hash of the previous block in addition to its own hash.

As a result, any fraud and manipulation are prevented, because it is only enough to compare the hashes with each other. This is an interesting and exciting feature of blockchain because everyone on the network only needs to agree on one hash, confirm that the transactions of the new block are correct, and finally consent to the creation of that block. If the hashes do not match, they can easily detect the fraud and reject it. In this way and in this simple way, hashing makes the blockchain immutable, a feature that is considered to be the most prominent feature of this technology.

6.18 Mining

Mining is the process of finding and creating a new block and adding it to the blockchain. As you know, the supply of many digital currencies is limited to a certain number, which is 21 million units for Bitcoin. If miners can mine new blocks very quickly and with minimal computing power, all the coins would be mined in a short period of time. In addition, in the blockchain, there is a time limit for creating each block. If the blocks are mined in less time than the intended time, the following problems arise:

- **More collisions:** More hash functions will be generated, resulting in more collisions.
- **More orphan blocks:** If a large number of miners are able to mine a new block, then they may arrive at a new block at the same time. But as you know, only one of these blocks will be part of the chain and the rest will be abandoned, which are called orphan blocks.

Therefore, in order to limit block production, a certain difficulty level is considered for calculating and mining new blocks. Mining is like a game. You solve the puzzle and get a prize. Considering the difficulty makes solving the puzzle more difficult and time-consuming. For example, Bitcoin's target difficulty is a 64-character string that starts with a few zeros. As the level of extraction difficulty increases, the number of these zeros also increases. The difficulty level changes after every 2016 block.

When mining software wants to add a new block to the blockchain, they do it like this. When a new block is added to the network for mining, everything in the block is turned into a code. If the hash number is smaller than the target level of difficulty, it gets added to the blockchain, and everyone on the network needs to agree on the new block.

Getting a new block isn't easy. This is where details matter. A nonce is a random string of letters and numbers that is added to a hash block. This new linked string is then turned into a code and checked against how hard it is. If the value isn't less than the target difficulty level, the nonce needs to be changed, and this can be done millions of times until the right conditions are achieved. When this happens, the new block gets added to the blockchain.

6.19 Determinant Criteria in the Quality of Hash Functions

1. **Determinism:** The result of the function must be unchanging in any situation. Regardless of when the function is used, who uses it, or how many times the function is used, if the original message is the same, the result must be the same.
2. **Efficiency:** The hash function must make it possible to obtain an immediate result, otherwise it risks slowing down the computer systems that depend on it too much.

3. **Resistance to attacks:** It is not possible to discover the original message through reverse engineering unless all possible outcomes are tried one by one.
4. **Difference and differentiation:** Two very similar messages should produce two very different results.
5. **The uniqueness of the results:** It should be impossible to produce two identical results with two different messages (Khalili et al., 2020).

6.20 Updated Encryption in Functions

Of course, these criteria change over time. Yesterday's ultimate security is nothing but mediocre in today's systems. And some computational tasks that were once considered enormous have become ridiculously easy for modern processors.

For example, the salting technique slightly slows down the performance of the hash function. In this technique, instead of directly calculating the hash of the original message, a sequence of characters is randomly generated and mixed with the message text. This method does not allow the attacker to calculate the message by viewing the fingerprint.

There is another similar method whose code is called pepper. This method is actually a sequence of secret and non-random characters, like some kind of additional password that is added to the message before calculating the hash of the message. However, the principle is the same. This is also an additional step in the hash function that adds complexity and slows down the performance of the computer that executes it. Fortunately, the power of today's computers has increased to the point where these two functions have negligible impact on the hash functions that use them. While these operations were very time-consuming in 1980, today they are only minor obstacles and it would be a shame not to resort to them, as they greatly increase the safety of the operation.

Of course, there are not only these two methods. Merkel trees, key stretching, and many other techniques allow cryptographers to improve security in any number of ways, but at the cost of slowing down the performance of the hash function. Finally, it should be noted that the criteria listed above do not all apply to hash functions in the same way, depending on the intended goals. Most hash functions should produce the result immediately, but in some contexts where security is a priority, a delay of ten seconds is perfectly acceptable. In short, it all depends on what you want to do with it. These criteria are only ideals, not rules.

6.21 Conclusion

Hash functions are the heart of digital currencies. Functions are constantly being updated and evolved. They are the pillar on which the entire ecosystem rests. It is therefore essential that you at least know the name of the function used by the digital assets you are investing in because an error in a hash function is equivalent to an error in the cryptocurrency that uses that function. And if you have the misfortune to discover that there is a flaw in the performance of a particular function, consider all digital assets directly or indirectly related to it. Because such an error means the loss of the full value of a digital currency.

References

1. Chen, Q., Shen, Z. and Chen, Z. (2022). A penalty function semi-continuous thresholding methods for constraints of hashing problems. *Journal of Visual Communication and Image Representation, 87*, 103552.
2. Eslamian, S. and Eslamian, F. (2025). *Blockchain Technology in Water and Environmental Systems*. CRC Press, Taylor and Francis, USA, p. 282.
3. Eslamian, S., Fallah, A.E. and Sabzevari, Y. (2025). Optimizing Rain Gauges with the cuckoo Algorithm and Entropy of Information Transfer: A Case Study on the Gavkhouni Basin in Iran. *Environmental Earth Sciences, 84*(15), 436.
4. Hu, S., Zhang, R., Wang, F., Chen, K., Lian, B. and Chen, G. (2022). A sanitizable signcryption scheme with public verifiability via chameleon hash function. *Journal of Information Security and Applications, 71*, 103371.
5. Hu, Y., Cheng, G., Tang, Y. and Wang, F. (2020). A practical design of hash functions for IPv6 using multi-objective genetic programming. *Computer Communications, 162*, 160–168.
6. Karthik, P. and Bala, P.S. (2022). A new design paradigm for provably secure keyless hash function with subsets and two variables polynomial function. *Journal of King Saud University-Computer and Information Sciences, 34*(5), 1933–1949.
7. Khalili, M., Dakhilalian, M. and Susilo, W. (2020). Efficient chameleon hash functions in the enhanced collision resistant model. *Information Sciences, 510*, 155–164.
8. Mishra, N., Islam, S.H. and Zeadally, S. (2021). A comprehensive review on collision-resistant hash functions on lattices. *Journal of Information Security and Applications, 58*, 102782.
9. Orea-Flores, I.Y., Rivero-Angeles, M.E., Onofre-Soto, A.L., Azpeitia-Rebollar, E.G., Torres-Cruz, N., Villordo-Jiménez, I. and & Menchaca-Mendez, R. (2022). Teletraffic analysis of energy-efficient intruder detection using hash function techniques in images for remote monitoring in Wireless Sensor Networks. *Computers and Electrical Engineering, 103*, 108373.
10. Rashidi, B. (2021). Efficient full data-path width and serialized hardware structures of SPONGENT lightweight hash function. *Microelectronics Journal, 115*, 105167.
11. Sabzevari, Y., Eslamian, S., Pamula, A.S.P. and Bazrkar, M.H. (2025). Drought Trend Analysis Using Standardized Precipitation Evapotranspiration Index in Cold-Climate Regions. *Atmosphere, 16*(4), 482.
12. Sateesan, A., Biesmans, J., Claesen, T., Vliegen, J. and Mentens, N. (2023). Optimized algorithms and architectures for fast non-cryptographic hash functions in hardware. *Microprocessors and Microsystems*, 104782.
13. Tutueva, A.V., Karimov, A.I., Moysis, L., Volos, C. and Butusov, D.N. (2020). Construction of one-way hash functions with increased key space using adaptive chaotic maps. *Chaos, Solitons, and Fractals, 141*, 110344.
14. Wang, Y., Chen, L., Wang, X., Wu, G., Yu, K. and Lu, T. (2021). The design of keyed hash function based on CNN-MD structure. *Chaos, Solitons, and Fractals, 152*, 111443.

7 Consensus Algorithm and Blockchain

Tanuj Surve[1*] and Amit Tyagi[2]

Consensus algorithms are the foundational mechanisms that enable blockchain networks to operate securely, transparently, and without centralized control. This chapter explores the deep interconnection between consensus mechanisms and blockchain architecture, highlighting how decentralized agreement is achieved across distributed nodes. Core concepts such as trustless coordination, ledger immutability, and fault tolerance are introduced to illustrate why consensus is critical for blockchain reliability. Major consensus models, including Proof of Work (PoW), Proof of Stake (PoS), Delegated Proof of Stake (DPoS), and Byzantine Fault Tolerant (BFT) protocols, are analyzed in terms of performance, security assumptions, energy consumption, and scalability. Real-world applications and blockchain platforms are examined to demonstrate how different consensus choices influence system behavior and design trade-offs. The chapter also discusses emerging innovations in hybrid, energy-efficient, and scalable consensus systems. By linking theory with practical implementation, this chapter provides a clear understanding of how consensus algorithms ensure integrity and trust within blockchain ecosystems.

7.1 Introduction

Blockchain technology has appeared as a transformative power, reforming how we theorize and achieve digital dealings (Mwandosya & Luhanga, 2021). Blockchain is founded on the ideas of decentralization, transparency, and security (Puthal et al., 2018; Sabzevari et al., 2025). This chapter presents an impression of blockchain and dispersed ledger technology (DLT), highlighting the vigorous role of consensus algorithms in the operation of blockchain networks.

7.1.1 *Background on Blockchain and Distributed Ledger Technology*

Blockchain, which was initially imagined as the supporting innovation for cryptocurrencies such as Bitcoin, has advanced into a flexible instrument with

[1] University of California, Berkeley.
[2] National Institute of Fashion Technology.
* Corresponding author: tanujsurve@gmail.com

DOI: 10.1201/9781003686026-7

employment in an assortment of businesses (Morabito, 2017). A blockchain is basically a decentralized and conveyed record that records exchanges over an organization of computers (Sarmah, 2018; Sabzevari and Eslamian, 2025). Since, of this decentralized nature, no single substance has control, cultivating belief and straightforwardness is important.

The overarching concept that covers blockchain is known as Dispersed Record Innovation (DLT). A decentralized database that's shared, replicated, and synchronized among a few members is alluded to as a DLT (Aviv, 2021). Blockchain could be a sort of DLT that organizes information into squares that are connected together utilizing cryptographic hashes, coming about in a permanent and obvious chain of exchanges (Laroiya et al., 2020).

7.1.2 Importance of Consensus Algorithms in Blockchain Networks

Agreement calculations are at the heart of blockchain systems, permitting for exchange assertion and approval in a decentralized setting (Yao et al., 2021). A central specialist approves and records exchanges in conventional centralized frameworks. In any case, the nonattendance of a central specialist in a decentralized blockchain organization needs an instrument to preserve ascension among members on the state of the record (Pereira, et al., 2019).

The significance of agreement calculations can be paralleled to how a gathering of individuals make a collective choice. Within the nonattendance of a central decision-maker, agreement calculations serve as the convention that coordinates how members concur on exchange lawfulness and record state. This decentralized agreement secures the blockchain's judgment and security.

7.1.3 Overview of Consensus Algorithms Covered

This chapter burrows into a few agreement calculations, each of which is implied to manage with distinctive challenges and cater to particular utilize cases in blockchain systems. Understanding these calculations is fundamental for understanding the complex scene of blockchain innovation. This chapter incorporates the taking after calculations:

Proof of Work (PoW): Originally introduced by Bitcoin, PoW requires network participants—known as miners—to solve complex computational problems to confirm transactions and append new blocks to the chain. Although this mechanism offers strong security, it is highly energy-intensive (Lasla et al., 2022).

Proof of Stake (PoS): In this model, validators are selected according to the quantity of cryptocurrency they commit as a stake. By replacing energy-heavy computations with economic commitment, PoS presents a more environmentally sustainable alternative to PoW (Gola & Sedlmeir, 2022).

Delegated Proof of Stake (DPoS): DPoS allows token holders to vote for a limited set of representatives who are responsible for validating transactions and producing blocks. This delegation-based structure enhances both transaction throughput and operational efficiency (Saad & Radzi, 2020).

Practical Byzantine Fault Tolerance (PBFT): PBFT aims to achieve consensus in systems where some nodes may behave maliciously or fail. It provides agreement despite a predefined proportion of faulty participants (Meshcheryakov et al., 2021).

Directed Acyclic Graphs (DAGs): DAG-based systems offer an alternative to traditional blockchain architectures. Instead of storing transactions in sequential blocks, each transaction references multiple previous ones, forming a directed acyclic network that supports faster and more scalable validation (He et al., 2021).

Hybrid Consensus Mechanisms: These systems combine features from multiple consensus protocols to leverage their respective advantages. By blending different approaches, hybrid models seek an optimal balance of decentralization, security, and performance (Aggarwal et al., 2021).

Exploring these consensus mechanisms in depth reveals the distinct benefits, limitations, and practical applications that make each of them an essential element of the broader blockchain landscape.

7.2 Proof of Work

PoW represents one of the earliest and most influential consensus protocols within blockchain technology, first rising to prominence through its use in Bitcoin (Romano & Schmid, 2021). This section provides an in-depth exploration of PoW, examining how the mechanism operates, the key elements that underpin its process, and the broader implications it holds for the performance and security of blockchain systems.

7.2.1 How Proof of Work Functions

At its fundamental level, PoW serves as a decentralized consensus method that requires participants—referred to as miners—to expend computational effort by solving complex cryptographic puzzles (Kumar et al., 2019). The central purpose of this mechanism is to demonstrate a tangible investment of resources, primarily in the form of processing power and energy.

The process begins when a transaction is disseminated throughout the network. Miners then attempt to solve a cryptographic challenge associated with the transaction. This challenge—commonly called a hash puzzle—requires miners to identify a specific value (a nonce) which, when combined with the block's data, produces a hash output meeting pre-determined conditions. The miner who first discovers a valid solution broadcasts it to the network, where other nodes independently verify its correctness.

Once the solution is validated, the corresponding block is appended to the blockchain, and the successful miner receives a reward consisting of newly generated cryptocurrency along with transaction fees. This competitive, computation-heavy mechanism protects the blockchain by ensuring that no single actor can feasibly accumulate the majority of the network's processing power, thereby preserving its security and integrity.

7.2.2 *Mining, Hash Puzzles, and Block Difficulty*

Mining: Within the PoW framework, mining refers to the competitive process in which participants attempt to resolve cryptographic puzzles in order to confirm transactions and append new blocks to the blockchain (Han et al., 2019). To perform these intensive computations, miners rely on specialized machines known as mining rigs. As the number of miners increases, the level of competition also rises, making it progressively more difficult and resource-demanding to solve puzzles and validate blocks.

Hash Puzzles: Hash puzzles are computational challenges that require miners to identify a particular input value—referred to as a nonce—that produces a hash output meeting predefined condition, such as containing a specified number of leading zeros (Mardiansyah & Sari, 2022). The difficulty of these puzzles is adjustable to ensure that new blocks are generated at a predictable pace; for instance, in Bitcoin, approximately every ten minutes.

Block Difficulty: Block difficulty is a dynamic metric that determines how hard the cryptographic puzzle will be to solve. This parameter automatically scales based on the aggregate computational power (hash rate) of the network. As the combined hash rate grows, the difficulty level increases accordingly to maintain a stable and consistent interval between block creations (Dahiya et al., 2022).

7.2.3 *Pros and Cons of Proof of Work*

Advantages

High Security: PoW is widely recognized for providing strong security guarantees. Because validating a block demands substantial computational effort, attempting to alter the blockchain becomes prohibitively costly for attackers.

Promotes Decentralization: The PoW model supports a decentralized network structure by enabling any individual with appropriate hardware to join the mining process and contribute to consensus.

Disadvantages

Excessive Energy Use: A major drawback of PoW is its significant energy requirement. The continuous computation needed to solve complex cryptographic puzzles consumes large amounts of electricity, raising concerns about environmental sustainability.

Limited Scalability: PoW-based systems also face challenges in scaling. As more participants engage in mining, the network may experience congestion, resulting in slower transaction confirmation and reduced overall throughput.

7.2.4 Examples: Bitcoin and Ethereum (PoS Transition)

Bitcoin: Bitcoin—the first and most widely recognized cryptocurrency—employs the Proof of Work consensus mechanism (Bach et al., 2018). Its long-standing effectiveness has established PoW as a dependable and robust method for securing decentralized networks.

Ethereum and the Move Toward PoS: Ethereum, the second-largest blockchain platform, initially operated under the PoW model as well. However, concerns over energy demand and limited scalability motivated the development of Ethereum 2.0, which marks a transition to a PoSe framework. This shift aims to mitigate PoW's environmental footprint and enhance system performance.

PoWdecentralized means of achieving agreement across a distributed network. Yet, as the industry advances, the shortcomings of PoW have become more pronounced, encouraging the exploration of alternative mechanisms such as PoS. Ethereum's ongoing migration to PoS reflects the sector's broader commitment to improved scalability and environmental sustainability, demonstrating the evolving and adaptive nature of blockchain technology.

7.3 Proof of Stake

PoS was introduced as a promising successor to PoW, designed to address its high energy demands and limited scalability while still preserving the core values of decentralization and network security.

7.3.1 How Proof of Stake Functions

In contrast to Proof of Work—where miners compete using computational power—Proof of Stake selects validators according to the amount of cryptocurrency they commit as collateral, or "stake" (Saleh, 2021). The greater the amount staked, the higher the probability a participant will be chosen to propose the next block. This approach enables decentralized networks to achieve consensus in a far more energy-efficient and economically sustainable manner.

Validator Nodes: In PoS-based systems, validator nodes take on the role previously held by miners. Their responsibility is to verify transactions and produce new blocks. Selection is determined by the quantity of cryptocurrency they have locked as collateral, signaling their commitment to the network's security.

Staking: Staking requires users to deposit a specified amount of cryptocurrency into a designated wallet to become eligible for participation in the consensus mechanism. These locked funds serve as collateral, ensuring that validators' incentives are aligned with maintaining the system's reliability and trustworthiness.

Slashing: To discourage harmful or dishonest behavior, PoS networks employ a penalty mechanism known as slashing. Validators that attempt to validate invalid or malicious transactions risk losing a portion—or potentially all—of their staked assets. This enforcement tool helps preserve the integrity of the consensus process.

Unlocking: The assets committed as stake are typically restricted for a predefined period during which they cannot be accessed or withdrawn. After fulfilling their validation responsibilities, participants regain access to their staked funds through an unlocking procedure.

Governance: Many PoS networks incorporate governance frameworks that enable token holders to take part in shaping protocol updates and parameter changes. This participatory model enhances decentralization by giving stakeholders a direct voice in system evolution.

7.3.2 Pros and Cons of Proof of Stake

Pros

Energy Efficiency: PoS consumes far less energy compared to PoW because it eliminates the need for heavy computational operations and complex cryptographic puzzle-solving.

Decentralization: PoS supports a decentralized environment by enabling any token holder to engage in the validation process, creating a more open and widely distributed network of participants.

Scalability: Because validator selection requires fewer resources, PoS systems often achieve higher scalability, enabling quicker transaction confirmation and smoother network performance.

Cons

Wealth Concentration: A common criticism of PoS is that it may encourage an uneven distribution of influence, as individuals or entities with larger stakes gain greater control over network decisions, potentially weakening the system's decentralized foundation.

Security Issues: Some observers raise concerns about PoS's security—particularly in scenarios where a large portion of the cryptocurrency supply is controlled by a single validator or group, increasing the risk of manipulation.

7.3.3 Examples: Ethereum (Beacon Chain), Solana, Cardano

Ethereum (Beacon Chain)

Ethereum—one of the earliest and most influential blockchain platforms—has taken a leading role in moving from a PoW system to a PoS model through its Ethereum 2.0 roadmap (Leonardos et al., 2020). A major catalyst in this transformation is the Beacon Chain. Introduced as a separate PoS network running alongside the original Ethereum chain, the Beacon Chain establishes the structural foundation needed to create a more energy-efficient and scalable ecosystem.

Transitioning to PoS significantly cuts down the energy footprint associated with Ethereum's consensus process, effectively addressing long-standing environmental criticisms of PoW (Platt et al., 2021). This upgrade highlights Ethereum's commitment to continual improvement and its responsiveness to the evolving technological landscape of decentralized systems.

Solana: Solana distinguishes itself through its unique blend of PoS with an additional mechanism called Proof of History (PoH). PoH creates cryptographic timestamps for transactions before they are grouped into blocks, producing a verifiable sequence of events that boosts throughput and operational efficiency.

Within Solana's architecture, validator nodes are essential for upholding consensus. The combination of PoS and PoH allows Solana to achieve extremely rapid transaction processing, making it one of the high-performance networks in the blockchain sector. Its capacity to manage a substantial transaction load with minimal delay has positioned Solana as a strong candidate for decentralized applications requiring high-speed processing.

By integrating PoH into a PoS framework, Solana addresses typical scalability bottlenecks faced by many blockchain systems, demonstrating a focus on speed and mass-transaction capabilities.

Cardano

Cardano employs a PoS protocol designed to balance decentralization, security, and network expansion, and it is widely recognized for its research-oriented development philosophy. A distinguishing aspect of Cardano is its two-layer structure, which separates transaction settlement from computational operations.

Staking lies at the center of Cardano's economic and security model. ADA holders can delegate their tokens to staking pools, and the performance of these pools influences the rewards participants receive. This delegation system supports widespread participation and strengthens decentralization by allowing many users—not just large stakeholders—to engage in network validation.

Cardano's emphasis on formal methods, mathematical verification, and peer-reviewed scientific work underscores its pursuit of a rigorously engineered blockchain platform. This academically grounded strategy aims to ensure reliability and security, reinforcing Cardano's position as a major contributor to PoS innovation.

Collectively, these cases illustrate not only advances in consensus design but also the variety of approaches different blockchain platforms adopt when implementing PoS. As blockchain technology matures, these networks are helping shape a future that prioritizes sustainability, scalability, and enhanced security.

7.4 Delegated Proof of Stake

DPoS is a consensus mechanism that builds upon the foundational principles of PoS, offering a more efficient and streamlined approach for achieving agreement across blockchain networks.

7.4.1 Mechanism of Delegated Proof of Stake

DPoS introduces a democratic element by enabling cryptocurrency holders to elect a limited number of delegates or representatives to serve as validators (Calcaterra & Kaal, 2018). Key components of DPoS include:

Validator Nodes:
A fixed group of elected validators—also referred to as delegates or witnesses—conducts the consensus process. These validators take turns producing blocks and confirming transactions, ensuring orderly network operation.

Voting and Election of Validators:
In a DPoS system, token holders vote for delegates of their choice, with voting power generally proportional to the amount of cryptocurrency staked (Li et al., 2023). The delegates receiving the highest votes are selected as active validators, tasked with safeguarding the blockchain and validating transactions.

Block Production:
Unlike traditional PoS, where all validators can propose blocks, DPoS employs a rotational scheme (Liu et al., 2021). Elected validators generate blocks in a predetermined sequence, which increases efficiency and reduces the likelihood of forks.

Slashing Mechanism:
To deter misconduct, DPoS networks often implement slashing penalties. Validators who act dishonestly or attempt to compromise the network may forfeit a portion of their staked coins, reinforcing network security.

7.4.2 Voting, Governance, and Network Participation

Voting and Validator Selection:
Voting constitutes a central part of DPoS consensus. Token holders actively participate in network governance by electing validators, and they may adjust their votes over time based on validator performance or other criteria. This ongoing selection process ensures that validators with the most support form the active set responsible for block validation. By allowing stakeholders to influence the composition of validators, DPoS enhances network decentralization and accountability.

Governance: DPoS systems typically include governance structures enabling token holders to engage in decision-making regarding protocol upgrades, parameter modifications, and other network improvements. Participants can submit proposals or vote on existing ones, directly affecting the blockchain's operational direction. This participatory model strengthens decentralization by giving the community a meaningful role in shaping network development.

7.4.3 Pros and Cons of Delegated Proof of Stake

Advantages

High Efficiency:

DPoS is recognized for its operational efficiency, as limiting consensus to a fixed set of elected validators shortens decision-making cycles and accelerates block production. This structure allows for faster transaction confirmations compared with conventional PoW or PoS networks.

Enhanced Scalability:

The restricted number of validators minimizes the likelihood of network congestion, making DPoS highly scalable. Additionally, the rotational block production mechanism further supports consistent performance under increasing transaction loads.

Partial Decentralization:

Although DPoS does not achieve the same level of decentralization as some other consensus protocols, it still promotes participatory governance by allowing stakeholders to vote for validators.

Disadvantages

Potential for Centralization:

Critics highlight that DPoS may inadvertently favor centralization. Wealthier participants or highly supported validators can accumulate disproportionate influence over network decisions.

Risk of Vote Manipulation:

There exists a possibility of vote buying, where validators offer incentives to coin holders in exchange for their votes. Such practices can compromise the democratic nature of the system.

Security Vulnerabilities:

The limited number of validators can introduce security risks, particularly if a significant portion of them is compromised or colluded. This concentration of authority presents a potential weakness in the network's resilience.

7.4.4 Examples: EOS, Tron, Lisk

EOS

EOS is a blockchain platform designed for decentralized applications (DApps) that implements a DPoS consensus mechanism. Token holders in EOS vote for block producers, who are responsible for validating transactions and producing new blocks. The platform has gained recognition for its high throughput and ability to handle a large number of transactions per second. Through EOS's governance framework, token holders can also propose and vote on network upgrades, allowing the community to influence the platform's development. EOS illustrates how DPoS can be leveraged to build scalable and efficient blockchain infrastructures.

Tron
Tron is another blockchain network that relies on DPoS to achieve consensus. TRX holders elect a group of Super Representatives who validate transactions and generate blocks. Tron's implementation of DPoS emphasizes speed and scalability, making it suitable for applications that require high transaction volumes. Super Representatives also participate in governance by proposing and voting on protocol changes, reinforcing the participatory principles of DPoS.

Lisk
Lisk utilizes a variation of DPoS called Delegated Proof of Stake with Sidechains (DPoS+SC). This model allows developers to create sidechains linked to the main blockchain, while elected delegates secure the main chain and validate its transactions. Lisk's flexible fee and voting system enables users to participate in governance without locking their tokens, promoting wider engagement and involvement in decision-making processes.

These examples demonstrate the adaptability of DPoS to address specific network requirements, highlighting its potential for enhancing efficiency, scalability, and community governance. As with any consensus protocol, the effectiveness of DPoS depends on its implementation and the balance achieved between centralization and decentralization.

7.5 Byzantine Fault Tolerance

Byzantine Fault Tolerance (BFT) is a fundamental property of certain consensus protocols, designed to address challenges arising from the Byzantine Generals Problem in distributed computing environments.

7.5.1 Byzantine Generals Problem

The Byzantine Generals Problem is a well-known theoretical challenge in the study of fault-tolerant distributed systems. It describes a scenario in which multiple generals, each commanding a portion of an army, must coordinate a unified strategy—either to attack or to retreat (Dimitrios, 2022). The complication arises from the possibility that some generals may act as traitors, sending contradictory instructions to disrupt coordination and sow confusion.

Despite the presence of potentially deceitful participants and unreliable channels of communication, all generals must ultimately reach a consensus on the course of action. In distributed systems, this analogy reflects the difficulty nodes face in agreeing on a single system state when some components may behave maliciously or malfunction. Developing mechanisms to solve this problem is essential for ensuring reliability and consistency in decentralized networks.

7.5.2 BFT Algorithms: PBFT, RBFT, SBFT

Practical Byzantine Fault Tolerance (PBFT)

Practical Byzantine Fault Tolerance (PBFT), developed by Castro and Liskov, is a foundational consensus algorithm designed to achieve agreement in the presence of faulty or malicious nodes. PBFT employs a multi-phase protocol to establish consensus:

- **Request:** A client submits a transaction request to the network.
- **Pre-Preparation:** The primary node proposes a block and distributes it to the other nodes for review.
- **Preparation:** Nodes verify the proposed block and broadcast their approval to the network.
- **Commit:** Nodes finalize the block's validity, completing the consensus process.

PBFT guarantees both consistency and safety even when some nodes behave incorrectly. However, the algorithm requires at least a two-thirds majority of honest nodes to operate correctly.

Redundant Byzantine Fault Tolerance (RBFT)

RBFT is an enhancement of PBFT aimed at increasing fault tolerance and overall network security (Qin et al., 2022). RBFT introduces redundancy by allowing multiple nodes to act as leaders simultaneously, reducing the risk of single points of failure. This modification improves the system's resilience to Byzantine faults, ensuring that block proposals can continue even if one or more leaders fail.

Simplified Byzantine Fault Tolerance (SBFT)

SBFT is a lightweight adaptation of traditional BFT algorithms, designed to reduce the computational and communication overhead while maintaining the essential properties of fault tolerance (Wang & Nixon, 2023). SBFT is particularly suitable for scenarios demanding low latency and high throughput. Despite its simplified structure, SBFT preserves BFT, ensuring that consensus integrity is maintained under adverse conditions.

7.5.3 Pros and Cons of BFT Algorithms

Advantages

Resilience to Faults:

BFT algorithms are inherently designed to tolerate Byzantine faults, enabling networks to reach consensus even when some nodes act maliciously or experience failures.

Finality of Transactions:

These algorithms provide definitive consensus, ensuring that once an agreement is reached, it cannot be reversed. This property is crucial for applications that demand high levels of certainty and trust.

Enhanced Security:
BFT protocols are particularly suited to environments where security is paramount, making them ideal for financial systems and other mission-critical applications.

Disadvantages
Limited Scalability:
BFT mechanisms can struggle as network size increases. The associated communication overhead and the number of messages required for consensus can constrain performance and slow down processing.

Potential Centralization:
Depending on the specific implementation, BFT networks may display tendencies toward centralization, as a small subset of trusted nodes often play a decisive role in the consensus process.

7.5.4 Examples: Hyperledger Fabric and Stellar

Hyperledger Fabric
Hyperledger Fabric is a permissioned blockchain platform designed with a modular and scalable architecture. It implements PBFT to ensure that all nodes reach a consistent state of the ledger. Within Fabric, the network is composed of endorsing peers, validating peers, and ordering nodes. The ordering service employs PBFT to coordinate the sequence of transactions, enhancing the platform's security and reliability, making it suitable for enterprise-level applications.

Stellar
Stellar is a decentralized network tailored for fast, low-cost cross-border payments. It utilizes the Stellar Consensus Protocol (SCP), a federated Byzantine agreement method. SCP combines BFTe with federated consensus to achieve agreement across the network. Validators are organized into quorums, and consensus is reached by verifying the validity of transactions. This structure allows Stellar to maintain decentralization while achieving rapid agreement, making it appealing for financial services and payment systems.

These real-world applications illustrate how BFT principles—from theoretical models like the Byzantine Generals Problem to practical implementations such as PBFT, RBFT, and SBFT—can be applied effectively in distributed systems.

7.6 Directed Acyclic Graphs (DAGs)

DAGs offer an alternative approach to consensus in decentralized systems. Unlike traditional blockchains with linear block structures, DAGs employ a graph-based architecture, improving scalability and efficiency (Wang et al., 2020).

7.6.1 DAG Mechanism

Transactions as Nodes: In a DAG network, each transaction is treated as a separate node containing information such as sender, recipient, amount, and unique identifier (Yang et al., 2020). Unlike blockchains, transactions are not grouped into blocks but exist as individual, interconnected nodes.

Transaction Linking: Nodes reference prior transactions directly, forming a web of connections rather than a sequential chain. This interlinking allows transactions to be confirmed without waiting for block creation.

Tips: Tips are unreferenced transactions, representing the most recent transactions that have not yet been validated by others (Pervez et al., 2018). New transactions reference tips to contribute to consensus, supporting continuous growth of the network.

Timestamps and Temporal Ordering: Timestamps are applied to transactions to maintain chronological order, allowing nodes to reach a coherent understanding of the transaction history.

Asynchronous Confirmation: Transactions in DAGs can be confirmed asynchronously. When a node issues a new transaction, it references existing tips, which are verified by other network participants. This approach reduces latency and enables faster processing compared with block-based systems.

Consensus Mechanism:
Approval by subsequent transactions ensures the validity of a transaction. A transaction is confirmed once it has been validated by a sufficient number of subsequent nodes, achieving distributed consensus without miners or centralized validators.

7.6.2 Pros and Cons of DAGs

Advantages

- **Scalability:** Transactions can occur in parallel, increasing throughput.
- **Reduced Latency:** Transactions are confirmed without waiting for block creation.
- **Low Fees:** Feeless or low-cost transactions due to absence of miners.
- **IoT Compatibility:** Ideal for Internet of Things applications, as exemplified by IOTA.

Disadvantages

- **Security Risks:** DAG integrity relies on the honesty of the majority; attacks by a dominant participant can compromise the network.
- **Consensus Complexity:** Maintaining a single valid transaction history without blocks requires sophisticated mechanisms.
- **Limited Adoption:** DAG networks are relatively new, with fewer developer resources and community support compared to traditional blockchains.

7.6.3 Examples: IOTA and Nano

IOTA: IOTA employs a DAG structure called the Tangle, designed for IoT applications. Each new transaction confirms two previous transactions, enabling efficient, feeless microtransactions. IOTA's Coordinator mechanism currently aids consensus, though the network aims for full decentralization in the future.

Nano: Nano uses a block-lattice DAG model where each account maintains its own blockchain. Transactions are added to individual account chains, reducing the need for network-wide consensus. Nano leverages a DPoS system to confirm transactions efficiently, providing a scalable and high-speed solution for peer-to-peer payments.

7.7 Hybrid Consensus Models

Hybrid consensus models integrate multiple consensus algorithms to combine their strengths and mitigate individual weaknesses, aiming to balance security, scalability, and decentralization (Pieroni et al., 2020).

7.7.1 Combining PoW and PoS

- **Proof of Work (PoW):** Provides security through computational effort, defending against attacks proportional to miners' computing power.
- **Proof of Stake (PoS):** Energy-efficient, selecting validators based on staked cryptocurrency and incentivizing honest participation.

7.7.2 Pros and Cons of Hybrid Models

Advantages

- **Enhanced Security:** Combines PoW's computational security with PoS's economic incentives.
- **Improved Scalability:** PoS layer reduces the bottlenecks of PoW.
- **Greater Decentralization:** Stakeholder participation enhances governance.
- **Flexibility:** Hybrid models can adapt dynamically to network conditions.

Disadvantages

- **Complexity:** Integrating multiple consensus mechanisms requires careful design and maintenance.
- **Centralization Risk:** Poorly balanced parameters may inadvertently favor certain participants.
- **Transition Challenges:** Migrating to hybrid models can introduce temporary risks and operational hurdles.

7.7.3 *Examples of Hybrid Models*

Decred: Decred combines PoW and PoS. Miners perform PoW to generate blocks, while PoS stakeholders participate in governance and block validation. This hybrid structure enhances security and decentralization simultaneously.

Casper (Ethereum):
Casper FFG (Friendly Finality Gadget) is a hybrid consensus mechanism designed for Ethereum's transition from PoW to PoS. Validators lock up cryptocurrency as collateral, providing economic incentives for honest behavior, while PoW continues to secure the network during the transition.

7.8 Conclusion

This chapter has explored the diverse array of consensus mechanisms that form the backbone of blockchain technology. Each method offers distinct advantages and faces specific limitations, ranging from traditional PoW to novel architectures like DAGs and hybrid systems. In this conclusion, we summarize the key insights, highlight prevailing challenges, consider future directions, and underscore the crucial role of consensus in maintaining the reliability and integrity of blockchain networks (Eslamian and Eslamian, 2025).

7.8.1 *Summary of Consensus Mechanisms*

The discussion began with **PoW,** the pioneering algorithm popularized by Bitcoin. PoW secures networks by requiring miners to solve complex computational puzzles, ensuring both the immutability and trustworthiness of the blockchain.

Next, **PoS** was examined as a more energy-efficient alternative. PoS selects validators according to the amount of cryptocurrency they stake, providing economic incentives for honest participation while enhancing scalability.

DAGs introduced a transformative approach, replacing linear block structures with interconnected transactions. This design improves throughput and efficiency, making DAGs especially suitable for high-frequency or IoT applications.

DPoS incorporates a governance layer where coin holders elect a limited number of delegates to validate transactions and produce blocks. This mechanism promotes community participation, while also improving transaction speed and overall network efficiency.

BFT protocols, including PBFT and RBFT, focus on achieving consensus in environments where nodes may act maliciously or fail. BFT ensures system integrity and reliability even under adverse conditions.

Finally, hybrid consensus models, as exemplified by Decred (PoW/PoS) and Ethereum's Casper protocol, combine multiple consensus mechanisms to achieve a balance of security, scalability, and decentralization. By leveraging the strengths of different algorithms, these hybrids aim to address the limitations inherent to single-method approaches.

In conclusion, consensus mechanisms are central to the functioning of blockchain systems. The choice of algorithm profoundly impacts security, efficiency, and network resilience, and ongoing innovations in this area continue to shape the evolution of decentralized technologies.

7.8.2 Current Challenges in Blockchain Consensus

Scalability: As blockchain adoption grows, the ability of consensus algorithms to efficiently handle larger transaction volumes and expanding network sizes becomes a critical concern. Some consensus mechanisms may experience bottlenecks or slower processing times as the system scales.

Energy Consumption: Energy-intensive protocols, particularly PoW, have raised concerns regarding environmental sustainability. Balancing the demands of security with the need to reduce energy usage remains a pressing challenge for the blockchain community.

Interoperability: Integrating diverse blockchain networks and consensus protocols is essential for a truly interconnected decentralized ecosystem (Belchior et al., 2022). Standards and protocols for cross-chain communication are still evolving, limiting seamless interaction between different systems.

Security and Governance: Maintaining robust security while promoting decentralization is a delicate task. Governance structures must be carefully designed to ensure fair decision-making without inadvertently concentrating power or undermining network integrity.

7.8.3 Future Directions and Concluding Remarks on Consensus

Layer 2 Scaling Solutions: To address scalability limitations, Layer 2 technologies such as Bitcoin's Lightning Network and Ethereum's state channels have been developed. These solutions process transactions off-chain and periodically settle them on the main blockchain, reducing congestion and increasing transaction throughput.

Sustainability Measures: The blockchain industry is actively exploring energy-efficient consensus mechanisms and environmentally conscious design strategies to mitigate the carbon footprint of decentralized networks. Transitioning to less energy-intensive algorithms or integrating sustainability considerations into consensus models represents a promising path forward.

Cross-Chain Compatibility: Achieving interoperability through standardized protocols and frameworks will allow different blockchain networks to communicate and collaborate seamlessly. Enhanced cross-chain compatibility is essential for building a more cohesive and versatile decentralized ecosystem.

Advanced Security Protocols: Ongoing research in cryptography and advanced security techniques aims to strengthen the resilience of consensus algorithms against emerging threats. These innovations will ensure the continued integrity and reliability of blockchain networks.

Concluding Remarks

Consensus algorithms form the backbone of blockchain networks, enabling trust, security, and decentralization. They establish the integrity of distributed ledgers, guarantee transaction immutability, and underpin the overall functionality of blockchain systems.

Beyond maintaining agreement on ledger states, consensus mechanisms allow blockchain networks to evolve and adapt to diverse application requirements. They serve as the cornerstone of trust in decentralized environments, fostering innovation, collaboration, and democratization across industries.

As the blockchain ecosystem continues to mature, the development of optimal consensus mechanisms remains a critical pursuit. Challenges such as scalability, sustainability, and interoperability are driving the evolution of more robust, efficient, and inclusive systems. The ongoing exploration of consensus solutions will shape the future of decentralized technologies, laying the foundation for a new era of trust, transparency, and global collaboration. In essence, the study and advancement of consensus protocols is an ever-evolving journey, with each innovation paving the way toward more resilient and adaptive decentralized networks.

References

1. Aggarwal, A., Gaba, S. and Mittal, M. (2021). A Comparative Investigation of Consensus Algorithms in Collaboration with IoT and Blockchain. *Transforming Cybersecurity Solutions Using Blockchain*. Springer Nature Link. pp.115–140.
2. Aviv, I. (2021). The Distributed Ledger Technology as Development Platform for Distributed Information Systems. In: *Proceedings of the International Conference on Intelligent Vision and Computing* (*ICIVC 2021*). pp. 344–355.
3. Bach, L.M., Mihaljevic, B. and Zagar, M. (2018). Comparative analysis of blockchain consensus algorithms. *IEEE*,1545–1550.
4. Belchior, R., Vasconcelos, A., Guerreiro, S. and Correia, M. (2022). A Survey on Blockchain Interoperability: Past, Present, and Future Trends. *ACM Computing Surveys*, *54*(8), pp.1–41.
5. Calcaterra, C. and Kaal, W.A. (2018). Secure Proof of Stake Protocol. *SSRN Electronic Journal*.
6. Dahiya, A., Gupta, B.B., Alhalabi, W. and Ulrichd, K. (2022). A comprehensive analysis of blockchain and its applications in intelligent systems based on IoT, cloud, and social media. *International Journal of Intelligent Systems*, *37*(12).
7. Dimitrios, P. (2022). *Distributed Consensus Inference and Blockchain*. Thesis. Aristotle University of Thessaloniki.
8. Eslamian, S. and Eslamian, F. (2025). *Blockchain Technology in Water and Environmental Systems*. CRC Press, Taylor and Francis, USA, p. 282.
9. Gola, C. and Sedlmeir, J. (2022). Addressing the Sustainability of Distributed Ledger Technology. *SSRN Electronic Journal*.

10. Han, R., Foutris, N. and Kotselidis, C. (2019). Demystifying Crypto-Mining: Analysis and Optimizations of Memory-Hard PoW Algorithms. In: *2019 IEEE International Symposium on Performance Analysis of Systems and Software (ISPASS)*.
11. He, J., Wang, G., Zhang, G. and Zhang, J. (2021). Consensus Mechanism Design based on Structured Directed Acyclic Graphs. *Blockchain: Research and Applications*, *2*(1).
12. Kumar, G., Saha, R., Rai, M.K., Thomas, R. and Kim, T.-H. (2019). Proof-of-Work Consensus Approach in Blockchain Technology for Cloud and Fog Computing Using Maximization-Factorization Statistics. *IEEE Internet of Things Journal*, *6*(4).
13. Laroiya, C., Saxena, D. and Komalavalli, C. (2020). Applications of Blockchain Technology. In: *Handbook of Research on Blockchain Technology*. Academic Press, pp. 213–243.
14. Lasla, N., Al-Sahan, L., Abdallah, M. and Younis, M. (2022). Green-PoW: An energy-efficient blockchain Proof-of-Work consensus algorithm. *Computer Networks*, *214*.
15. Leonardos, S., Reijsbergen, D. and Piliouras, G. (2020). PREStO: A Systematic Framework for Blockchain Consensus Protocols. *IEEE Transactions on Engineering Management*, *67*(4).
16. Li, C., Xu, R. and Duan, L. (2023). Liquid Democracy in DPoS Blockchains. In: *BSCI '23: Proceedings of the 5th ACM International Symposium on Blockchain and Secure Critical Infrastructure*.
17. Liu, Y., Wang, B., Ye, W., Ning, X. and Gu, B. (2021). Global Estimation Method Based on Spatial–Temporal Kalman Filter for DPOS. *IEEE Sensors Journal*, *21*(3).
18. Mardiansyah, V. and Sari, R.F. (2022). SimBlock Simulator Enhancement with Difficulty Level Algorithm Based on Proof-of-Work Consensus for Lightweight Blockchain. *Sensors*, *22*(23).
19. Meshcheryakov, Y., Melman, A., Evsutin, O. and Morozov, V. (2021). On Performance of PBFT Blockchain Consensus Algorithm for IoT-Applications with Constrained Devices. *IEEE Access*, *9*.
20. Morabito, V. (2017). *Business Innovation Through Blockchain*. ResearchGate.
21. Mwandosya, M.J. and Luhanga, M.L. (2021). Blockchain: A Disruptive and Transformative Technology of the Fourth Industrial Revolution. *Business Management Review*, *23*(2), pp.16–31.
22. Pereira, J., Tavalaei, M.M. and Ozalp, H. (2019). Blockchain-based platforms: Decentralized infrastructures and its boundary conditions. *Technological Forecasting and Social Change*, *146*, 94–102.
23. Pervez, H., Muneeb, M., Irfan, M.U. and Haq, I.U. (2018). A Comparative Analysis of DAG-Based Blockchain Architectures. In: *2018 12th International Conference on Open Source Systems and Technologies (ICOSST)*.
24. Pieroni, A., Scarpato, N. and Felli, L. (2020). Blockchain and IoT Convergence: A Systematic Survey on Technologies, Protocols, and Security. *Applied Sciences*, *10*(19).
25. Platt, M., Sedlmeir, J., Platt, D., Xu, J., Tasca, P. and Vadgama, N. (2021). The Energy Footprint of Blockchain Consensus Mechanisms beyond Proof-of-Work. In: *2021 IEEE 21st International Conference on Software Quality, Reliability, and Security Companion (QRS-C)*.
26. Puthal, D., Malik, N., Mohanty, S.P., Kougianos, E. and Yang, C. (2018). The Blockchain as a Decentralized Security Framework [Future Directions]. *IEEE Consumer Electronics Magazine*, *7*(2), 18–21.
27. Qin, H., Cheng, Y., Ma, X., Li, F. and Abawajy, J. (2022). Weighted Byzantine Fault Tolerance consensus algorithm for enhancing consortium blockchain efficiency and security. *Journal of King Saud University - Computer and Information Sciences*, *34*(10, Part A).
28. Romano, D. and Schmid, G. (2021). Beyond Bitcoin: Recent Trends and Perspectives in Distributed Ledger Technology. *Cryptography*, *5*(4).
29. Sabzevari, Y., Eslamian, S., Okhravi, S. and Bazrkar, M.H. (2025). Evaluating the Performance of Soil and Water Assessment Tool (SWAT) in a Snow-Dominated Climate (Case Study: Azna–Aligoudarz Basin, Iran). *Atmosphere*, *16*(4), 382.

30. Sabzevari, Y. and Eslamian, S. (2025). Application and evaluation of modern irrigation. In: *Handbook of Nature-Based Drought Solutions*. Elsevier, pp. 155–186.
31. Saleh, F. (2021). Blockchain without Waste: Proof-of-Stake. *The Review of Financial Studies*, *34*(4).
32. Sarmah, S.S. (2018). Understanding Blockchain Technology. *Computer Science and Engineering*, *8*(2), 23–29.
33. Skh Saad, S.M., and Raja Mohd Radzi, R.Z. (2020). Comparative Review of the Blockchain Consensus Algorithm Between Proof of Stake (POS) and Delegated Proof of Stake (DPOS). *International Journal of Innovative Computing*, *10*(2).
34. Wang, G. and Nixon, M. (2023). X-assisted BFT Consensus Protocols. In: *International Conference on Blockchain* (pp.54–71).
35. Wang, Q., Yu, J., Chen, S. and Xiang, Y. (2020). Diving into DAG-based Blockchain Systems. *Computer Science*.
36. Yang, W., Dai, X., Xiao, J. and Jin, H. (2020). LDV: A Lightweight DAG-Based Blockchain for Vehicular Social Networks. *IEEE Transactions on Vehicular Technology*, *69*(6).
37. Yao, W., Ye, J., Murimi, R. and Wang, G. (2021). A Survey on Consortium Blockchain Consensus Mechanisms. *IEEE Access*, *11*.

8 | SoK in Blockchain

Paven Thuat D.O.[1*]

A Systematization of Knowledge (SoK) in blockchain aims to consolidate, categorize, and critically evaluate the rapidly expanding body of research in decentralized systems. This chapter provides a structured and comprehensive systematization of blockchain knowledge, covering foundational concepts, architectural components, consensus mechanisms, cryptographic primitives, smart contract platforms, and major application domains. By synthesizing findings across academic literature, technical reports, and real-world implementations, the chapter identifies common patterns, research gaps, methodological trends, and evolving challenges within the blockchain ecosystem. Particular attention is given to security vulnerabilities, scalability limitations, interoperability issues, and governance models. The SoK approach not only maps the state of the art but also highlights future research directions and emerging innovations that could shape the next generation of blockchain technologies. This systematic perspective provides researchers, practitioners, and policymakers with a coherent understanding of blockchain's evolution, capabilities, and open problems.

8.1 Introduction

This part offers a review of literature on categorizing and organizing blockchain, encompassing scholarly works and non-academic pieces. It also outlines the main significance of the paper and the methodology used in the research. Section 2 briefly outlines the history of Blockchain technologies, including the creation of Bitcoin, the world's first operational blockchain network developed by Satoshi Nakamoto. Section 3 showcases various database layouts for distributed ledger technologies (DLTs). Section 4 provides a categorization of DLTs and blockchains based on governance and accessibility perspectives. In Section 5, a thorough examination is given regarding the significant issues and obstacles faced by public blockchains, along with their advancements and changes, followed by the introduction of a classification system comprising four generations of public blockchains. Section 6 provides an overview of the layers (0, 1, 2, 3) concepts in blockchains. Section 7 consists of a brief dialogue on Web3 and the forthcoming evolution of the Internet's next era, along with a summary of the project. Within that context, individuals can

[1] Hong Kong University of Science and Technology.
* Corresponding author: thuat86@gmail.com

DOI: 10.1201/9781003686026-8

discover the implementation of our categorization system on well-known public networks.

8.1.1 A Literature Review

Blockchain has been widely acknowledged as a revolutionary technology globally. Web3 describes upcoming Internet applications relying on blockchain protocols, envisioning *the Internet of Money* to exchange and hold value, as stated by Antonopoulos in his book on the topic. Alternatively, the concepts of *Internet of Value*, were introduced and defined by the authors in (Vadgama, Xu, and Tasca 2022) and (Treiblmaier 2022), who also discussed how blockchain technologies connect and change businesses.

In the academic realm, numerous researchers have thoroughly studied Blockchain (Sabzevari et al., 2025; Sabevari and Eslamian, 2025). According to system architects, Xu et al. (2017) introduced a classification outlining important architectural features of blockchains and how their key designs affect the performance and quality attributes of blockchain-based systems, which is helpful for architectural analysis. (Tasca & Tessone, 2019) broke down blockchains into their basic components starting from the bottom, then categorized them into main and subcomponents in a hierarchical way for identification and comparison. Next, from a technical perspective, a taxonomy tree is implemented for various blockchain architectural setups.

Sarkintudu et al. (2018) employed established techniques in the field of information systems to create a categorization of blockchain platforms. Olga, Tobias, and Ali (2019), along with Goundar et al.(2020) categorized how Blockchain technologies are utilized in different industries and areas. In their work, Cangir, Cankur, and Ozsoy (2021) put forward a classification system for distributed storage technologies based on blockchain, which they then applied to assess and compare different solutions.

Bano et al. (2019) introduced a structured framework for analyzing blockchain consensus mechanisms, focusing on their security and performance characteristics with consensus at the core. Bouraga (2021) examined and evaluated 28 agreement protocols, and then systematically classified them based on their origins, designs, performance, and security. Nijsse (2020) and Garay and Kiayias (2020) both suggested a classification system for consensus mechanisms.

In the works of Stephan Cummings (Feb. 2019), Ruchika Dubey (Sep. 2019), Kirsty Moreland (May 2021), Nathan Reiff (Sep. 2021), Willigut (Oct. 2022), and The Nation Thailand (Nov. 2022), the authors made different efforts to categorize three generations of public blockchains. They mostly reached a consensus on the 1st and 2nd generations, but had a dispute regarding the 3rd and subsequent ones. Nevertheless, the articles did not include proper scientific research methods.

8.1.2 Chapter Contribution

Readers are unable to locate a condensed and thorough overview of the major milestones and advancements in Blockchain's evolution history. In the academic field and scientific literature, there is a gap in DLT classification and taxonomy for the development of public Blockchain generations. In this study, the author's goal is to address the deficiencies.

8.1.3 Research Method

- The writer examines the historical perspective of Blockchain Evolution to identify important problems and challenges that drive innovation, as well as groundbreaking inventions that are shaping a new era for advancements in Blockchain Technology. The author doesn't focus on consensus mechanisms, application perspectives, or deep technology designs of blockchain protocols. Instead, criteria are used to categorize DLTs and blockchains into four evolutionary generations of public chains.
- Development history
- Significant issues targeted for resolution (such as digital currency system, distributed settlement, ability to grow, Blockchain Trilemma, superior efficiency, linking different chains, mix-and-match capability)
- System architecture:
  - Types of ledger (database): UTXO vs account, linear vs DAG
  - Governance models: permissioned vs permissionless
  - Access modes: public vs private
- Other significant technological designs: consensus, virtual machine, application-oriented modularity.

8.2 The Early History of Blockchain

The idea of a series of interconnected blocks using a hash function was introduced by Ralph Merkle in his dissertation from 1979. The design Merkle connecting data is widely recognized as the Merkle hash tree (Eslamian and Eslamian, 2025; Merkle, 1979). The initial protocol resembling a blockchain was introduced in David Chaum[1] dissertation from 1982 (Chaum, 1982). In 1991, Stuart Haber and W. Scott Stornetta (1991) suggested a cryptographic secure chain of blocks layout for ensuring the integrity of document timestamps. Dave Bayer et al. (1993) introduced Merkle trees to enhance the efficiency of the chain design (Bayer, 1993). The research in Sherman et al. (2019) examines the historical development and different forms of blockchain technology.

Basic elements like hash function, Merkle tree, block-chaining design, and public key cryptography such as the Elliptic Curve Digital Signature Algorithm were essential for Satoshi Nakamoto[2] in developing Bitcoin, the first functional blockchain network globally. The Bitcoin whitepaper, titled *Bitcoin: A Peer-to-Peer Electronic Cash System* by Nakamoto (2008), was released on the website

[1] David Chaum is credited as the godfather of cryptocurrency. He introduced the first digital currency in 1995 but failed in practice.

[2] Pseudonym of an anonymous programmer or an unknown group of cryptographers.

www.bitcoin.org, and the Bitcoin mainnet was launched on 3 January 2009. The initial coinbase transaction in the first block of Bitcoin included the statement, “The Times 03/Jan/2009 Chancellor on brink of second bailout for banks”, referring to an article in *The Times of London* about the 2008-2009 financial crisis. The Great Recession had a significant global impact for almost ten years.

In the author’s opinion, by inventing Bitcoin, Satoshi Nakamoto made critical dedications to the world.

i. He coordinated the creation of a single software (Bitcoin core) that combines distributed system designs, immutable data structures, and cryptography techniques.
ii. He brought in Nakamoto Consensus, the initial and most widely used Proof of Work (PoW) consensus algorithm in the Blockchain industry.
iii. Shocked by the news, she couldn’t believe what she was hearing. He presented a secure ledger structure (the world’s first replicated state machine).
iv. He understood the idea of the electronic money system, specifically cryptocurrency, along with token economics rooted in game theory.
v. Satoshi Nakamoto is credited with originating the concept of decentralized systems and governance, in opposition to centralized entities.

Bitcoin, being the original blockchain network, serves as the cornerstone technology for all subsequent blockchain platforms, such as Ethereum. Being the original cryptocurrency, Bitcoin has emerged as the digital equivalent of gold worldwide, a marvel in the 21st century, and possibly ushering in a new era of global financial and monetary structures. More information on Bitcoin can be located in the study by Antonopoulous et al. (2016).

8.3 Varieties of Distributed Ledgers: UTXO Versus Account, Linear Versus DAG

Blockchain, also referred to as DLT, relates to the technological structure and protocols that allow multiple entities or locations to simultaneously access, validate, and update records in an immutable way. DLT requires a decentralized network and a consensus algorithm to maintain the distributed database, eliminating the need for a central authority to manage the ledger.

Within distributed ledgers, there exist two types of models for keeping records of balances. The initial technique is known as UTXO (Unspent Transaction Output) and resembles physical cash (bank notes). Bitcoin, Cardano, Dogecoin, IOTA, among others, utilize the UTXO model (refer to Figure 8.1). Another approach is the account model, which monitors the account balances (i.e., the system’s state changes) on each block, resembling internet banking systems. Ethereum, Polygon, BNB Chain, and most smart contract platforms utilize the account model, which also provides Turing completeness. Although UTXO systems are easier to implement and deploy, they have limited applications, whereas account models offer greater programming capabilities despite their complexity.

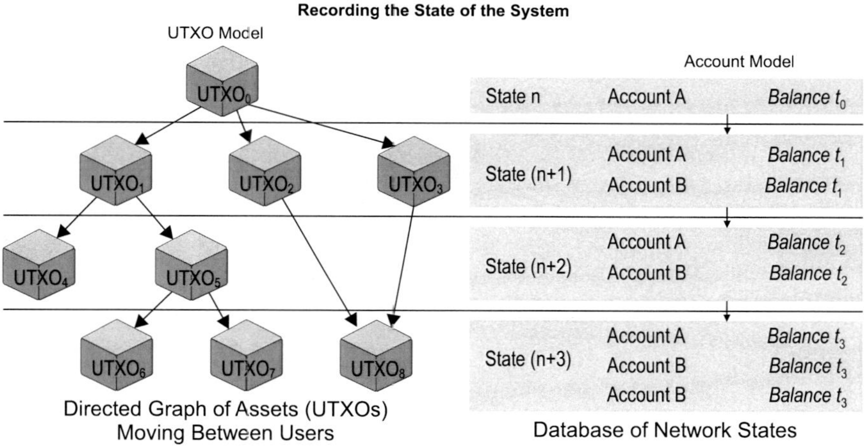

Fig. 8.1 UTXO vs Account model, by Horizon Academy.

Distributed ledgers are either a linear chain or a Directed Acyclic Graph (DAG) of transactions when considering data structure. The linear format, known as blockchain, is a total ordering method that connects transactional data blocks using hash pointers, widely utilized by Bitcoin, Ethereum, and other platforms. DAG is a partial order, a more intricate and nonlinear structure of incoming transactions connected to previous ones. DAG is utilized by IOTA, Avalanche, Hedera, as shown in Figure 8.2. Linear design offers synchronicity with latency variability, yet has constraints on throughput and scalability. Linear ledgers can also be easily implemented for virtual machines (VMs) and smart contract programs operating on linear ledgers. On the flip side, DAG enables the construction of distributed systems that have increased throughput and horizontal scalability. Nevertheless, DAG architecture proves to be challenging for constructing VMs and smart contracts due to its asynchronous nature. IOTA is a well-known distributed system based on DAG technology. Mainnet was released in July 2016, with the smart contract beta-devnet being released five years later. Avalanche constructs a decentralized network of linear chains (for hosting smart contracts) interconnected with a DAG-based ledger (for issuing assets, managing accounts, and facilitating exchanges). Hedera, Avalanche, and Cardano have tried to create their Virtual Machines on Unspent Transaction Outputs (UTXOs), but they have had limited success compared to what was initially hoped for.

Either UTXO or account model can be linked to linear chains or nonlinear ledgers. Yet, in the realm of DLT, the model of an account and linear structure (with complete ordering) are the most commonly used and thriving smart contract platforms. Readers can refer to Table 8.1 for a brief overview.

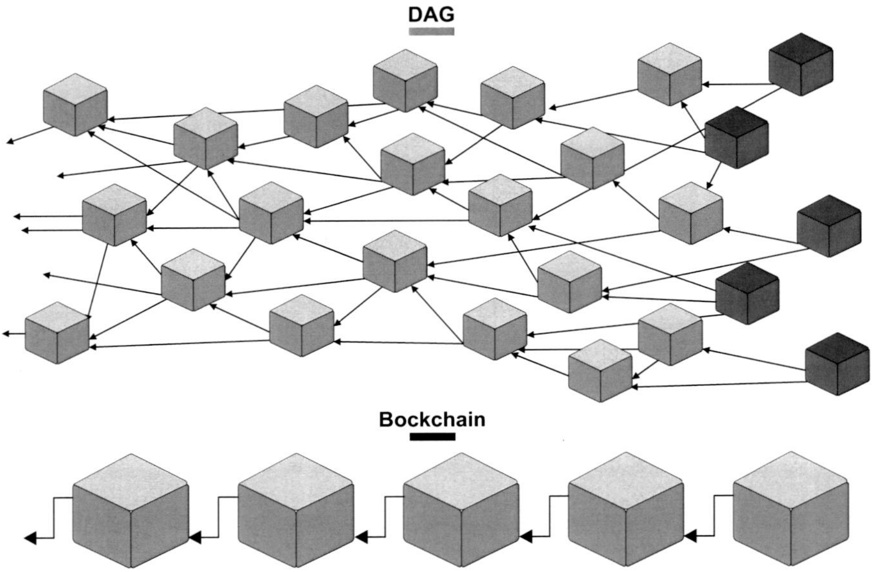

Fig. 8.2 DAG vs linear block-chain structure, by Horizon Academy.

Table 8.1 A summary on ledger types.

	UTXO	*Account*
Linear	Bitcoin, Cardano	Ethereum
DAG	IOTA, Avalanche	Hedera

8.4 Categories of DLTS and Blockchains

As Blockchains, DLTs, and cryptocurrencies continue to advance, numerous researchers have tried to classify the industry. The categorization of Blockchain into public, private, and consortium classifications was put forward by Puthal et al. (2018). Included in the author's suggestions is a classification system for DLTs and the Blockchain sector, taking into account various characteristics.

- Different forms of governing or authorization to become part of the network as operators and overseers.
- Different ways end users can access the DLT.We can categorize DLTs into public and private based on their access mode. The first option is completely open and available for all to view and utilize the ledgers, whereas the second option limits access to specific registered accounts. We can classify governance into two main categories: permissioned and permission-less. Readers also observe Table 8.2.
- **Permissioned blockchains** are specifically created for business purposes, requiring new nodes to be approved by authorities or current operators to

join the network. This implies that a trust model, either complete or partial, is necessary. They can be classified as either personal chains utilized for sharing data in a single organization, or consortium chains (also known as federation chains) used for cooperation among multiple companies. The first one is shut, completely reliable, and doesn't necessarily need agreement, whereas the second one is partly open, trusted, and involves a consensus mechanism between operating and governing nodes. Quorum, Hyperledger Fabric, and Corda are the top choices for enterprise blockchain frameworks. Numerous companies utilize their public sources to create their own exclusive or collaborative chains for different reasons. The majority of permissioned DLT platforms have various layers of privacy and accessibility.

- **Open-access blockchains** allow anyone to participate as operators, miners, or validators without requiring permission, hence operating on a trustless basis. At times, this group mistakenly implies that public blockchains rule the Blockchain industry with numerous active networks and platforms such as Bitcoin, Ethereum, BNB Chain, Dogecoin, Cardano, Polygon, Polkadot, and more.

Some public blockchains such as Hedera, POA Network, XRP Ledger, Stellar, VeChain are still permissioned. However, according to the author, there are no permissionless private chains as far as they know (refer to Table 8.2).

Table 8.2 Blockchain and DLT categories.

	Permissionless	*Permissioned*
Public	Bitcoin, Ethereum, Cardano *Fully transparent and fully trustless*	Hedera, Stellar, XRP Ledger *Fully transparent and partially trustless*
Private		Hyperledger, Corda, Quorum *Partially transparent and partially trustless*

8.5 Taxonomy on Public Blockchain Generations

One of the reasons for the author's decision to trace the historical development of Blockchain Technology is to create a classification system for public DLTs and blockchains based on various criteria and characteristics, making it easier for readers to understand.

- Evolution of development.
- Some of the key problems under consideration for solving include digital money system, distributed settlement, scalability, Blockchain Trilemma, high efficiency, cross-chain compatibility, and composability.
- Different aspects of the system's structure and technological plans: network layer, agreement system, virtual machine, application-focused modularity, and so on.

8.5.1 The 1st Blockchain Generation

The first generation celebrated Bitcoin's birth on January 3, 2009, recognizing the genesis block as its mainnet launch date. The fans include XRP[3] (released on June 2, 2012), Dogecoin (released on December 6, 2013), Stellar - XLM (released on July 31, 2014), which used Byzantine Fault Tolerance (BFT) consensus to create digital currency for the banking and financial industry. Litecoin, launched on 13 October 2011, Bitcoin Cash, launched on 1 August 2017, and Dogecoin, launched on 6 December 2013, are all trying to create their own digital currencies using distributed ledger technologies inspired by Bitcoin and its creator, Satoshi Nakamoto. These digital currencies strive to address the issue of centralized, nontransparent control present in fiat currencies and banking systems, moving towards a money system in the future that is decentralized, transparent, and resistant to censorship, supported and created by the community.

8.5.2 The 2nd Blockchain Generation

The 2nd generation of blockchains aims to address two primary issues: **private transactions and programmable money**. Dash[4] was released on January 18, 2014, followed by Monero on April 18, 2014, and Zcash on October 28, 2016. These cryptocurrencies were created with the intention of providing transaction privacy by concealing the link between senders and recipients, unlike Bitcoin and the initial generation of blockchains.

Scripting languages are not supported on Bitcoin and cryptocurrency networks, preventing programmable money and decentralized applications from being executed. Vitalik Buterin, born in 1994, showed a strong understanding of Blockchain Technology when he introduced and established Ethereum in 2013 (read more about his vision in the Ethereum Whitepaper). Ethereum mainnet went live on 30 July 2015, providing a platform that allowed for the implementation of any smart contract and application programming on blockchain, a capability that was not possible with Bitcoin, Dash, XRP, Stellar, and other cryptocurrencies at that time. Vitalik Buterin and the Ethereum Foundation's innovative ideas revolutionized the Blockchain industry, surpassing Bitcoin Core and introducing groundbreaking concepts like smart contract platforms, tokenization, tokenomics, DeFi, stablecoins, DAOs, decentralized file storage, oracle services, etc. The introduction of Turing-completeness on the blockchain, the Ethereum Virtual Machine (EVM), and the Solidity language by Ethereum developers are notable contributions that have significantly impacted the world of Blockchain technology. Moreover, Ethereum utilized the Ethash-PoW algorithm, to safeguard the network from ASIC-mining, a problem present on Bitcoin and its hard-forked chains.

So far, the majority of public blockchains are considered smart contract platforms (or layer-1s in Section 6) totaling in the hundreds. Some common

---

[3] XRP, previously known as XNS, is a digital currency created by Ripple Labs.

[4] At the time of its release, it was called Xcoin, and later renamed Darkcoin.

examples following Ethereum include Cardano (September 2017), EOS (June 2018), Tron Network (July 2018), Polygon (June 2020), BNB Chain[5] (September 2020).

At present, Ethereum remains the top smart contract platform when it comes to developer and user community, variety of applications, DeFi growth, and composability. The EVM is widely used for smart contracts, while Solidity is the top programming language for smart contracts.

IOTA, which was released on July 11, 2016, implemented a unique strategy by introducing a Tangle protocol utilizing DAG. This was designed specifically for IoT devices and their applications, aiming for a cryptocurrency network with limitless scalability in theory. Nevertheless, IOTA's applications are severely restricted because of its lack of synchronicity and absence of a virtual machine.

Essentially, Ethereum took note of the 2nd generation and the rise of Turing-complete blockchains, such as smart contract platforms. Privacy-supported chains like Dash, Monero, and Zcash have not gained the same level of adoption and applications as other blockchains. The decreased usage and acceptance of privacy coins is attributed to the absence of practical uses and regulatory worries.

8.5.3 The 3rd Blockchain Generation

The third generation sought to resolve scalability issues and the Blockchain Trilemma in Bitcoin and Ethereum's smart contract platform. In the past, Ethereum, with PoW, managed around 14 transactions per second, causing regular congestion and gas price increases on the network. Between 2016 and 2018, scalability and gas efficiency became the main concern for blockchain developers. Scalability is the ability of a blockchain to manage higher transaction levels through the addition of additional computing resources. Achieving this is very difficult (maybe even close to impossible) for Bitcoin and Ethereum.

Several initiatives like Cardano mainnet (2017), EOS (2018), and Tron Network (2018) aim to develop blockchains with Turing-complete capabilities, smart contract platforms that surpass Ethereum in terms of performance and scalability. They developed cutting-edge consensus mechanisms that utilize Proof of Stake (PoS) instead of resource-heavy Proof of Work (PoW), enabling scalability, along with progress in architectural designs and virtual machines. Cardano developed the PoS Ouroboros and introduced a Turing-complete language with UTXO assets, whereas EOS and Tron opted for Delegated Proof of Stake (DPOS) with 21 and 27 validators, offering superior performance compared to Ethereum, respectively. Hedera revealed a public, permissioned ledger protocol based on Hashgraph architecture and asynchronous BFT consensus, in place of blockchain. The Hedera mainnet, launched on August 14, 2018, claimed to support

[5] The rebranding took place in February 2022, changing the name from Binance Chain to Binance Smart Chain - BSC. Binance Chain is constructed using Tendermint Core, but lacks a runtime environment (VM). Currently, BNB Chain serves as the guiding light, with BSC functioning as the EVM-chain, transforming BNB Chain into a multi-chain system.

around 10,000 transactions per second and is capable of additional scaling. In June 2020, Polygon, which was formerly called Matic Network, was launched to enhance Ethereum's capabilities through the use of Plasma Bridge as a sidechain solution for scaling. At present, Tron and Polygon have many applications and are widely used, able to process thousands of transactions per second. Still, the EOS platform boasts strong technological capabilities, but its utilization is hindered by a dearth of applications due to the perpetual discord between the EOS community and Blockone, the platform's creator. Even though Cardano has made significant progress with the PoS-based Ouroboros consensus protocol (David et al., 2018), it faces a shortage of resources for developing and deploying smart contracts.

The Blockchain Trilemma, identified by Vitalik Buterin from 2018 to 2020, highlighted the challenge of balancing decentralization, security, and scalability in the blockchain space. Hedera, EOS, Tron, and other projects were noted for their superior performance compared to Ethereum, although they were found to compromise on decentralization. The Triangle comprises decentralization, safety, and scalability, as described by Leonardos, Reijsbergen, and Piliouras (2020). In their paper, the authors introduced PREStO, a systematic framework for assessing blockchain consensus protocols based on five categories: optimality, stability, efficiency, robustness, and persistence. Figure 8.3 illustrates a visual representation of the Blockchain Trilemma in the framework of the PREStO system. The different subcategories of optimality, efficiency, and stability accurately reflect the balance of safety, scalability, and decentralization. Both resilience and endurance offer separate ways to address the Trilemma over time.

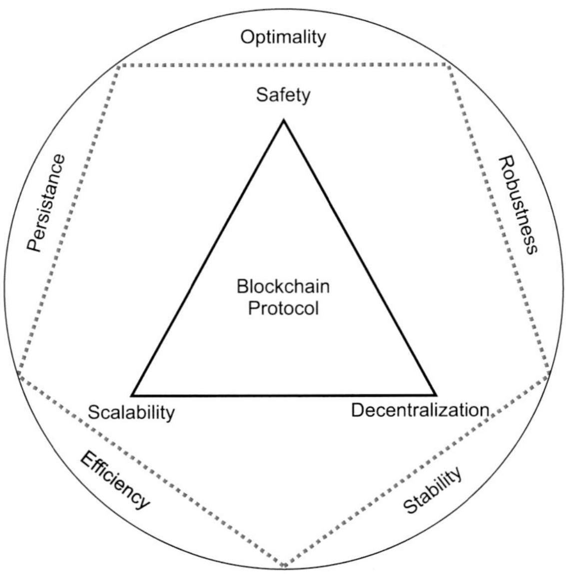

Fig. 8.3 The Blockchain Trilemma and its connection to the PREStO framework.

- Figure 8.4 depicts how the Blockchain Trilemma is solved: The green dot symbolizes an ideal protocol that achieves a harmonious balance among safety, decentralization, and scalability. The protocol symbolized by the blue dot consistently pursues the optimal solution, balancing conflicting traits with a degree of persistence.
- Safety can impede scalability and decentralization, but any attempt to increase scalability may compromise safety on the network.
- Projects can either concentrate on two aspects of the trilemma or commit to solving all three components once and for all.
- It is extremely challenging to find an equilibrium among the three attributes. Yet, finding a satisfactory answer to the issue may result in increased acceptance of cryptocurrency and Blockchain, as well as widespread implementation of the technology in various sectors worldwide.

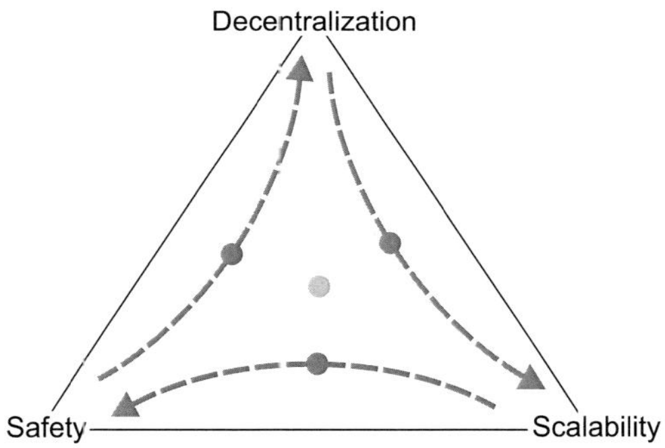

Fig. 8.4 Dealing with the Blockchain Trilemma.

MIT scientist Silvio Micali established Algorand, which launched its mainnet in June 2019, and is said to have resolved the Blockchain Trilemma through the creation of a Pure Proof of Stake (PPoS) consensus and a verifiable random function (VRF). Despite Silvio Micali claiming his team could increase Algorand's transaction speed to 46,000 TPS, the current throughput of Algorand (averaging 1,300 with a maximum of 6,000 TPS) as per the Algorand 2021 Performance report is comparable to that of Tron, EOS, and Polygon. Audiences have the option of looking at (Mauro Conti, 2019) discussing a security analysis of Algorand and a potential attack by exploiting a security vulnerability in the Byzantine Agreement's message validation process.

Utilizing sharding technology is a crucial solution for scalability, as it involves dividing a large database and blockchain network into multiple shards to improve performance. Readers looking for in-depth analysis of sharding techniques and protocols can refer to (Wang et al., 2019). Leading sharding protocols, Zilliqa, Near Protocol, and TON - Telegram Open Network, launched their mainnets in January 2019, April 2020, and May 2021. Sharding protocols demonstrate the possibility of achieving high TPS levels (such as 100k TPS or more). After The Merge was finished on 15 September 2022, moving Ethereum to a PoS consensus system, the Blockchain groups are now excitedly waiting for Ethereum 2.0 and its sharding enhancements. Every sharding chain requires a main chain to manage and coordinate all additional shards. They can be seen as consistent multichain systems, as opposed to varied multichain systems referred to as the 4th generation in Section 5.4.

In general, while The Blockchain Trilemma candidate isn't widely recognized, the revolutionary developments of the 3rd generation play a crucial role in advancing Blockchain Technology and expanding its applications and uses in the cryptocurrency field, ultimately leading to mainstream adoption down the line. The third generation is characterized by advancements in PoS and DPoS consensus mechanisms, hashgraph design, and sharding technologies, all suggesting superior performance to Ethereum.

8.5.4 The 4th Blockchain Generation

The fourth era is still focused on scalability, aiming at high-performance blockchains, and presents a new major difficulty in cross-chain communication and custom application flexibility.

Solana implemented a unique approach, known as high performance blockchain, to address the scaling problem, being the first to do so in commercial use in March 2020. The Parallel Chain initiatives by Aptos strive to develop monolithic chains with high throughput (measured by TPS) and fast finality (i.e., low latency). The developers claim they can build individual, straight chains that can reach tens of thousands of TPS and achieve finality within a few seconds. Yakovenko (2020) states in the *Solana Whitepaper* that the protocol can reach 71,000 TPS thanks to a deterministic block producer selection mechanism, Proof of History, and Tower BFT consensus. Gelashvili et al. (2023) presents Block-STM, a parallel processing system for smart contracts, now live on Aptos, allowing the network to achieve speeds of 110k tps in Diem tests and 170k tps in Aptos tests. Unfortunately, they sacrifice security, particularly in terms of safety and liveness, as well as decentralization in certain areas. Solana has experienced seven periods of downtime since launching its mainnet on March 16, 2020. Aptos officially launched its mainnet on October 12, 2022, boasting 102 active validators and minimal transaction activity. Visit https://explorer.aptoslabs.com to find current details about the Aptos mainnet. In Little's (2020) view, Parallel Chain seeks to reach a high throughput of 100k TPS per node, with performance increasing as more nodes are added.

One of the significant issues and major hurdles faced by Blockchain technology is cross-chain interoperability. Each chain is kept apart and alone from the rest by sovereignty. In straightforward terms, transferring bitcoin from the Bitcoin Network to Ethereum is not possible (i.e., transferring assets from one blockchain to another). In a broader sense, how can various blockchains communicate and work together?

Cosmos (mainnet launched on 13 March, 2019) and Polkadot (released on 26 May, 2020) are at the forefront of introducing varied multi-chain systems with a central chain that supervises and coordinates all other chains within the system (built using a standardized framework). This framework allows for triple-addressing scalability, interoperability across different blockchains, and customization for various application-specific chains. Cosmos includes a Cosmos Hub and zones built with Tendermint Core and IBC, whereas Polkadot comprises a Relay Chain and parachains developed on BABE-GRANDPA. Avalanche made its debut on September 21, 2020.

The Internet Computer Protocol, also known as ICP, was introduced on May 7, 2021 and is known for its innovative inventions within the protocol. Avalanche is distinguished by its blend of DAG and linear ledgers, merging UTXO assets and account model, powered by Snowman consensus, and supervised by P-Chain at the core. Nonetheless, ICP employs Chain-key cryptography along with threshold BLS signatures, a multi-subnet architecture governed by a Network Nervous System, and the Motoko smart contract language. The ICP mainnet can handle up to 11,500 TPS under normal conditions and has the ability to expand infinitely by adding more data on the chain. Starting in early 2022, BNB Chain, Tron Network, and Klaytn have started implementing heterogeneous multichain models, also known as modular blockchain systems, aligning with the trend towards multichain.

Despite its highly intricate design and development, the multichain architecture is a significant advancement that aims to tackle scalability, interoperability, and application-custom modularity, with the potential to enhance composability as specified in Section 5.5 and (Walden, 2018).

Even though sharding chains are labeled as the 3rd generation in Section 5.3, we could view these uniform multichain systems as the 4th Blockchain generation, due to their resemblance to heterogeneous systems, such as intricate architecture and focus on solving scalability issues, ensuring secure cross-shared interaction simultaneously and potentially providing a higher level of scalable composability (refer to Section 5.5).

8.5.5 Regarding Blockchain Composability

As per Walden (2018), a blockchain platform is considered composable when its current resources can serve as foundational elements and be coded into more advanced applications. Composability is crucial as it enables developers to accomplish greater tasks with fewer resources, ultimately resulting in faster and more exponential innovation. In simpler terms, composability is a crucial feature of smart contract platforms (or Turing complete blockchains) that enables a smooth connection, interaction, and composition between contracts and applications. In the article, Jesse Walden explores the development of blockchain technology by creating a conceptual framework of four different time periods, each having different structures and focuses on composability.

- **Calculator Era:** Tailored for specific use (e.g., Bitcoin), with restricted ability to combine functions.
- **Mainframe Era:** Able to perform any calculation (e.g., Ethereum, BNB Chain), great ability to work together.
- **Server Era:** Focused on specific applications like Cosmos, Polkadot, and Avalanche, with a focus on avoiding composability.
- **Cloud Era:** Capable of executing any program (such as ICP, Ethereum 2.0, etc.), with the ability to easily combine different components.

As per Walden (2018), a blockchain platform is considered composable when its current resources can function as foundational elements that can be integrated into more advanced applications. The significance of composability lies in enabling developers to achieve greater results with minimal resources, ultimately driving accelerated and compounded innovation. Put simply, composability is a critical feature of smart contract platforms (or Turing complete blockchains) that allows for smooth connection, interaction, and integration between contracts and applications. Jesse Walden examines the development of blockchain technology in the article, outlining four different eras characterized by varying architectures and priorities in terms of composability.

- **Era of Calculators:** Designed for specific purposes (e.g., Bitcoin), with restricted ability to combine functions.
- **Mainframe Era:** Capable of Turing completeness (like Ethereum, BNB Chain), with high levels of composability.
- **Server Era:** Focusing on specific applications (such as Cosmos, Polkadot, Avalanche, and other multichain systems) instead of prioritizing composability.
- **Cloud Era:** Capable of performing any computational task (e.g., ICP, Ethereum 2.0, etc.), easily combined and expanded.

The significant achievement of Ethereum is its remarkable level of composability. Currently, integrated networks like Ethereum, Solana, Tron, BNB Chain, Polygon, Algorand are providing the highest level of interoperability in the Mainframe Era. We are enthusiastic about examining the same property in both the Server Era and Cloud Era of Blockchain computing, which involves diverse and similar multichain systems. This also corresponds with how we categorize our age group.

8.5.6 A Summary

While it may seem impossible to solve The Blockchain Trilemma, progress in the field is rapid due to numerous groundbreaking innovations and initiatives. We scrutinized and evaluated the key criteria and attributes to categorize public blockchain generations as follows.

Key issues tackled in the Blockchain industry include digital currencies, decentralized transactions, scalability (especially high-performance blockchain solutions), the Blockchain Trilemma, and interoperability between blockchains.

- Disrupting architectural blueprints (network layer, ledger database, custom modularization)
- Fresh agreement mechanisms
- Creative virtual machines (such as Turing-complete runtime environments) and intelligent contract language

Table 8.3 outlines four generations of Blockchain Evolution, including typical examples, significant issues, protocols solving them, consensus type, and key inventions. Performance metrics such as real TPS on mainnet, theoretical Max TPS, and time to transaction finality (TTF) or latency are also considered, despite challenges in measurement and distinguishing between performance and scalability (Bonneau, 2022).

Table 8.3 A summary on four generations of public blockchains.

Gens	*Problems*	*Protocols*	*Cons*	*Inventions*	*TPS*	*Max*	*TTF*
1st	P2P cash system Censorship resistant	Bitcoin 2009	PoW	Nakamoto consensus Decentralized ledger and economics	7	7	60m
2nd	Smartcontract Dapps	Ethereum 2015	PoS	Turing completeness EVM and Solidity	15	30	6m
3rd	Scalability	Hedera 2018	BFT	Hashgraph, a BFT	2k	10k	6s
	Scalability	Tron 2018	PoS	DpoS, Tron VM	2k	4k	36s
	Blockchain Trilemma	Algorand 2019	PoS	Pure PoS and VRF	1k	6k	4s
4th	Interoperability	Cosmos 2019	BFT	Tendermint and IBC	1k	10k	7s
	Interoperability	Polkadot 2019	NpoS	BABE-GRANDPA	1k	100k	60s
	High performance	Solana 2020	PoS	PoH and Tower BFT	4k	50k	1s
	High performance	Aptos 2022	BFT	MOVE and Move VM	2k	170k	1s
	Scalability and Interop	ICP 2021	PoS	ChainKey cryptography	11k	Infinite	1s
	Scalability and Interop	Avalanche 2021	PoS	Hybrid ledger, Snowman	1k	4500	3s

8.6 Layers in Blockchain by Application View: L0, L1, L2, L3

Blockchain communities are anticipating Ethereum upgrades progressing towards sharding phase and additional scaling solutions post the PoS-Merge. In the meantime, numerous scaling solutions like layer-2, side-chain, state-channel, rollups, etc., have been suggested to enhance the overall performance of Ethereum and can be adapted for other layer-1 blockchains too. rollup-based layer-2 solutions are gaining prominence as secure and efficient options for scaling Ethereum, with significant investments and development efforts. In a broader sense, we will explore the idea of four layers in Blockchain.

Layer-0 (L0) serves as the foundation for constructing various layer-1 blockchains. L0 is at the core of a sophisticated blockchain structure that addresses interoperability, scalability, and customizable blockchain SDKs concurrently. Common instances of Layer 0s include diverse multi-chain systems like Polkadot, Cosmos, Avalanche, and Internet Computers (ICP). Readers should check out the Binance article on Layer-0.

Layer-1 (L1) blockchains are the foundational protocols that support the network and its applications. An L1 can either be a standalone blockchain like Bitcoin, XRP, Ethereum, or Solana, or it can be created on the software development kits of another base layer blockchain like Cronos on Cosmos SDK, or Moonbeam on Polkadot Substrate. L1s can be categorized as either versatile smart contract platforms (like Ethereum, Solana, and Moonbeam) or specialized blockchains for specific applications (such as Bitcoin and XRP for cryptocurrency and settlements, Theta Network for video streaming, etc.).

- **Layer-2 (L2)** comprises various scaling solutions that handle transactions outside of layer-1 to lessen its load and lower gas expenses, while maintaining security through L1. Several layer-2 scaling solutions have been suggested and built for Ethereum using various technologies and goals before being utilized for other layer-1 blockchains.
- Polygon is a sidechain connected to Ethereum through a Plasma bridge in order to create a versatile smart contract platform. Polygon has a currency of its own specifically for paying for gas.
- Arbitrum One, Optimism, Boba Network, Base are generalized layer-2s based on optimistic rollup.
- Loopring, zkSpace, AzTec, Immutable, dYdX, all based on ZK-rollup, are application-specific layer-2s for payment, exchange, or NFTs.
- zkSync Era, Starknet, Consensys's Linea and Polygon's zkEVM are general-purpose L2s on Ethereum based on Zero-Knowledge Rollups (or ZK-rollups). ZK-rollups are emerging as the most secure scaling technology, extensively investigated and developed by many scientists and technologists.
  - **Layer-3 (L3)** is currently not well-liked, involving application-specific layers built on top of L2s for specific purposes such as games, wallets, and privacy. In September 2022, Vitalik Buterin talked about the importance and reasoning behind layer-3 scaling in an article, and also suggested some valuable strategies.
  - If L2 is designed for scaling in specific applications, then L3 is geared towards specific functions such as privacy.
  - L2 is used for general scaling while L3 is used for scaling tailored to specific applications.

- If L2 is for trustless scaling (ZK-rollup), layer-3 is for weakly-trusted scaling (e.g., validium).

Layer-2 (L2) is a set of scaling solutions that handle transactions from layer-1 to lessen its load and lower gas expenses, while still maintaining security through L1. Several layer-2 scaling solutions for Ethereum have been suggested and created for Ethereum utilizing various technologies and objectives, then implemented for other layer-1 blockchains.

- Polygon is a smart contract platform that is connected to Ethereum through a Plasma bridge, designed to be a general-purpose platform. Polygon has its own cryptocurrency for paying gas fees.
- Arbitrum One, Optimism, Boba Network, Base are all examples of layer-2 solutions that utilize optimistic rollup.technology.
- Loopring, zkSpace, AzTec, Immutable, dYdX are all applications that use ZK-rollup technology and are focused on payment, exchange, and NFTs on layer-2.
- zkSync Era, Starknet, Consensys's Linea and Polygon's zkEVM are L2s on Ethereum that are designed for general use and rely on Zero-Knowledge Rollups (ZK-rollups). ZK-rollups are becoming known as the most secure method for scaling, with substantial research and development efforts by numerous experts.

Layer-3 (L3) is currently out of favor and involves the creation of application-specific layers that are built on top of L2s, catering to specific applications like games, wallets, privacy, and more. In a September 2022 article on layer-3 scaling, Vitalik Buterin explored the importance and reasoning behind L3, and put forward several significant strategies.

- If L2 is used for scaling in specific applications, L3 is used for specific functions such as privacy.
- If L2 is used for universal scaling, then L3 is used for tailored scaling (e.g., for individual purposes).
- If Layer 2 focuses on trustless scaling through ZK-rollup, Layer 3 is dedicated to scaling with weaker trust through methods like validium.

ZK-rollups and zk-EVM are currently the most promising solutions for scaling Ethereum, particularly in Layer-2. Vitalik Buterin offers a tutorial on rollup technologies, which are increasing the scalability of Ethereum by hundreds of times for specific uses. He categorizes various general-purpose zk-EVMs with Turing-complete runtime environments that utilize Zero-Knowledge techniques. Still, for handling many thousands of transactions per second, he turns to upcoming sharding technology, specifically Ethereum 2.0. Readers have the opportunity to delve deeper into L2 scaling solutions in Gangwal, Gangavalli, and Thirupathi, (2023).

8.7 WEB3 and The Internet of Value

- Web2 is commonly used by the majority of people to access the information on the internet that we are familiar with and use on a daily basis. Big technology companies (such as Apple, Amazon, Google, Facebook, etc.) have a strong influence and authority across most of the internet, offering various platforms and services in return for user data. Web3, also known as the third generation of the Web, is gaining popularity as developers strive to create a new wave of Internet applications utilizing blockchain protocols. The next evolution of the World Wide Web, known as Web3, integrates decentralization, Blockchain technologies, and token-based economics. Web3 highlights user ownership and community governance, concepts not found in Web2, by leveraging Blockchain as its core infrastructure. Despite numerous obstacles that must be faced, Web3 aims to enable the Internet of Money to operate efficiently on the web, as described by Andreas M. Antonopoulos in his book of the same name, allowing users to have full control. On the other hand, Vadgama, Xu, and Tasca (2022) as well as Treiblmaier (2022) presented and explained the Internet of Value concepts, detailing how blockchain innovations impact and transform industries. During his presentation at the World Economic Forum in Davos 2023, Michael Gronager, who serves as the CEO and co-founder of Chainalysis, talked about the necessity of cryptocurrency for the Internet of Value and the potential of Web3 to assist the billions of people who are not served by traditional banking and finance systems. Web3 technologies have the potential to:
- Enable innovative applications in finance, such as decentralized finance, that were previously hindered by the lack of liquidity in traditional assets.
- Enhance visibility and promote closer connections between producers, sellers, and customers.
- Empower user ownership and community governance in the business world through the implementation of decentralization.
- Before observing actual demonstrations of Web3 applications in use, we are observing groups of blockchain protocols working towards developing decentralized assets and computing infrastructure, as well as decentralized applications for the future of the Internet. As decentralized finance (DeFi), NFTs (non-fungible tokens), and gamefi gain popularity in the crypto space, social and media platforms along with decentralized autonomous organizations (DAOs) are emerging on Web3, following the influx of billions of dollars and millions of users. People can expect future developments in socialfi, DAO, edutech, healthcare, insurance, and other areas to drive the adoption and integration of Blockchain and Web3. Numerous typical cases showcase the development of infrastructure and applications for Web3 across different projects.

- Filecoin functions as a decentralized network for storing files.
- Lens Protocol is in the process of creating a decentralized and composable social graph that allows multiple Web3 social platforms and applications to be built on top of it. Chainflix is a video sharing platform that utilizes blockchain technology.
- The Graph is an indexing protocol designed for Ethereum-compatible chains and IPFS networks in order to increase accessibility of on-chain data for all users.
- Ocean Protocol and Golem are developing platforms for data and computing power, each utilizing blockchain technology and token economics.
- Chainlink and DIA, decentralized systems, provide smart contracts with data from outside sources.
- Ethereum Name Services offers decentralized naming for domains and wallets, while Civic provides identity solutions on various blockchain networks.
- Singularity Net and Fetch Network are working on protocols that enable AI to be integrated, deployed, and run on Blockchain technology.
- Octan Network and Orange Protocol are working on distinct reputation systems to build trust and credibility on Web3.
- In the end, the author utilizes the proposed classification and arrangement to swiftly categorize well-known public blockchain platforms and L2-projects found in the Top list on Coinmarketcap. The characteristics and examples can be beneficial for readers to apply in their personal circumstances. It is crucial to note that some platforms are developing and have several upcoming updates, leading to potential convergence among different generations. Readers can view samples being categorized in Tables 8.3-8.7 and find definitions for abbreviations in the Appendix. We possess the comments below.
- Generation classification is only relevant for sovereign blockchains, not for individual applications or L2 solutions. It evaluates various system designs, such as monolithic chains, homogeneous sharding, and heterogeneous multichain systems.
- The original generation of developers and some in the second generation focus on creating platforms for cryptocurrencies and settlement layers that are not Turing-complete.
- The second generation is mainly defined by Turing-complete smart contract platforms, although it has notable performance limitations.
- Turing-complete smart contract platforms with improved performance and scalability are prevalent in the 3rd generation, surpassing the capabilities of the 2nd generation. The platforms are either independent L1s or constructed on top of L0s.
- The fourth generation stands out for its high-performance L1s and multichain systems, with the potential to provide increased scalability, customizable applications, and expandable composability.

Table 8.4 First and second blockchain generations, including Generation L1. Ethereum, Ethereum Classic, and EthereumPoW are smart contract platforms based on the EVM, all derived from a single blockchain through forking.

Platforms	*Gens*	*Ledger Types*		*Governance*	*Cons*	*Layer*
Bitcoin, Dogecoin, Litecoin, Bitcoin Cash	1st	UTXO	Linear	Permissionless	PoW	L1
XRP, Stellar	1st	Account	Linear	Permissioned	BFT	L1
Monero, Zcash, Dash	2nd	UTXO	Linear	Permissionless	PoW	L1
IOTA (Approval Weight consensus)	2nd	UTXO	Dag	Permissionless	AW	L1
Ethereum (PoS from Sep. 2022)	2nd	Account	Linear	Permissionless	PoS	L1
Ethereum PoW, Ethereum Classic	2nd	Account	Linear	Permissionless	PoW	L1

Table 8.5 Outlines the layer-1 smart contract platforms of the 3rd generation, with some transitioning to heterogeneous multi-chain systems.

Platforms	*Ledger Types*		*Governance*	*Cons*	*VM*	*Architecture*
BNB Chain	Account	Linear	Permissioned	PoSA	EVM	Heterogeneous
Tron	Account	Linear	Permissionless	DpoS	TVM	Heterogeneous
Cardano	UTXO	Linear	Permissionless	PoS		Monolithic
Algorand	Account	Linear	Permissionless	PoS	AVM-TEAL	Monolithic
Hedera	Account	Dag	Permissioned	BFT		Monolithic
VeChain	Account	Linear	Permissioned	PoA	EVM	Monolithic
Fantom	Account	Linear	Permissionless	BFT-PoS	EVM	Monolithic
Theta	Account	Linear	Permissionless	PoS	EVM	Monolithic
EOS	Account	Linear	Permissionless	DPoS	WASM	Monolithic
Tezos	Account	Linear	Permissionless	PoS	TezosVM	Monolithic
Flow	Account	Linear	Permissionless	PoS	FlowVM	Monolithic
Klaytn	Account	Linear	Permissionless	PoS	EVM	Heterogeneous
Casper	Account	Linear	Permissionless	PoS	CasperVM	Monolithic
Oasis	Account	Linear	Permissionless	PoS	EVM, Eswam	Monolithic
Chiliz	Account	Linear	Permissioned	PoSA	EVM	Monolithic

Table 8.6 Shows first languages created from base languages and second languages built on top of first languages. Certain L1s lack virtual machines but function as application-oriented platforms with integrated contracts, classifying them as third generation. Certain L2s are specialized applications and therefore do not fit into a particular blockchain generation.

Platforms	*Gens*	*Governance*	*Cons*	*Layer*	*VM*	*Underlying Technology*
Polygon	3rd	Permissionless	PoS	L2	EVM	L1-Plasma
Arbitrum, Base, Optimism		Centralized		L2	EVM	L1 Optimistic Rollups
ImmutableX, Loopring, dYdX		Centralized		L2	no	L1 ZK-rollups
zkSync Era, Linea, Scroll		Centralized		L2	EVM	L1 ZK-rollups
Starknet		Centralized		L2	CairoVM	L1 ZK-rollups
Cronos	3rd	Permissioned	PoA	L1	EVM	L0-CosmosSDK
Osmosis, THORChain, Terra	3rd	Permissionless	Tendermint	L1	no	L0-CosmosSDK
Moonbeam	3rd	Permissionless	DPoS	L1	EVM	L0-Polkadot Substrate
BitTorrent	3rd	Permissionless	PoS	L2	EVM	L1-Tron

Table 8.7 Showcases the 4th Gen, which consists of efficient L1s with high performance and uniform sharding, as well as diverse multichain systems. All of them have no need for permission.

Platforms	*Ledger Types*		*Cons*	*Layer*	*VM*	*Architecture*
Solana	Account	Linear	PoH+BFT	L1	LLVM	Monolithic
Aptos, Sui	Account	Linear	BFT	L1	MoveVM	Monolithic
Cosmos	Account	Linear	Tendermint BFT	L0	EVM, WASM	Heterogeneous
Polkadot	Account	Linear	NpoS-BABE	L0	EVM, WASM	Heterogeneous
Avalanche	Hybrid	Hybrid	PoS	L0	EVM, WASM	Heterogeneous
ICP	Account	Linear	PoS	L0	WASM	Heterogeneous
Near, MultiversX	Account	Linear	PoS	L1	EVM, WASM	Homogeneous
TON	Account	Linear	PoS	L1	TonVM	Homogeneous
Zilliqa	Account	Linear	PoS	L1	EVM	Homogeneous
Ethereum 2.0	Account	Linear	PoS	L1	EVM, WASM	Homogeneous

8.8 Appendix

Explanation of terms and abbreviations found in Tables 8.3–8.7.

- **BABE-GRANDPA:** BABE refers to Blind Assignment for Blockchain Extension and serves as the protocol for selecting block authors among validator nodes. BABE (Alper, 2023) can be likened to Ouroboros Praos (David et al., 2018), which is the consensus algorithm utilized by Cardano. GRANDPA (Stewart and Kokoris-Kogia, 2020) is the conclusive tool designed for the Polkadot Relay Chain to settle blocks (see, Polkadot Consensus Wiki).
- **NPoS:** Nominated Proof of Stake, similar to DPoS, used in Polkadot.
- **PoSA:** Proof of Staked Authority by BNB Chain.
- **MOVE and MoveVM:** MOVE is a recently developed programming language for smart contracts in Libra-Diem, a consortium blockchain project financially backed by Facebook (Meta). On June 18, 2019, an official announcement was made to develop a stable-coin system that spans borders and is supported by a basket reserve of various fiat currencies. The project was ended and all Diem assets were sold as announced by the Diem Association announcement on 31 January 2022. After Libra-Diem was shut down, certain ex-developers are working on creating Aptos and Sui, L1s that utilize the MOVE language and Move Virtual Machine (MoveVM).
- **IBC** is short for Inter-Blockchain Communication, an interchain protocol created by Cosmos developers for cross-chain interoperation, with the goal of constructing an interconnected network of blockchains.
- **Tendermint Core** serves as a BFT consensus engine implemented in Cosmos and CosmosSDK.
- **WASM**, short for WebAssembly, is a format for binary instructions designed for a virtual machine that is stack-based.

References

1. Alper, A.M., og Handan Kilinc. (2023). BABE protocol. web3.foundation. https://research.web3.foundation/en/latest/polkadot/block-production/Babe.html.
2. Antonopoulos, A.M., Hariry S.H.E., Lords, M.K., Morgan, P., Scothorn, M., og Zolt-Gilburne, S. (2016). *The Internet of Money: Talks by Andreas M.* Antonopoulos. Merkle Bloom LLC.
3. Bano, Shehar, Alberto Sonnino, Mustafa Al-Bassam, Sarah Azouvi, Patrick McCorry, Sarah Meiklejohn, og George Danezis. (2019). SoK: Consensus in the Age of Blockchains. In: *Proceedings of the 1st ACM Conference on Advances in Financial Technologies* (*AFT'19*). New York, USA: ACM. 183–198. https://doi.org/10.1145/3318041.3355458.
4. Bonneau, Joseph. (2022). Why blockchain performance is hard to measure. a16z.com. https://a16zcrypto.com/why-blockchain-performance-is-hard-to-measure/.
5. Bouraga, Sarah. (2021). A taxonomy of blockchain consensus protocols: A survey and classification framework. *Expert Systems with Applications*. ISSN 0957-4174 168. https://doi.org/10.1016/j.eswa.2020.114384.
6. Cangir, Omer F., Onur Cankur, og Adnan Ozsoy. (2021). A taxonomy for Blockchain based distributed storage technologies. *Information Processing and Management*, *58*(5). https://doi.org/10.1016/j.ipm.2021.102627.

7. Chaum, David Lee. (1982). *Computer Systems Established, Maintained, and Trusted by Mutually Suspicious Groups*. PhD Thesis. University California, Berkely. https://nakamotoinstitute.org/static/docs/computer-systems-by-mutually-suspicious-groups.pdf.
8. David, Bernardo, Peter Gazi, Aggelos Kiayias, og Alexander Russell. (2018). Ouroboros Praos: An Adaptively-Secure, Semi-synchronous Proof-of-Stake Blockchain. B. 10821, í *Advances in Cryptology – EUROCRYPT 2018, Lecture Notes in Computer Science*, eftir: J. Nielsen, og V. Rijmen (Eds.). Springer Cham. https://doi.org/10.1007/978-3-319-78375-8_3 2017.
9. Eslamian, & Eslamian (2025). *Blockchain Technology in Water and Environmental Systems*. CRC Press, Taylor and Francis, USA, p. 282.
10. Gangwal, Ankit, Haripriya Ravali Gangavalli, og Apoorva Thirupathi. (2023). A survey of Layer-two blockchain protocols. *Journal of Network and Computer Applications, 209*. https://doi.org/10.1016/j.jnca.2022.103539.
11. Garay, J., og Kiayias A. (2020). SoK: A Consensus Taxonomy in the Blockchain Era. Í *Topics in Cryptology, Lecture Notes in Computer Science*, eftir: Jarecki S. (Ed.). Springer Cham. https://doi.org/10.1007/978-3-030-40186-3_13.
12. Gelashvili, Rati, Alexander Spiegelman, Zhuolun Xiang, George Danezis, Zekun Li, Dahlia Malkhi, Yu Xia, og Runtian Zhou. (2023). Block-STM: Scaling Blockchain Execution by Turning Ordering Curse to a Performance Blessing. In: *Proceedings of the 28th ACM SIGPLAN Annual Symposium on Principles and Practice of Parallel Programming* (*PPoPP '23*). New York, NY, USA: ACM. pp. 232–244. https://doi.org/10.1145/3572848.3577524.
13. Goundar, Sam, Shalvin Chand, Jalpa Chandra, Akash Bhardwaj, og Fatemeh Saber. (2020). A Taxonomy of Blockchain Applications. Ín: *Blockchain Technologies, Applications, and Cryptocurrencies* World Scientific, pp. 49–-71. https://doi.org/10.1142/9789811205279_0003.
14. Haber, S., og Stornetta W.S. (1991). How to time-stamp a digital document. *J. Cryptology, 3*, 99–111.
15. Leonardos, S., Reijsbergen, D., Piliouras, og G. (2020). PREStO: A Systematic Framework for Blockchain Consensus Protocols. *IEEE Transactions on Engineering Management, 67*(4): 1028–1044. doi:doi: 10.1109/TEM.2020.2981286.
16. Little, Arthur D. (2020). *Market Opportunity Report*. https://parallelchain.
17. Mauro Conti, Ankit Gangwal and Michele Todero. (2019). Blockchain Trilemma Solver Algorand has Dilemma over Undecidable Messages. In: *Proceedings of the 14th International Conference on Availability, Reliability and Security* (*ARES'19*). New York: ACM. 1–8. https://doi.org/10.1145/3339252.3339255.
18. Merkle, R.C. (1979). *Secrecy, Authentication, and Public-key Systems.*, PhD Thesis. Stanford University. https://cir.nii.ac.jp/crid/1571417125520618240?lang=en.
19. Nakamoto, Satoshi. (2008). *Bitcoin: A Peer-to-Peer Electronic Cash System*. https://bitcoin.org/bitcoin.pdf.
20. Nijsse, Jeff. (2020). A Taxonomy of Blockchain Consensus Methods. *Cryptography, 4* (4), 32. https://doi.org/10.3390/cryptography4040032.
21. Olga, Labazova, Dehling Tobias, og Sunyaev Ali. (2019). From Hype to Reality: A Taxonomy of Blockchain Applications. In: *Proceedings of the 52nd Hawaii International Conference on System Sciences*. Wailea, Maui, HI, USA: IEEE.
22. Puthal, D., Malik, N., Mohanty, S.P., Kougianos, E., Das, og G. (2018). Everything You Wanted to Know About the Blockchain: Its Promise, Components, Processes, and Problems. *IEEE Consumer Electronics Magazine, 7*(4), 6–14. doi: 10.1109/MCE.2018.281629.
23. Sabzevari, Y., Eslamian, S., Okhravi, S. and Bazrkar, M.H. (2025). Evaluating the Performance of Soil and Water Assessment Tool (SWAT) in a Snow-Dominated Climate (Case Study: Azna–Aligoudarz Basin, Iran). *Atmosphere, 16*(4), 382.
24. Sabzevari, Y. and Eslamian, S. (2025). Application and evaluation of modern irrigation. In: *Handbook of Nature-Based Drought Solutions*. Elsevier, pp. 155–186.

25. Sarkintudu, Shehu M., Huda H. Ibrahim, og Alawiyah Bt Abdwahab. (2018). Taxonomy development of Blockchain platforms: Information systems perspectives. In: *AIP Conference Proceedings, 2016*, 020130. https://doi.org/10.1063/1.5055532.
26. Sherman Alan, T., Javani Farid, Zhang Haibin, og Golaszewski Enis. (2019). On the Origins and Variations of Blockchain Technologies. *IEEE Security Privacy*, *17* (1), 72–77. doi:10.1109/MSEC.2019.2893730.
27. Stewart, Alistair, og Eleftherios Kokoris-Kogia. (2020(). *GRANDPA: A Byzantine Finality Gadget*. web3.foundation. https://github.com/w3f/consensus/blob/master/pdf/grandpa.pdf.
28. Tasca, P., og Tessone C.J. (2019). A Taxonomy of Blockchain Technologies: Principles of Identification and Classification. *Ledger*, *4*. https://doi.org/10.5195/ledger.2019.140.
29. Treiblmaier, H. (2022). Defining the Internet of Value. Í Enabling the Internet of Value: Future of Business and Finance. eftir: Vadgama, N., Xu, J., og Tasca P. (Eds.). Springer Cham. https://doi.org/10.1007/978-3-030-78184-2_1.
30. Vadgama, Nikhil, Jiahua Xu, og Paolo Tasca. (2022). *Enabling the Internet of Value: How Blockchain Connects Global Businesses*. Springer Cham. https://doi.org/10.1007/978-3-030-78184-2.
31. Walden, Jesse. (2018). *4 Eras of Blockchain Computing: Degrees of Composability*. a16z.com. {https://a16z.com/2018/12/16/4-eras-of-blockchain-computing-degrees-of-composability/.
32. Wang, Gang, Zhijie Jerry Shi, Mark Nixon, og Song Han. 2019. SoK: Sharding on Blockchain. In: *Proceedings of the 1st ACM Conference on Advances in Financial Technologies* (*AFT'19*). New York: ACM. 41–61. https://doi.org/10.1145/3318041.3355457.
33. Xu, X., og et al. (2017). A Taxonomy of Blockchain-Based Systems for Architecture Design. 2017. *IEEE International Conference on Software Architecture* (*ICSA*). Gothenburg, Sweden: IEEE, pp. 243–252. doi: 10.1109/ICSA.2017.33.
34. Yakovenko, Anatoly. 2020. *Solana Whitepaper*. https://solana.com/.

9 | Capabilities of Blockchain and Outlook of Blockchain Technology

Bharat Ramdas Pawar[1], G. Sasikala[2*] and B. Sathyasri[3]

Blockchain technology has evolved from a niche innovation supporting digital currencies into a versatile infrastructure with transformative potential across numerous sectors. This chapter provides a comprehensive review of the core capabilities that make blockchain a powerful technological paradigm, including decentralization, transparency, immutability, security, programmability through smart contracts, and resilient data integrity. These capabilities are examined in the context of real-world applications such as finance, supply chains, energy systems, healthcare, and digital identity. The chapter also presents a forward-looking outlook on blockchain technology, exploring key trends shaping its future development. These include advancements in scalability, privacy-preserving techniques, interoperability frameworks, regulatory integration, and the convergence of blockchain with emerging technologies like artificial intelligence, edge computing, and the Internet of Things. By combining current capabilities with prospective innovations, this chapter provides readers with a holistic understanding of blockchain's present strengths and its long-term technological trajectory.

9.1 Introduction

"Blockchain Technology" is a word that many people use to refer to a wide range of topics. They could be discussing Bitcoin, cryptocurrencies, digital tokens, Ethereum Blockchain, smart contracts, or any combination of these at any given time. The truth is that many people misuse the term "blockchain", which has led to confusion among many people regarding both the phrase "blockchain technology" and all of the associated terminology and terminologies. However, the distributed ledger technology (DLT) that underpins all of these applications unites them. Instead of being saved on a centralized server, transactions with distributed ledger technology are duplicated and stored among several computers connected to the network.

[1] Assistant Professor, Department of Electronics & Telecommunication, VPM MPCOE, Velneshwar.
[2] Professor, Department of ECE, Vel Tech Rangarajan Dr. Sagunthala R&D Institute of Science and Technology, Chennai, India.
[3] Associate Professor, Department of Electronics and Communication Engineering, Vel Tech Rangarajan Dr. Sagunthala R&D Institute of Science and Technology.
* Corresponding author: gsasikala@veltech.edu.in

DOI: 10.1201/9781003686026-9

In the present era, when trust is a crucial consideration when designing any new technology, blockchain technology leads the cryptographic approach to technology when considering encryption from decades ago. The main focus of technological advancement is on data that needs to be created, transferred, and simply and securely delivered to the end user in a way that preserves all of its original form and eliminates lag and loss.

This chapter's primary goal is to introduce all of the essential elements of blockchain technology, as using it in any sector requires having these elements in place. The whole transaction execution process, as well as the block creation technique and structure, have all been thoroughly addressed in this chapter. This chapter goes into detail on the many uses of blockchain technology that are addressed, as its use is becoming more and more widespread. Furthermore, there are issues or challenges associated with applying blockchain technology in other sectors, which can be very beneficial to researchers, professors, and business professionals.

A distributed database powered by blockchain technology keeps an unchangeable public ledger of every transaction. Blockchain enables all transactions to be recorded with a time stamp. Every node in the network is in charge of upkeep and ongoing transaction verification. Digital tokens are created using blockchain technology for digital data, including transactions or documents. You may think of these digital tokens as the files' digital fingerprints. These fingerprints are stored in what are known as "block" groups. Subsequently, the separate blocks are joined to create a chain of blocks, with each new block containing a digital token from the previous block. It becomes challenging to update an older block in the chain without simultaneously changing the blocks that came before it. The primary goal of blockchain technology is to eliminate the need for middlemen in the registration, confirmation, and transfer of all types of agreements and properties (Swan, 2015).

The most widely used and well-known technology is inextricably linked to bitcoin. It is also one of the most talked about phenomena in the recent past because it makes it easier for multibillion dollar transactions on the international market to occur without interference from the government. As a result, there are many regulatory issues involving national governments as well as other financial institutions. But the concept of blockchain technology has emerged as a major topic recently, is completely uncontroversial, and has found effective applications in both the financial and non-financial sectors. Marc Andreessen, the doyen of Silicon Valley's capitalists, claims that the "Blockchain's distributed consensus process is the most significant invention in Internet history". This economy runs on the digital system, and some of nature's most reliable authorities depend on it. Any person conducting business must have confidence in some sort of system. Examples of such systems include email services that deliver information, Facebook posts that share information with the public, and bank transactions that provide confirmation that funds are being transferred to a recipient anywhere in the world (Frame & White, 2004; Nakamoto, 2018). We can examine the fact that we live in a digital age where security and privacy concerns must be handled by a third party. But the truth is that it is also possible to operate or hack this third source. This is where blockchain technology enters the picture. Currently, the mechanism for transactions between two people or two businesses is mostly centralized and managed by a third party. A third party is always involved when we make a digital payment in order for the transaction to be completed. In addition, a credit card company or bank may levy additional fees. A comparable tendency is also observed in other domains, such as software, gaming, and music. Thanks to the introduction of Blockchain technology, this problem has been fixed. This technology's primary goal is to establish a decentralized environment where transactions and data are handled without the need for a third party (Lerner, 2006). With the use of distributed consensus, this technology has the ability to track every transaction made in the past and present, which can be confirmed in the future. This is accomplished even without compromising the privacy of complex and digital assets belonging to third parties. Anonymity and distributed consensus are two of blockchain technology's key characteristics.

With the aid of distributed database solutions, blockchain makes it possible to continuously increase the number of data entries that are contributed by the various nodes that make up the network. Every completed transaction is entered onto a public ledger. The decentralized approach offered by blockchain technology eliminates the need for third-party mediation. All of the nodes involved in the Blockchain exchange information about every transaction that is completed. Because of this feature, blockchain technology offers greater transparency than centralized transactions mediated by a third party. Transactions are safer for other nodes since every node in the Blockchain is anonymous. Bitcoin was the first use of blockchain technology to be introduced. Bitcoin is the digital payment network that facilitates the exchange of products in a decentralized environment (Frame & White, 2004).

Even while this technology appears to be ideal for using cryptocurrencies to execute transactions, it still has certain technical limitations and difficulties that need to be thoroughly examined and handled. Nodes in the Blockchain must remain secret in order to fend off attacks, preserve a high degree of security, and facilitate transactions (Swan, 2015). Furthermore, with the Blockchain, processing power is needed for transaction confirmation. In this situation, it becomes crucial to comprehend the subjects that various writers have studied as well as the subjects and areas that require additional attention and are posing risks and difficulties for future research. We have employed methodical mapping to comprehend the research methodology and identify the many works pertaining to the Blockchain sector in order to investigate and comprehend these problems.

Figure 9.1 shows the block structure of the Blockchain.

9.2 Blockchain Technology Inside

It's said that Bitcoin combines already-existing technologies to generate new uses. Implementing safeguards against data fabrication and duplicate payments, along

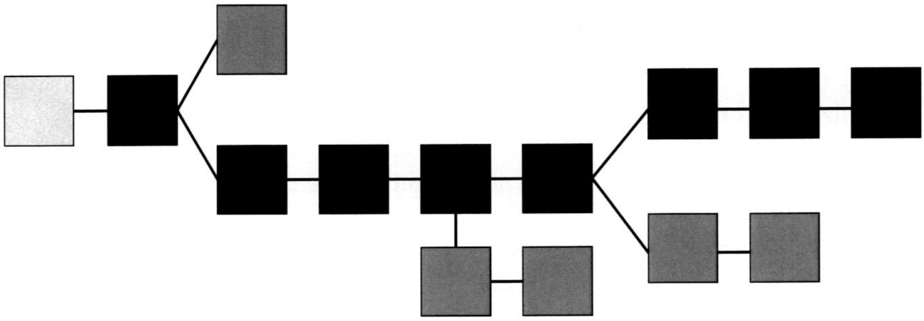

Fig. 9. 1 Structure of Blockchain.

with a means to keep the system safe from malicious user attacks, are essential for the smooth operation of a decentralized electronic money system. The following is a list of major blockchain technologies that enable the proper operation of apps like Bitcoin: P2P, hash, public-key cryptography and digital signatures, Proof of Work, and Proof of Stake (Bheemaiah, 2021).

DLT, which is unchangeable by nature, is the foundation of blockchain technology. In a blockchain network, when a user wants to perform a transaction, they record their request in the ledger, which is accessible to all other users in the network, openly. A node is added to the network once every user has confirmed that it is authentic and has verified the required transaction in that node. It is nearly impossible for more than 51% of the network's nodes to be hacked in order for someone to change or hack the transaction. Every technology underlying blockchain technology is covered in this section.

Block Structure

There are two components to a block: the block body and the block header. As was previously indicated, every block creates a chain when it is connected to its parent or preceding block. The parent block is absent from the initial block, often known as the genesis block. In this case, the header of each block contains the hash of that block.

Peer-to-Peer Network

The roles of the two parties in a client-server network are predetermined since the client requests data from the server and is granted access to it. The server is responsible for maintaining and providing data. On the other hand, in peer-to-peer (P2P) networks, every node involved stores data independently and forms a self-contained network in which nodes can exchange requests and information. The corresponding nodes' role (server or client) is flexible in such a network. P2P networks must take data transfer and search strategies into account. Nodes can be located using search methods, and data can be transmitted between nodes (direct or relayed) using data transmission methods.

Block Version	02000000
Parent Block Hash	B6ff0b2b1680a2862a30 ca44d346d9e910d334b eb48ca0000000000000 00
Merkle Tree Root	9d10aa52ee94936ca93 85695f04ede270dda208 10decd12bc9b048aaa3 1417
Timestamp	24d95a54
nBits	30c31b18
Nonce	Fe9f0864

Transaction Counter

TX 1 TX 2 TX n

Fig. 9.2 Block structure.

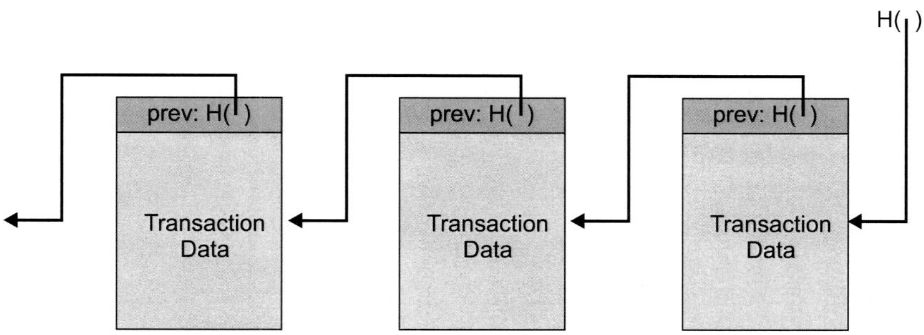

Fig. 9.3 Blockchain nodes.

Hash

Hashing and hash functions are heavily used in blockchain technology. A hash function is a mathematical operation that transforms an input into an output. To create a hash, the original data given as input is altered. Collision resistance, or the difficulty of reconstructing the input data from the hash value alone, is the main characteristic of the hash function. This procedure generates the same hash value from the same data, hence any modification to the original data will result in a completely different hash value. This technique is used to detect data manipulation by taking advantage of such traits (Wagner, 2014).

Public Key Cryptography Digital Signature

A form of cryptography known as public-key cryptography uses separate keys for encryption and decryption. By splitting the key into two versions—a public key that is accessible to all users and a private key that is kept for personal use—the issue of key distribution was resolved. Symmetric-key cryptography requires a number of security precautions to ensure that the key is only exchanged with the appropriate counterparty for encryption and decryption. Public-key cryptography, however, can only be used to securely transfer and receive data if the recipient has already created a set of private and public keys and given the sender the public key. As long as the recipient appropriately handles his or her private key, safety can be preserved even while other people utilize the public key. Blockchain verifies transaction authentication by an asymmetric cryptographic technique. A digital signature, which is similar to public-key cryptography, verifies the legitimacy of data transmitted over a network using two keys. A digital signature created by encrypting the file's hash value with the sender's private key is delivered to the recipient along with the pertinent file. Using the same hash function as the sender, the recipient independently calculates the file's hash value. It then compares this value to the hash value obtained by decrypting the sender's digital signature using the sender's public key. Asymmetric cryptography is used by Blockchain to validate transaction authentication. A digital signature is a method of using two keys to prove the authenticity of data sent over a network, much to public-key cryptography. A digital signature is produced and sent to the receiver with the pertinent file by encrypting the hash value of the file using the sender's private key. The recipient uses the same hash function as the sender to independently produce the file's hash value, which it then compares to the hash value obtained by decrypting the sender's digital signature using the sender's public key.

Consensus Mechanism

Ensuring that all peer-to-peer network users reach the consensus protocol is a significant challenge when using blockchain technology with public ledgers (Namasudra et al., 2021; Williams, 2016; Tran, 2020). In order to ensure that transactions are validated in the correct order and that network members abide by its rules, a consensus protocol is employed. It is also used to prevent problems like the double spending problem, ensure that the information within a block is accurate, and ensure that the nodes, or miners, receive appropriate reward.

There must be some sort of vote among a known group of peers in order for algorithms for reaching consensus with arbitrary flaws. There are two primary methods that are in use: Byzantine Fault Tolerance (BFT) and the "Nakamoto consensus". In the first method, a leader is chosen by a "lottery" of sorts, after which they put forward a block that can be added to a series of blocks that have already been committed to. The second method relies on BFT algorithms and reaches the required consensus through several rounds of explicit voting. The most popular consensus methods are introduced in this part, including Proof of Work (PoW) and Proof of Stake (PoS), which are mostly tested with cryptocurrencies.

Proof of Work (PoW)

Anyone can take on the PoW mining challenge. The miners fight with each other to add the next block. A specific reward is given to the miner who solves the puzzle the quickest. Actually, the winner of the race is usually the node with higher computational efficiency. The PoW algorithm is applied to Bitcoin.

The general term for a system that assigns a simple but challenging work that can be easily verified as done to prove someone's innocence is PoW. One way to ensure authenticity is to give each member of the network one vote to decide which transaction belongs in the next block. Which set of transactions should be included is determined by the total number of votes. This type of consensus-process is susceptible to Sybil assaults, in which a single person pretends to be numerous, in order to gain more clout in the network. The Bitcoin inventor Nakamoto found a solution to this influence problem by making voting more expensive. The quantity of processing power possessed by each individual determines their level of influence (Wagner, 2014). Higher processing power results in higher energy requirements and higher hardware prices. The PoW consensus protocol is conceptualized as follows. The PoWk consensus algorithm used by bitcoins allows the network to compile all of the transactions that have been made during a given time frame into a block. The nodes' job is to validate the transactions and record them in a blockchain, hashing the data to keep it safe from hackers. Nodes have financial incentives to continue hashing and mining; the more blocks generated, the more bitcoins are awarded to the nodes. Neighboring nodes receive blocks that are created by a node. The nearby nodes independently confirm that the regulations have been observed and that the data within the block is accurate. It is advised to wait at least six blocks on a bitcoin network to ensure that the transaction is final. A blockchain fork occurs when many nodes attempt to work on the same transaction at the same time in competition to be the first to produce a block The node that creates the block with the longest blockchain first wins and is paid. Although PoW requires a lot of hardware and energy, which increases mining prices, it has been empirically shown to be reliable and safe.

Among the drawbacks of a proof-of-work consensus method are the high energy expenditures associated with generating a single block and the possibility of a 51% attack. According to Courteous, "the proof-of-work process is going to self-destruct". The mining industry is becoming more specialized and smaller, and large organizations possessing superior resources are able to outmaneuver an individual miner. A 51% attack is becoming more likely as a result of the system's increased centralization to a few numbers of powerful companies due to this mining specialization.

Proof of Stake (PoS)

The Blockchain community adopted PoS, a new consensus mechanism, to lower the possibility of a 51% attack and to use less energy. Rather than demonstrating that it has completed a computationally challenging activity, as required by the PoW protocol, a node could instead demonstrate that it possesses a specific quantity of coins. In a PoS system, a new block is created using coins rather than processing power, and the node with the most coins has the most sway. Game theory states that the majority owner will not attack a network that is secure and stable because it serves their interests. In the event of an attack, the digital currency's value will only drop and it will become unstable. Forking is one problem with the PoS protocol. Without incurring any additional computational costs, multiple nodes may begin mining on a transaction at the same time. Comparing a PoW protocol to a blockchain increases the risk of malicious nodes forking the blockchain. This raises the possibility of greedy nodes mining on all forks in order to avoid missing out on block rewards, as well as double spending assaults. The problem of double spending attacks can be resolved by employing check-point blocks, which prevent blocks prior to a check-point from being changed. The arguments put out by Manning and the bitcoin community are unfounded since a PoS technology is still susceptible to a 51% attack. He claims that by acquiring 50% of a PoW cryptocurrency network's money base for free, a miner can take over the network (Namasudra et al., 2021).

Three Main Categories of BCT:
Public blockchain: A distributed, decentralized ledger that is accessible to anybody wishing to participate is called a public blockchain. Without the need for a central authority, it enables the transparent and impenetrable recording of transactions over a network of computers (nodes).

Permissioned or private blockchain: A private blockchain is a type of blockchain where network access is restricted to a chosen subset of users. Private blockchains are usually centralized and permissioned, in contrast to public blockchains, which are decentralized and accessible to everybody. This implies that users must have authorization in order to connect to the network, which is frequently run by a single organization or a group of organizations.

Federated or consortium blockchain: A consortium blockchain refers to a type of blockchain network that is not controlled by a single organization, but rather is maintained and administered by several. Consortium blockchains achieve a balance by incorporating several organizations, in contrast to private blockchains, where one entity has exclusive control, and public blockchains, like Bitcoin and Ethereum, where anybody can participate and contribute to the network.

9.3 Compute, Build, and Deploy Blockchain

At first, it was all about Bitcoin, but as the movement spread, a lot more cryptocurrencies entered the market. Some of them succeeded in making money, while other cryptocurrencies trailed behind. But before long, blockchain technology realized its true potential and expanded into a wide range of unexpected fields. Healthcare, enterprise software development, banking, insurance, and other financial sectors are among the domains where blockchain technology is presently radically changing the existing frameworks of technology.

Banking and Payments
The future of payment and financial systems is now blockchain. Cryptocurrencies that resemble Bitcoin have the ability to govern payment systems independently of national boundaries. Remittances based on bitcoin are exemplified by BRA.

Cyber Security
Cryptography is used in blockchain to safeguard and verify data. This will prevent any unapproved modifications and system hacks. By doing so, intermediaries are eliminated from the system, preventing any unauthorized alterations.

Supply Chain
Better accountability, transparency, and feedback mechanisms along the supply chain are just a few of the ways that the blockchain may transform the supply chain. Complete product tracking is possible with blockchain-based supply chain management. IoT sensors enable the blockchain to record every action and state of a product. A supply chain management system built on the blockchain is called Block Verify and Provenance.

Online Data Storage
A single point of failure can occur with OneDrive, Google Drive, and other centralized systems that hold data. Blockchain technology increases the security and dependability of distributed data storage. One such encrypted cloud storage platform is called Storj.

Networking and IoT: A decentralized network of IoT devices can be established by utilizing blockchain technology in networking and IoT. As a result, managing the Internet of Things devices no longer requires a central location.

Insurance: The foundation of the global insurance market is trusting management. The new paradigm in trust management is blockchain. Blockchain relies on participant mistrust to maintain confidence.

Government: Blockchain technology will reduce red tape and bureaucratic barriers while increasing operational efficiency and transparency in government processes.

Crowdfunding: It is a well-liked means of financing new ventures and initiatives. Smart contracts and online reputation systems are used in blockchain-based crowdfunding platforms to establish trust, doing away with the need for a middleman who would otherwise charge exorbitant fees for this service. It is possible for new enterprises to issue their own tokens, which may then be traded for goods, services, or money.

Multimedia and entertainment: Blockchain is now present in the entertainment industry, where excessive third-party influence is a problem. The middleman will no longer be present in this scenario thanks to the blockchain's implementation. One of the entertainment industries where blockchain is already being implemented is online music.

For instance, Ujo and Mycelia's music
Real estate: This is another crucial area where the use of blockchain technology will bring about significant change. There are numerous ownership and transfer

problems in the current real estate system. The complete real estate system with shared ledgers can be controlled by the blockchain implementation in this industry.

For instance, the state government of Andhra Pradesh in India has begun using blockchain technology to handle land registration in its entirety.

The list includes a lot more other locations, such as energy management, voting, healthcare, forecasting, and transportation. Along with blockchain came industry-specific blockchain development frameworks, DApp and digital asset management software, blockchain management software, and other blockchain-related products and services. And as it expands, a ton of more tools are added. Blockchain creation and management are now easier than ever thanks to all of these frameworks and tools. Thus, it's now simpler than ever to build and implement blockchain-applied solutions.

Features of Blockchain

These days, blockchain is utilized in a wide range of applications across numerous industries and use cases thanks to its distinctive features. Below is a discussion of some of the key characteristics of blockchain:

(1) **Enhanced security:** Increased security is possible with blockchain technology since all data is stored in a decentralized network. Since each node's hash is linked to the one before it, changes to one node's hash impact the hashes of all nodes. In this scenario, data cannot be changed. As was previously said, there are various types of blockchain networks. However, because permissioned blockchain enables specific users to add new blocks to the network, it is usually utilized in industries.

(2) **Decentralized network:** Decentralization is one of the main characteristics of a blockchain network. It suggests that a blockchain network lacks a central authority. In this case, the network is decentralized, with all system control being exercised by a group of peer nodes. Users on a blockchain network are assigned specific positions. Because a blockchain network doesn't require a central authority, people can utilize it directly through the internet and keep their data there.

(3) **Immutable:** A blockchain network's data is viewed as an endless transaction log. A block cannot be removed or changed once it has been uploaded to the network. In this instance, each blockchain node maintains a copy of the web ledger and verifies each transaction's authenticity before allowing it to join the network. A transaction is added to the public ledger if it is validated by the majority of nodes in a blockchain network. This guarantees a blockchain network's dependability.

(4) **Transparent:** Data control over users is intended to be transparent with blockchain technology. Every peer node in this scenario has access to data blocks, which they can use anyway they see fit. This encourages openness, which dismantles boundaries across systems, organizations, and individuals.

(5) **Distributed ledger:** Distributed ledgers store, share, and synchronize transactions in their individual electronic ledgers using independent computers,

or nodes. Unlike a standard ledger, this one does not store data centrally. A public ledger is typically used to make all transaction data, including the identities of the participants, available to the public. Federated and private blockchain networks function substantially differently. Nonetheless, a lot of users and administrators might find a lot of ledger facts in these networks. This is primarily because everyone who uses the blockchain network contributes to maintaining the ledger.

(6) Consensus: The foundation of a blockchain network's architecture is its consensus algorithms. The network uses these consensus algorithms to help it decide what to do in each situation. A consensus method is necessary to validate a specific transaction that enables the seamless operation of the entire blockchain network, even with millions of nodes connected. As a result, every choice the network makes is a means—directly or indirectly—of accomplishing its objectives. One of the main advantages of DLT is its consensus method.

(7) Faster settlement: Conventional financial processes move slowly; a transaction may take days to complete. The quicker settlement method supported by blockchain technology is in contrast to these conventional approaches. As a result, users can finish transactions quickly, keeping the system from becoming sluggish overall. Even if there are situations when the network cannot handle a high volume of users or quick settlement, blockchain is now used for quick transactions and peer-to-peer money transfers.

(8) Chronological data: Blockchain technology is a network of blocks, as was previously explained. Here, each block contains transaction-related data and uses a cryptographic hash to connect to the block before it. The block that comes after with the same hash value is linked to the block that came before it. By joining together, these networked bricks create a chronological chain of data that documents ownership transfers and establishes provenance.

(9) Smart contracts: One of the key elements of a blockchain network is a smart contract. These programmers, scripts, or logics are used in a blockchain network for terms, conditions, or network-wide automated chores. Environments with permissioned blockchains are ideal for smart contract operation. It expedites the completion of any settlement. Many of the operations that take place in a blockchain network can be automated by smart contracts.

9.4 Risks and Adoption

New techniques for managing dispersed and decentralized systems are needed as cutting-edge technologies like data business models, on cloud technologies like Azure, AWS or GCP, and the Internet of Things (IoT). Since all of these technologies operate online, there are more and more criminal users and attackers, which leads to a constant rise in security concerns. Furthermore, due to the growing volume of data and susceptible gadgets on the market, it is critical that verified and trustworthy services are enforced today. Blockchain is a cutting-edge and exciting technology. As we know before, blockchain-based technology can be used to tackle a wide range of problems or applications. This covers both financial and non-financial uses, such as remittance and investment banking, as well as notary services. Most of them are extreme innovations. Adoption is fraught with peril, much like dramatic inventions.

Shift in behavior: Although resistance to change exists, change is a constant. In the non-tangible world that BlockChain presents, users must become used to the notion that their electronic transactions are full, safe, and secure. The roles of the contemporary middlemen, like Visa or Mastercard (for credit cards), will likewise change. We expect them to invest as well as migrate to a blockchain-based system. They will continue to provide customer relationship-related services. Scaling: One of the issues with the newly developed BlockChain services at the moment is scaling. Imagine that this is your first time executing a BlockChain transaction. You have to download and validate all BlockChains that are currently in existence before you can finish your first transaction. Considering the exponential increase in the quantity of blocks, it can take several hours or even longer (Eslamian & Eslamian, 2025).

Bootstrapping: To convert the current contracts, business papers, or frameworks to the new BlockChain based approach, a significant set of migration procedures needs to be finished. For example, present records maintained by counties or escrow services on real estate ownerships or liens need to be converted to the appropriate BlockChain format. This could include expenses and time. The fact that these agencies are trusted by their clients may help adoption in the United States. Strong opposition to adoption is expected in economies like China that have more stringent rules.

Sutardja Center for Entrepreneurship and Technology Technical Report
The bad guys can utilize this for fraudulent operations like money trafficking because it is easy to transfer assets and because BlockChain transactions are pseudonymous. However, if there are enough regulations and technical assistance, law enforcement agencies will be able to monitor them and file charges.

Quantum computing: The foundation of blockchain technology is the logical impossibility of a single party gaming the system due to insufficient processing power. However, if quantum computers are available, it would be feasible to crack cryptographic keys quickly enough using merely brute force. The system will collapse as a result. The counterargument is that keys ought to become stronger in order to make them harder to crack (vcresearch.berkeley.edu).

9.5 Conclusion

In conclusion, we always use a blockchain approach to build microservices with different versions, design the artifact, run on different platforms, compute, and deploy successful microservice applications in a best suited and adoptive framework with all reliability and comprehensiveness, whether on premises or off. To comply with GDPR regulations, this method uses a number of novel adaptive encryption technologies, including consensus algorithms and Merkle trees. The core technology of Bitcoin is blockchain. The immutable and fast adopting distributed ledger feature, security algorithms, and protocols of Blockchain technology make it an appealing solution for contemporary financial and non-financial company issues worldwide. Due to the great amount of interest in blockchain-based commercial applications, many start-ups are developing them. Adoption certainly faces several difficulties, as was previously indicated. PayPal, Visa, and other e-payment corporations are investing in the research of integrating blockchain technology into their current business strategies. In fact, several of them are searching the blockchain space for fresh company ideas.

Given how quickly blockchain technology is being developed and used as a lead technology, it will soon become a commonplace piece of technology. The primary obstacle to the acceptance and deployment of smart contracts and related technologies is the significant challenges that must be resolved in the near future. Very few blockchain frameworks are ideal for resolving issues in the real world. Technology is more portable and easier to adopt when it is deployed via an integrated approach thanks to consensus algorithms. Continuous integration with digital signatures, NFT, and smart contracts gives us a wealth of options for end-to-end user deployments. The Blockchain approach provides a thorough evaluation of innovative and comprehensive business ideas, leading to the discovery and fulfillment of several technological problem solutions.

References

1. Bheemaiah, K. (2015). *Blockchain 2.0: The Renaissance of Money*. Wired. https://www.wired.com/insights/2015/01/block-chain-2-0/
2. Eslamian, S. and Eslamian, F. (2025). *Blockchain Technology in Water and Environmental Systems*. CRC Press.
3. Frame, W. and White, L. (2004). Empirical studies of financial innovation: Lots of talk, little action? *Journal of Economic Literature, 42*(1), 116–144. http://www.jstor.org/stable/3217038.
4. Lerner, J. (2006). The new financial thing: The origins of financial innovations. *Journal of Financial Economics, 79*(2), 223–255. http://search.proquest.com/docview/231721046.
5. Nakamoto, S. (2018). *Bitcoin: A Peer-to-peer Electronic Cash System*. https://www.ussc.gov/sites/default/files/pdf/training/annual-national-training-seminar/2018/Emerging_Tech_Bitcoin_Crypto.pdf.
6. Namasudra, S., Deka, G.C., Johri, P., Hosseinpour, M. and Gandomi, A.H. (2021). The revolution of blockchain: State-of-the-art and research challenges. *Archives of Computational Methods in Engineering, 28*, 1497–1515.
7. Swan, M. (2015). *Blockchain: Blueprint for a New Economy*. O'Reilly Media.
8. Tran, S. (2020). IBM and Oracle collaborate on interoperability work for their blockchains to communicate with each other. https://blockchain.news/news/ibm-oracle-collaborate-interoperability-work-blockchains-to-communicate.

9. University of California, Berkeley. (n.d.). *Sutardja Center for Entrepreneurship and Technology*. https://vcresearch.berkeley.edu/research-unit/sutardja-center-entrepreneurship-technology.
10. Wagner, A. (2014). Ensuring network scalability: How to fight blockchain bloat. https://bitcoinmagazine.com/technical/how-to-ensure-network-scalibility-fighting-blockchain-bloat-1415304056.
11. Williams, A. (2016). *IBM to Open First Blockchain Innovation Centre in Singapore, to Create Applications and Grow New Markets in Finance and Trade. The Straits Times.*

10 Broad Applications of Blockchain Technology

Mohammad Karrabi[1], Farkhondeh Jabari[2] and
Asghar Akbari Foroud[3*]

Blockchain technology has evolved far beyond its initial role in supporting cryptocurrencies, developing into a versatile platform capable of revolutionizing data management, security, and multi-party coordination across a wide range of industries. This chapter presents an in-depth analysis of blockchain applications, emphasizing its capacity to enhance transparency, trust, and operational efficiency within complex systems. Key sectors explored include finance, supply chain management, healthcare, energy, digital identity, governance, and the Internet of Things, illustrating how blockchain can address persistent challenges such as data manipulation, inefficiencies caused by intermediaries, and fragmented information flows. The discussion is supported with real-world case studies and examples of blockchain platforms to highlight tangible benefits and emerging implementation trends. Furthermore, the chapter examines barriers to adoption, including scalability limitations, regulatory uncertainties, and difficulties integrating blockchain with legacy systems. By providing a comprehensive overview of its applications, this chapter underscores blockchain's growing potential to reshape digital ecosystems and enable next-generation innovation.

10.1 Introduction

In recent years, blockchain technology has garnered significant global attention, particularly for its potential to digitalize the power and energy sector. This technology enables the use of smart contracts, secure communication platforms, active producers and prosumers, and smart metering infrastructures, among others, from multiple perspectives including economic, environmental, energy, social, political, and legal considerations. Initially introduced in 2008 with Bitcoin for digital payment systems, blockchain has since expanded to applications in smart contracts, banking, financial services, healthcare, energy systems, and other

[1] Electrical and Computer Engineering Faculty, Semnan University, Semnan, Iran.
[2] Power Systems Operation and Planning Research Department, Niroo Research Institute (NRI), Tehran, Iran.
[3] Electrical and Computer Engineering Faculty, Semnan University, Semnan, Iran.
E-mail: mkarrabi@semnan.ac.ir[1]; fjabbari@nri.ac.ir[2]
* Corresponding author: aakbari@semnan.ac.ir

DOI: 10.1201/9781003686026-10

sectors—often in combination with artificial intelligence (Eslamian & Eslamian, 2025).

As a distributed digital ledger, blockchain eliminates the need for centralized authorities or trusted intermediaries in peer-to-peer (P2P) transactions across business, industrial, and commercial domains. It ensures that data is recorded in immutable, time-stamped blocks that are securely shared across a network of computers and accessible to all end-users in a cyber-physical-social framework (Tripathi et al., 2023).

Through P2P interactions, blockchain functions as both a ledger and a database for recording transactions. In the context of distributed energy networks and smart cities, one of blockchain's key benefits is the decentralization of energy and financial exchanges, eliminating intermediaries and enabling direct peer-to-peer trading.

Several notable advantages of blockchain technology include:

- **Data Immutability:** Once recorded, blockchain data cannot be altered, enhancing trust in financial transactions, asset ownership, and identity management. The decentralized P2P network structure mitigates the risk of single-point failures and cyberattacks, which are common in centralized server-based systems.
- **Smart Contracts:** These self-executing contracts allow users to carry out exchanges and transfer value—including assets, currency, or digital content—directly on the blockchain platform without intermediaries.
- **Consensus Mechanisms:** Blockchain networks enable all participants to reach agreement on the validity and integrity of data, ensuring that the network maintains a consistent and reliable ledger.

A schematic representation of the primary applications of blockchain technology is illustrated in Figure 10.1.

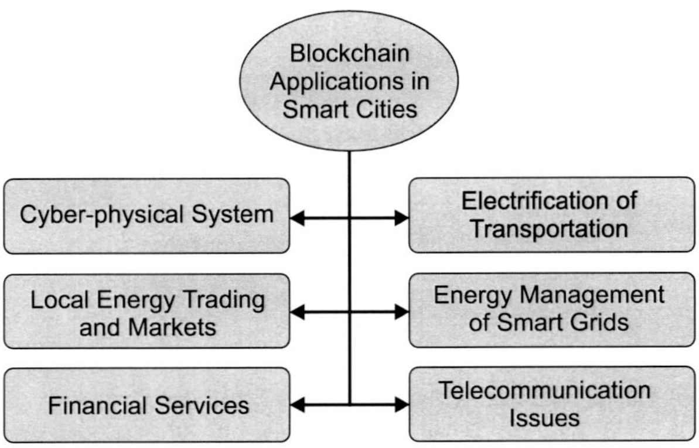

Fig. 10.1 Different applications of blockchain technology in cyber-physical power and energy systems.

Blockchain Integration in Smart Energy Systems

Recent advancements in smart grid technology aim to meet growing energy demand through optimized energy networks, grid automation, and enhanced energy management. Integrating blockchain systems into smart grids offers a pathway for fast, secure, and efficient energy exchanges. Blockchain enables features such as peer-to-peer (P2P) energy trading, smart metering, transparent operations, and automated energy management (Uddin et al., 2023).

10.2 Energy Management in Smart Grids

Blockchain plays a critical role in managing smart energy grids, which can be implemented through three main strategies: centralized platforms, multiagent systems, and decentralized frameworks (Miglani et al., 2020; Alladi et al., 2019).

10.2.1 Centralized Energy Management

In a centralized framework, a central authority or trusted third party oversees generation, transmission, distribution, storage, and consumption. While this approach allows for coordinated control, it suffers from low resiliency and flexibility due to the high risk of cyber-physical attacks on a single data center.

10.2.2 Multiagent Management Infrastructure

Multi-agent systems distribute management tasks among several agents. While effective, these infrastructures are often costly to operate, particularly in developing countries.

10.2.3 Peer-to-Peer Energy Trading Platforms

Blockchain-based P2P energy trading enables direct data exchange among all microgrid participants, including consumers, prosumers, producers, aggregators, retailers, and storage unit operators. Compared to centralized or multiagent approaches, blockchain-enabled P2P platforms offer enhanced reliability, security, cost-efficiency, flexibility, and resilience in microgrid energy management (Wang et al., 2021).

For example, in blockchain-based water-energy hybrid trading models, transaction information is stored in distributed blocks, forming an immutable ledger for P2P transactions (Li et al., 2023).

10.3 Blockchain Types in Energy Networks

Technically, blockchain can be classified into three main types (Yang et al., 2020; Cantillo-L. et al., 2022):

- **Public Blockchain:** All nodes are visible and open, allowing collaboration on transactions and computations.

- **Private Blockchain:** Access is restricted to authorized nodes, providing controlled and secure operations.
- **Consortium Blockchain:** Nodes must be registered and verified to participate in data exchange, offering a balance between decentralization and security.

The primary goal of blockchain is to reduce the reliance on centralized intermediaries, particularly in closed networks.

10.4 Permissioned vs. Permissionless Blockchain in Microgrids

Smart microgrids comprise multiple zones and equipment, which can lead to losses or inconsistencies during data exchanges. To ensure secure and reliable transactions, blockchain can be implemented as either permissioned or permissionless:

- **Permissioned Blockchain:** Uses secure channels for data exchange and transaction validation based on identity access. Shared consensus mechanisms enhance computational efficiency. Private blockchains and certain consortium blockchains fall under this category, providing security and privacy while enabling decentralized energy management without third-party intervention (Wang et al., 2021).
- **Smart Contracts and Cryptocurrency:** In P2P energy trading platforms, smart contracts automate predefined scenarios, ensuring secure, transparent, and immutable exchanges between producers and consumers. Blockchain facilitates local energy trading in microgrids, energy hubs, vehicle charging, and distribution network management while guaranteeing data integrity, encryption, and user authentication (Khan et al., 2021; Yapa et al., 2021).

10.5 Key Features of Blockchain-Enabled Smart Grids

- Integration of renewable energy sources and storage systems
- Community participation in energy production and trading
- Transparent real-time billing and energy pricing
- Fast decision-making protocols for effective grid management
- Secure storage and analysis of energy data while maintaining privacy
- Self-healing and resilient network performance

10.6 Challenges of Blockchain Integration in Smart Grids

Despite its advantages—reliability, transparency, immutability, and distributed operation—integrating blockchain into smart grids faces several challenges:

- Large-scale integration of distributed energy sources at different voltage levels and maintaining supply-demand balance
- Limited network connectivity and unreliable communication technologies
- Scalability concerns for handling numerous transactions
- High computational requirements for blockchain operations

- Privacy and security issues in grid integration
- Lack of standardized frameworks for blockchain-grid compatibility and cooperation

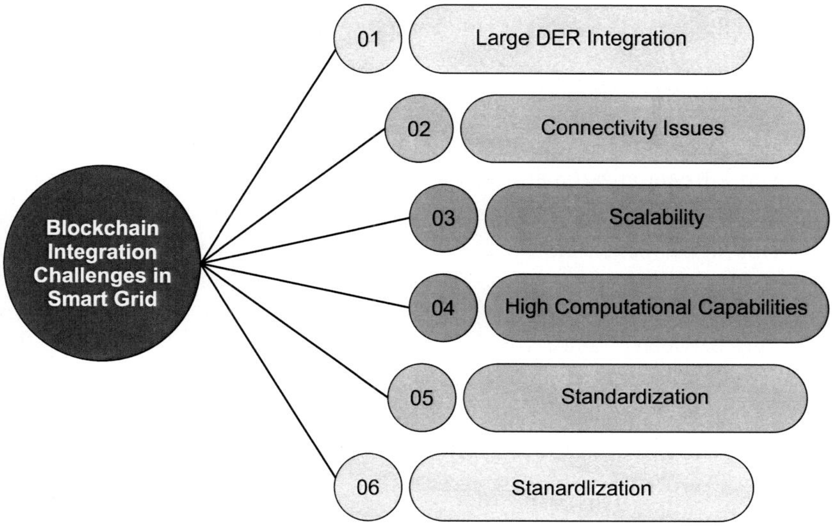

Fig. 10.2 Blockchain integration challenges in smart grids.

10.7 Local Energy Trading and Markets

Local energy markets rely on multiple actors, including smart homes, load response aggregators, distribution network operators, distributed generation units (both fossil-fuel and renewable), energy storage systems, parking lots for hybrid and electric vehicles, and smart meters enabling two-way communication with data concentrators in demand aggregation centers (Johanning & Brucknar, 2019).

In these markets, buyers and sellers of electricity—and other energy forms such as water, heat, cooling, hydrogen, and natural gas—participate to increase flexibility in managing technical and financial constraints while balancing supply and demand. Consumers and prosumers, who generate part or all of their energy needs, can exchange energy with microgrids or nearby consumers (Honarmund et al., 2021).

Active users with flexible or shiftable loads, such as EVs and air conditioning systems, may respond to signals from load response programs or market controllers, while passive users do not engage in consumption management. Surplus energy can be sold to other consumers, and demand-side management participants may shift usage in response to tariff signals and network load forecasts. All transactions are executed in a decentralized framework, maintaining security and privacy (Siano et al., 2019).

Distributed energy sources—including solar panels, wind turbines, storage systems, heat and refrigeration pumps, and controllable loads—are increasingly participating in energy exchanges. Emerging blockchain-based P2P platforms provide a suitable infrastructure for such markets, but several challenges remain:

- Ensuring stability and security of energy exchanges while respecting personal ownership rights.
- Coordinating large-scale P2P energy trading to provide flexibility and support new market participants.
- Integrating households, corporate managers, and market operators into hybrid decentralized markets, requiring coordination with traditional market structures (Wu et al., 2022).

With higher penetration of small-scale renewable energy, consumers may act as both producers and consumers depending on grid and market conditions. Blockchain-enabled smart contracts facilitate automated, transparent energy exchanges between producers, consumers, and storage systems (Afzal et al., 2022; Aloqaily et al., 2020).

10.8 Sustainable Development of Smart Distribution Networks

Sustainable development in energy systems integrates economic growth, environmental protection, and social inclusion. While industrialization and economic activities have historically caused environmental degradation, blockchain technology can help coordinate financial transactions, promote clean energy, and support equitable energy distribution (Mbaidin et al., 2023).

However, technical challenges remain, particularly in developing countries. For instance, blockchain scalability is limited: traditional block creation times (~19 minutes) and block sizes (1 MB) constrain transaction throughput (~8 transactions per second). Privacy risks and potential data loss can affect both users and asset owners. Despite these challenges, blockchain enables transparent, decentralized tracking of energy network characteristics, improving stability, reliability, monitoring, protection, and control. Such infrastructures can reduce greenhouse gas emissions and promote the integration of clean energy (Sahoo et al., 2022).

10.9 Electric Vehicles and Blockchain Integration

Electric vehicles (EVs) play a transformative role in modern electricity systems but face barriers such as limited charging infrastructure. Blockchain technology facilitates fast, secure, and cost-efficient transactions for charging and discharging, reducing adoption barriers.

Smart contracts enable automated and real-time energy management for EVs, including wireless and inductive charging at public points, traffic signals, or dedicated stations (Livingston et al., 2018). EVs can operate bidirectionally with the grid: vehicle-to-grid (V2G) and grid-to-vehicle (G2V). This flexibility allows charging during off-peak hours and discharging during peak demand, providing valuable grid support (Jabari et al., 2019).

The unpredictable behavior of EVs, including variable location and energy use, presents technical challenges for energy management. Blockchain provides a platform for real-time monitoring, control, and coordination of energy storage resources, including aggregated plug-in hybrid or pure electric vehicles (Al-Saif et al., 2021). By optimizing energy usage across locations and times, blockchain integration enhances energy efficiency, economic performance, and environmental sustainability (Tonev & Nikolev, 2020).

10.10 Blockchain Applications in Cyber-Physical Systems

Smart grids face multiple challenges and vulnerabilities, as many components within these systems can be manipulated or attacked. The potential threats can be broadly categorized into physical and cyber-attacks, each varying in type and impact (Zhao et al., 2020). Some common cyber-physical attacks include:

1. Time synchronization (or time delay) attacks
2. Global Positioning System (GPS) spoofing attacks
3. Denial of Service (DoS) attacks

A particularly critical threat is the False Data Injection (FDI) attack, in which an attacker manipulates or injects false data into measurements or control signals, aiming to alter the operating state of the system. FDI attacks pose serious risks to the stability and operation of smart energy networks. Conventional detection methods often rely on redundant measurement sets and signals; however, these approaches may have limited effectiveness across varying operational states and require significant measurement redundancy.

By introducing a cyber-layer using blockchain, smart grids can enhance security, reliability, and operational resilience. Blockchain provides a decentralized infrastructure to facilitate secure energy trading, microgrid control, and interactions between diverse entities, including consumers, producers, prosumers, storage units, aggregators, and electric vehicle charging stations. This cyber-layer ensures that all operations—from energy trading to system control—are verifiable, immutable, and resistant to manipulation.

For example, blockchain-based electricity markets connect energy traders—consumers, producers, and retailers—through a decentralized framework, allowing microgrids and distributed generation units to operate securely. Additional players such as prosumers, storage technologies, and EV aggregators can also participate in this decentralized ecosystem.

10.11 Protection and Control of Smart Systems

Blockchain technology significantly enhances security and flexibility in electric power systems. Its cryptographic security, combined with consensus mechanisms, ensures that once energy data or transactions are recorded on the blockchain, they become immutable, making unauthorized modifications extremely difficult.

Additionally, blockchain decentralizes the ledger, giving every network participant a copy of the data. This structure eliminates single points of failure, reduces vulnerability to attacks, and improves system flexibility compared to centralized architectures. Decentralized computing across blockchain nodes enhances network resilience, while smart grid layers can be managed by blockchain-based aggregators to ensure secure and reliable information storage. Moreover, blockchain ensures pseudonymity, disconnecting a participant's real identity from their blockchain identity and further reducing attack risks (Musleh et al., 2019).

Smart grids incorporate advanced technologies such as:

- Advanced Metering Infrastructure (AMI)
- Load response programs
- Micro Phasor Measurement Units (μPMUs)
- Renewable energy-based generation units
- Hybrid and electric vehicles
- Energy storage systems

These technologies improve monitoring, protection, and control. AMI and μPMUs transfer data between control centers and system entities—including producers, consumers, prosumers, and retailers—via multiple communication protocols. Cyber-attacks may target these communication channels or computing infrastructures by exploiting software vulnerabilities, misconfigurations, or insider threats (Kabachi, 2016).

The AMI, through smart meters at the consumer side, provides two-way communication with distribution companies, enabling:

- Real-time electricity pricing
- Consumption data collection
- Advanced failure analysis

Although compromising a few meters may have minimal impact, **coordinated cyber-attacks** targeting multiple meters or communication channels can seriously degrade distribution system reliability (Li et al., 2019).

Figure 10.3 illustrates potential cyber-attack points in a smart distribution network. Branch currents and node voltages are measured by μPMUs and transmitted to controllers and protection devices through high-speed communication links. Controllers make decisions based on system conditions and issue commands to specific protection elements. Cyber-attacks can introduce delays in direct or feedback paths, manipulate sensor data, or compromise protective and control devices. These attacks are commonly classified into three categories:

1. Time attacks
2. Data integrity attacks
3. Response attacks

Integrating blockchain into smart grids offers a robust, decentralized approach to mitigating these cyber-physical threats while enabling secure energy management and trading.

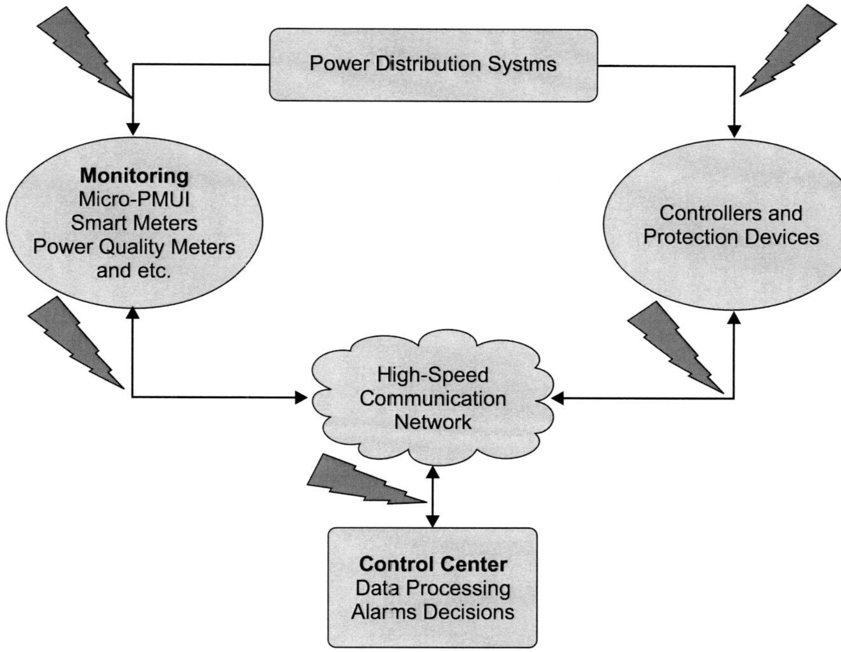

Fig. 10.3 Possible cyber-attacks to monitoring, protection, and control of power distribution systems.

10.11.1 Cyber-Attacks on Smart Grids

Smart grids, as dynamic cyber-physical systems, are highly sensitive to cyber-attacks that target time synchronization, data integrity, and system responses. Blockchain integration provides resilience, but understanding potential attack vectors remains critical. Cyber-attacks on smart grids can be broadly classified as follows:

10.11.1.1 Time Cyber-Attacks

In dynamic systems such as control and protection schemes, time is a critical component. Systems cannot tolerate communication delays; therefore, temporal cyber-attacks pose significant risks. These attacks may affect the communication network by injecting delayed packets or causing intermittent interruptions, potentially resulting in Denial of Service (DoS) scenarios that disrupt the normal operation of the grid.

10.11.1.2 Data Integrity Attacks

Data integrity attacks involve the corruption of sensor or control data, either in the forward or reverse communication path. For instance, attackers may manipulate μPMU (micro-Phasor Measurement Unit) measurements or block trip signals from controllers to protective devices. Such attacks compromise the accuracy of system state estimation and may prevent proper protection or control actions.

10.11.1.3 Response Attacks

Response attacks are similar to data integrity attacks, wherein attackers hijack transmission packets between metering devices and the control center. This manipulation can distort μPMU measurements or control signals, potentially causing incorrect automated responses in protection and control systems.

10.11.1.4 Cyber-Attack Vectors

Cyber-attacks in smart grids typically exploit malware (viruses, worms, Trojans) but may also target physical layers, as seen in attacks on transmission lines. The cybersecurity of measurement and aggregation systems, such as micro-synchrophasors, can be analyzed at three levels (Nejabakkhah et al., 2020):

1. Device and Aggregator Level
Attacks at this level compromise individual devices or aggregators, including μPMUs and data concentrators. Common attack methods include:

- **Signal manipulation:** Interfering with measurements or altering device calibration to report false readings.
- **Data falsification:** Injecting incorrect measurements to misrepresent system conditions.
- **GPS spoofing:** Broadcasting fake GPS signals to desynchronize phasor calculations. Mitigation strategies include:
  - Predicting satellite visibility for GPS receivers.
  - Comparing received GPS signals with estimated signals to detect anomalies. IEEE C37.118 standards recommend error rates below 40 ns at 95% confidence.

2. Communications Level
Smart meters rely on two-way communication between devices and control centers. Vulnerabilities in communication protocols can compromise data integrity, access, and confidentiality. Attack types include:

- **DoS attacks:** Target μPMUs, data concentrators, or aggregators to cause signal loss, attenuation, or delay.
- **Man-in-the-Middle (MITM) attacks:** Impersonating devices or aggregators to send false commands.

- **False Data Injection (FDI):** Intercepting and altering transmitted data frames before propagating them to higher levels.
- **Router delays:** Compromising routers to intentionally delay data, negatively affecting situational awareness.

3. Control Center Applications

Even with organizational cybersecurity measures, attacks on control center applications can significantly impact system performance. Since monitoring data drives analysis of power quality, reliability, fault detection, and grid topology, successful attacks can compromise the self-healing capabilities of smart grids. Solutions to mitigate these threats include:

- Machine learning-based anomaly detection
- Game theory approaches for strategic defense
- Data visualization tools for situational awareness.

10.11.1.5 Mitigation Strategies in Blockchain-Oriented Smart Grids

Blockchain-enabled networks facilitate direct, secure information exchange between devices, but the interconnected nature of these systems means that an attack on one component can propagate throughout the network. Cyber-attacks generally affect the acceptability, completeness, accuracy, and consistency of data. Strategies to mitigate these impacts include:

- Using dedicated communication channels between smart meters and aggregation centers
- Encrypting data before transmission
- Increasing redundancy in data exchange points
- Implementing firewalls and secure routers.

By integrating blockchain and these mitigation strategies, smart grids can enhance data integrity, privacy, and resilience against sophisticated cyber-attacks.

10.12 Renewable Energy and Demand Response

Blockchain technology enables enhanced coordination and management in renewable energy integration, grid control, and distributed energy resources (DERs). It facilitates production certification, participation of consumers in electricity markets, and demand-side management programs. Key operational challenges in transmission and distribution networks include (Di Silvestre et al., 2020):

- Cooperation of active and flexible customers in demand response schemes.
- Aggregation of local energy providers to participate in energy markets.
- Provision of ancillary services required for distributed systems.
- Coordination of the above processes for stable operation.

The decentralized nature of blockchain allows secure exchange of data, control signals, and commands between distributed energy resources and system operators. However, due to the diversity of resources, direct control over generators and devices is nearly impossible, necessitating consumer participation in demand response strategies through energy management systems.

From recent studies and practical implementations by aggregators (Junaidi et al., 2023), blockchain-based advantages in demand response and renewable integration include:

- **Peer-to-peer decentralized demand response schemes:** Allowing customers and microgrids to autonomously manage energy resources.
- **Aggregation of distributed generation units:** Coordinating multiple local energy providers for market participation.
- **Incentive-based demand-side management:** Enabling demand bidding, load reduction during emergencies, interruptible loads, and direct load control.

By implementing decentralized demand-side management in blockchain-enabled microgrids, operators can collect accurate information from smart meters, store produced energy data, and enforce smart contracts that automatically evaluate rewards or penalties. This ensures balanced energy flows between consumption and production sides while maintaining transparency and security.

10.13 Improving Energy Efficiency

Advances in digital technologies and consumer engagement present significant opportunities for energy efficiency and emission reduction. Blockchain and distributed ledger technologies (DLTs) enhance security, transparency, and sustainability in the energy sector, yet their application in energy efficiency markets remains underexplored.

Energy service companies (ESCOs) are exploring blockchain for the implementation of Energy Performance Contracts (EPCs). EPCs involve multiple stakeholders recording energy data, project costs, technology implementation, and energy savings. Traditionally, this process can cause delays and increased administrative costs. Smart contracts on blockchain platforms mitigate transaction inefficiencies, reduce costs, and streamline implementation.

Blockchain also enhances the value of energy efficiency by enabling consumers to exchange surplus energy. This creates incentives for energy savings, financial benefits, and greater participation in efficiency programs. Key advantages of blockchain for energy efficiency include (Khatoon et al., 2019; Mao et al., 2021):

- **Secure energy saving data:** Encryption ensures that energy savings records are protected from unauthorized access, increasing reliability and privacy.
- **Facilitated energy exchange:** Consumers can trade excess energy securely within local microgrids, encouraging participation in energy-saving schemes.
- **Higher transparency:** Distributed ledgers allow transparent, tamper-proof data sharing, ensuring trust in energy efficiency programs.
- **Cost savings in transactions:** Blockchain eliminates intermediaries, reducing the complexity and cost of energy efficiency transactions between ESCOs, distribution companies, and consumers.

- **Increased reliability:** Decentralized storage across multiple nodes improves data tracking, auditing, and system resilience.
- **Enhanced trust and security:** Blockchain provides secure verification of energy savings, permits, and payments, improving the credibility and effectiveness of energy efficiency markets.

By integrating blockchain into energy efficiency strategies, stakeholders gain a robust platform for secure, transparent, and efficient energy management, fostering both environmental sustainability and economic benefits.

10.14 Financial Applications

Blockchain technology has proven to be highly efficient in financial services, enhancing credit, security, and risk management. Numerous financial institutions are deploying blockchain platforms for customers to improve transparency, operational performance, and to create new revenue opportunities. Key capabilities of blockchain in the financial sector include **asset** settlement, ownership verification, data measurement, and analysis, which are particularly valuable for auditors and accountants. By clearly defining asset and liability ownership, blockchain improves productivity and accountability.

Blockchain-based financial systems enable the issuance of digital securities that are faster, more customized, and cost-effective. This expansion facilitates broader investor participation, reduces the cost for issuers, and lowers counterparty risk. Additionally, blockchain reduces transaction complexity, streamlining business operations and promoting market efficiency through mutual cooperation (Javaid et al., 2022).

Beyond traditional finance, blockchain has been widely explored in peer-to-peer (P2P) energy markets. In these systems, a decentralized paradigm is integrated with multi-factor models and two-way communication, ensuring that each user has equal privileges and shares in the electricity market. Traditional users are replaced by electricity price predictors, collectively determining a shared energy price without transmitting private load data to a central authority.

However, as more companies enter the electricity market, challenges regarding trust and anonymity emerge. Blockchain technology, with its decentralized, transparent, and privacy-preserving features, provides a robust solution for secure and reliable energy trading. Consortium blockchain techniques allow for safe P2P energy sharing, particularly among electric vehicles, while multi-crypto digital identity schemes protect users' privacy within neighborhood microgrids (Wen et al., 2021).

In building-level energy management, each user's energy load forecast contributes to price determination, enabling active participation in purchase strategies without the need for a trusted intermediary. Pseudo-digital identities ensure privacy during information exchange, while the immutable, decentralized, and distributed nature of blockchain smart contracts enforces pre-defined operations automatically and reliably. These smart contracts mitigate errors and streamline transactions, addressing multiple challenges in both financial and energy sectors.

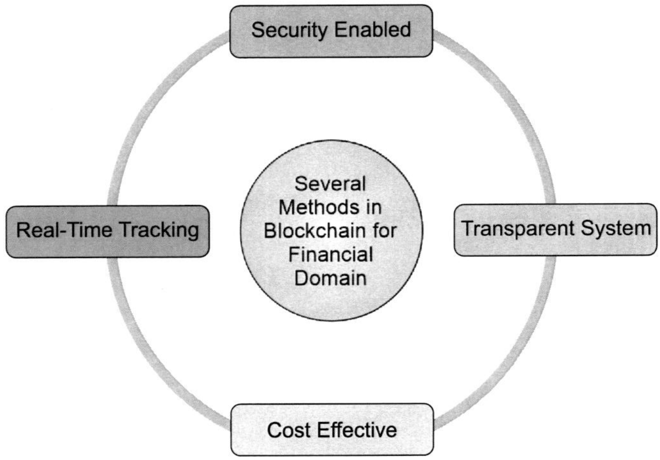

Fig. 10.4 Several methods in blockchain for the financial domain.

10.15 Identity Management and Access Control

Access control and identity management are critical services in any organization. Traditional identity management systems are typically centralized and rely on expensive hardware, such as smart cards and hardware security modules, to connect users to the system. Centralized systems introduce single points of failure, requiring strong backup, recovery, and disaster management strategies. Moreover, scalability limitations in these systems can cause delays when connecting to large numbers of Internet of Things (IoT) devices.

Blockchain-based access control and smart contract systems offer a next-generation solution, providing accuracy, fault tolerance, and high availability. Personal identity information stored in smart devices—including location and other sensitive data—is increasingly valuable, yet centralized platforms often limit users' control over their data, leaving it vulnerable to misuse. By leveraging blockchain's decentralized nature, smart contracts enable data owners to manage access to their personal information transparently and securely, with all access transactions recorded on a distributed ledger (Hewa et al., 2021; AlBadi et al., 2023).

Classical systems like Lightweight Directory Access Protocol (LDAP), Identity and Access Management (IAM), and Public Key Infrastructure (PKI) are optimized for centralized architectures, but they carry risks such as vulnerability of private keys stored on central servers. Blockchain-based systems eliminate the need for costly hardware, reduce reliance on central authorities, and ensure availability while maintaining security through decentralized enforcement (Pal et al., 2022).

10.16 Security Policy in Access Control

Security policies define required controls and access privileges based on organizational rules and strategic objectives. Traditional centralized policy management can be cumbersome and costly to customize, especially across distributed locations. Additionally, centralized systems are susceptible to tampering or unauthorized modification of policies without user consent (Hewa et al., 2021; Patil et al., 2021).

Smart contracts offer an ideal solution for decentralized security policy management. Policies and controls can be encoded programmatically, ensuring automatic execution without human intervention. The distributed nature of blockchain ensures that no single administrator or user can alter the execution logic, enhancing transparency, availability, and trustworthiness of the security management system.

10.17 Smart Contracts and the Internet of Things (IoT)

The IoT ecosystem is rapidly expanding, with billions of connected devices expected across industries. Future IoT systems require automation, security, and efficient resource sharing—areas where blockchain-based smart contracts can play a critical role (Khan & Salah, 2018).

Smart contracts operate on an "if–then" logic, distributed across the blockchain network. Once a transaction triggers the contract, all network nodes validate and execute it independently. Errors in contract code, however, are irreversible, emphasizing the need for careful programming (Islam, 2023).

In IoT networks, peer-to-peer resource sharing enabled by blockchain reduces computational overhead and network traffic compared to cloud-based solutions. This is especially beneficial for resource-constrained IoT devices with limited memory and computing capacity. By embedding access control policies in smart contracts, stakeholders can ensure secure, decentralized enforcement of resource permissions while minimizing centralized vulnerabilities (Minoli & Occhiogrosso, 2018).

10.17.1 Trust Management in IoT

Trust management is critical for applications like smart grids, smart cities, healthcare, and electric vehicle networks, where reliability is paramount. Blockchain enables decentralized trust management, which can handle real-time trust evaluation, mitigation of malicious behavior, and secure operation of billions of connected devices.

A blockchain-based IoT transaction platform can be conceptualized in multiple layers:

1. **Physical Layer:** Manages energy production, storage, consumption, and conversion at the customer side.

2. **Perceptive and Active Layer:** Incorporates sensors, smart meters, inverters, relays, and other devices for energy measurement and control.
3. **Connection and Network Layer:** Coordinates wired and wireless communications, translating sensor data for monitoring and decision-making.
4. **Middleware Layer:** Provides integration and coordination between IoT devices and higher-level services, managing the complexities of lower layers.
5. **Application Layer:** Implements energy-aware applications, providing data for consumers, utilities, and other stakeholders. This layer also supports P2P energy management, demand-side participation, and carbon reduction initiatives (Wu et al., 2022).

By combining blockchain, smart contracts, and IoT infrastructure, decentralized networks can achieve secure, transparent, and efficient management of identity, access, energy, and trust in next-generation smart systems.

10.18 Smart Communications

Smart cities represent a major innovation of the Internet of Things (IoT), providing the foundation for future urban and rural infrastructure development. The integration of millions of interconnected devices and communication systems aims to enhance the quality of life, improve services, and enable applications in diverse domains such as entertainment, transportation, and autonomous vehicles. However, scalability of operating systems remains a critical factor due to the large number of devices and the complexity of interactions. Centralized architectures can result in significant overheads and introduce additional security and operational risks.

Blockchain-based smart contracts and digital networks provide an effective solution for addressing these challenges. Their decentralized execution model ensures high service availability, reduces network resource consumption, and increases operational efficiency. In the communications sector, where subscriber volumes and service complexity continue to rise, blockchain-enabled smart contracts can autonomously implement customized, scalable service-level agreements (SLAs) while maintaining security and trustworthiness (Aggarwal et al., 2019; Liu et al., 2020).

By embedding operational logic into smart contracts, blockchain allows the efficient allocation and sharing of communication resources, ensures transparency in subscriber agreements, and enhances the reliability of network services. As a result, smart contracts become a core technology for supporting the scalable, secure, and optimized operation of modern communication infrastructures.

10.19 Concluding Remarks and Future Trends

This chapter has provided a comprehensive review of blockchain technology and its applications in power and energy systems, highlighting both the advantages and limitations. Blockchain's decentralized structure, immutability, individual data control, and tokenization capabilities offer significant potential for transforming energy management, peer-to-peer trading, demand response, and smart city infrastructure.

Despite these advantages, several challenges remain. High capital investment requirements, substantial energy consumption for mining processes, environmental concerns, the potential misuse of blockchain platforms for money laundering due to anonymity features, and time-consuming validation processes caused by limited block sizes are notable limitations. These challenges must be addressed to fully realize the digitalization and optimization of urban energy systems.

Future trends indicate a growing focus on Layer 2 scaling solutions, energy-efficient consensus mechanisms, cross-chain interoperability, and the integration of blockchain with IoT, smart contracts, and AI. As these technologies evolve, blockchain is expected to enhance not only energy efficiency and sustainability but also the security, transparency, and resilience of complex urban and industrial infrastructures. Ultimately, blockchain will continue to shape the future of decentralized digital ecosystems, offering new opportunities for innovation, collaboration, and sustainable development.

References

1. Afzal, et al. (2022). Role of blockchain technology in transactive energy market: A review, *Sustainable Energy Technologies and Assessments, 53*, 102646.
2. Aggarwal, S., Chaudhary, R., Aujla, G.S., Kumar, N., Choo, K.-K.R. and Zomaya, A.Y. (2019). Blockchain for smart communities: Applications, challenges, and opportunities, *Journal of Network and Computer Applications, 144*, 13–48.
3. AlBadi, A., Hajamohideen, F. and AlSaqri, D. (2023). A Review on Blockchain Techniques Used for Identity Management System: Privacy and Access Control. In: *International Conference On Systems Engineering*, Springer, pp. 361–375.
4. Al-Saif, N., Ahmad, R.W., Salah, K., Yaqoob, I., Jayaraman, R. and Omar, M. (2021). Blockchain for electric vehicles energy trading: Requirements, opportunities, and challenges, *IEEE Access, 9*, 156947–156961.
5. Alladi, T.,Chamola, V., Rodrigues, J.J., and Kozlov, S.A. (2019).Blockchain in smart grids: A review on different use cases. *Sensors, 19*(22), 4862.6. Aloqaily, M., A. Boukerche, O. Bouachir, F. Khalid and S. Jangsher. (2020). An energy trade framework using smart contracts: Overview and challenges. *IEEE Network, 34*(4), 119–125.
7. Cantillo-Luna, S., Moreno-Chuquen, R., Chamorro, H.R., Sood, V.K., Badsha, S. and Konstantinou, C. (2022). Blockchain for distributed energy resources management and integration. *IEEE Access.*
8. Di Silvestre, M.L., et al. (2020). Blockchain for power systems: Current trends and future applications. *Renewable and Sustainable Energy Reviews, 119*, 109585.
9. Hewa, T., Ylianttila, M. and Liyanage, M. (2021). Survey on blockchain based smart contracts: Applications, opportunities, and challenges, *Journal of Network and Computer Applications, 177*, 102857.
10. Honarmand, M.E., Hosseinnezhad, V., Hayes, B. and Siano, P. (2021). Local energy trading in future distribution systems. *Energies, 14*(11), 3110.
11. Islam, M.D. (2023). A survey on the use of blockchains to achieve supply chain security. *Information Systems, 117*, 102232.

12. Jabari, F., Jabari, H., Mohammadi-ivatloo, B. and Ghafouri, J. (2019). Optimal short-term coordination of water-heat-power nexus incorporating plug-in electric vehicles and real-time demand response programs. *Energy, 174*, 708–723.
13. Javaid, M., Haleem, A., Singh, R.P., Suman, R. and Khan, S. (2022). A review of Blockchain Technology applications for financial services. *Bench Council Transactions on Benchmarks, Standards, and Evaluations, 2*(3), 100073.
14. Johanning S. and Bruckner, T. (2019). Blockchain-based peer-to-peer energy trade: A critical review of disruptive potential. In: *2019 16th International Conference on the European Energy Market* (*EEM*). IEEE, pp. 1–8.
15. Junaidi, N., Abdullah, M.P., Alharbi, B. and Shaaban, M. (2023). Blockchain-based management of demand response in electric energy grids: A systematic review, *Energy Reports, 9*, 5075–5100.
16. Kabalci, Y. (2016). A survey on smart metering and smart grid communication. *Renewable and Sustainable Energy Reviews, 57*, 302–318.
17. Khan, M.A. and Salah, K. (2018). IoT security: Review, blockchain solutions, and open challenges. *Future Generation Computer Systems, 82*, 395–411.
18. Khan, S.N., Loukil, F., Ghedira-Guegan, C., Benkhelifa, E. and Bani-Hani, A. (2021). Blockchain smart contracts: Applications, challenges, and future trends. *Peer-to-Peer Networking and Applications, 14*, 2901–2925.
19. Khatoon, A., Verma, P., Southernwood, J., Massey, B. and Corcoran, P. (2019). Blockchain in energy efficiency: Potential applications and benefits. *Energies, 12*(17), 3317.
20. Li, F., Yan, X., Xie, Y., Sang, Z. and Yuan, X. (2019). A review of cyber-attack methods in cyber-physical power system. In: *2019 IEEE 8th International Conference on Advanced Power System Automation and Protection* (*APAP*). IEEE, pp. 1335–1339.
21. Li, J., Herdem, M.S., Nathwani, J. and Wen, J.Z. (2023). Methods and applications for Artificial Intelligence, Big Data, Internet of Things, and Blockchain in smart energy management. *Energy and AI, 11*, 100208.
22. Liu, Y., Yu, F.R., Li, X., Ji, H. and Leung, V.C. (2020). Blockchain and machine learning for communications and networking systems. *IEEE Communications Surveys and Tutorials, 22*(2), 1392–1431.
23. Livingston, D., Sivaram, V., Freeman, M. and Fiege, M. (2018). Applying blockchain technology to electric power systems. Ed: *JSTOR*.
24. Mao, W., Zhao, Z., Chang, Z., Min, G. and Gao, W. (2021). Energy-efficient industrial internet of things: Overview and open issues. *IEEE Transactions on Industrial Informatics, 17*(11), 7225–7237.
25. Mbaidin, H.O., Alsmairat, M.A. and Al-Adaileh, R. (2023). Blockchain adoption for sustainable development in developing countries: Challenges and opportunities in the banking sector. *International Journal of Information Management Data Insights, 3*(2), 100199.
26. Miglani, A., Kumar, N., Chamola, V. and Zeadally, S. (2020). Blockchain for Internet of Energy management: Review, solutions, and challenges. *Computer Communications, 151*, 395–418.
27. Minoli D., Occhiogrosso, B. (2018). Blockchain mechanisms for IoT security. *Internet of Things, 1*, 1–13.
28. Musleh, A.S., Yao, G. and Muyeen, S. (2019). Blockchain applications in smart grid: Review and frameworks. *Ieee Access, 7*, 86746–86757.
29. Nejabatkhah, F., Li, Y.W., Liang, H. and Reza Ahrabi, R. (2020). Cyber-security of smart microgrids: A survey. *Energies, 14*(1), 27.
30. Pal, S., Dorri, A. and Jurdak, R. (2022). Blockchain for IoT access control: Recent trends and future research directions. *Journal of Network and Computer Applications, 203*, 103371.
31. Patil, P., Sangeetha, M. and Bhaskar, V. (2021). Blockchain for IoT access control, security, and privacy: A review. *Wireless Personal Communications, 117*, 1815–1834.

32. Sahoo, S., Kumar, S., Sivarajah, U., Lim, W.M., Westland, J.C. and Kumar, A. (2022). Blockchain for sustainable supply chain management: Trends and ways forward. *Electronic Commerce Research,* 1–56.
33. Siano, P., De Marco, G., Rolán, A. and Loia, V. (2019). A survey and evaluation of the potentials of distributed ledger technology for peer-to-peer transactive energy exchanges in local energy markets. *IEEE Systems Journal, 13*(3), 3454–3466.
34. Tonev T. and Nikolaev, N. (2020). Review on blockchain applications for electric power systems. In: *2020 21st International Symposium on Electrical Apparatus and Technologies* (*SIELA*). IEEE, pp. 1–4.
35. Tripathi, G., Ahad, M.A. and Casalino, G. (2023). A comprehensive review of blockchain technology: Underlying principles and historical background with future challenges. *Decision Analytics Journal,* 100344.
36. Uddin, S.S., et al. (2023). Next-generation blockchain enabled smart grid: Conceptual framework, key technologies and industry practices review. *Energy and AI, 12*, 100228.
37. Wang, L., et al. (2021). A permissioned blockchain-based energy management system for renewable energy microgrids. *Sustainability, 13*(3), 1317.
38. Wen, S., Xiong, W., Tan, J., Chen, S. and Li, Q. (2021). Blockchain enhanced price incentive demand response for building user energy network in sustainable society. *Sustainable Cities and Society, 68*, 102748.
39. Wu, Y., Wu, Y., Cimen, H., Vasquez, J.C. and Guerrero, J.M. (2022). P2P energy trading: Blockchain-enabled P2P energy society with multi-scale flexibility services. *Energy Reports, 8*, 3614–3628.
40. Yang, R., et al. (2020). Public and private blockchain in construction business process and information integration. *Automation in Construction, 118*, 103276.
41. Yapa, C., De Alwis, C., Liyanage, M. and Ekanayake, J. (2021). Survey on blockchain for future smart grids: Technical aspects, applications, integration challenges, and future research. *Energy Reports, 7*, 6530–65641.
42. Zhao, W., Jiang, C., Gao, H., Yang, S. and Luo, X. (2020). Blockchain-enabled cyber–physical systems: A review. *IEEE Internet of Things Journal, 8*(6), 4023–4034.

11 The Patent Boom in the Blockchain Frontier

Mohammad Karrabi[1*], Farkhondeh Jabari[2] and
Asghar Akbari Foroud[3]

The rapid expansion of blockchain technology has led to a surge in patent filings, reflecting intense innovation and competition in this emerging frontier. This chapter examines the "patent boom" in blockchain, analyzing trends, drivers, and implications for technology development, commercialization, and intellectual property management. Key areas of blockchain innovation, such as consensus mechanisms, smart contracts, cryptographic methods, decentralized finance (DeFi), supply chain solutions, and digital identity, are explored in the context of patent activity. The chapter also discusses the geographic and institutional distribution of blockchain patents, highlighting leading countries, companies, and research organizations. Additionally, legal, strategic, and technological challenges associated with patenting in the blockchain space, including overlaps, standardization issues, and open-source tensions, are considered. By systematically reviewing the patent landscape, this chapter provides insights into the dynamics of blockchain innovation, potential barriers to adoption, and opportunities for researchers, developers, and policymakers.

11.1 Introduction

The digital landscape is ablaze with innovation, and at the heart of this firestorm lies blockchain technology. But what's propelling this revolutionary tech forward? While code and community might be the cornerstones of blockchain's progress, another factor is quietly adding fuel to the flames: patents. In recent years, the number of patent applications filed in this field experienced exponential growth, climbing from 7 in 2014 to more to approximately 2,400 a year in 2018–2022.

I. Blockchain is Impacting Every Industry

Blockchain technology, which underpins cryptocurrencies like Bitcoin, is now making inroads into a variety of industries beyond finance. With its decentralized,

[1,3] Electrical and Computer Engineering Faculty, Semnan University, Semnan, Iran.
[2] Power Systems Operation and Planning Research Department, Niroo Research Institute (NRI), Tehran, Iran.
E-mail: fjabbari@nri.ac.ir[2]; aakbari@semnan.ac.ir[3]
* Corresponding author: mkarrabi@semnan.ac.ir

DOI: 10.1201/9781003686026-11

transparent, and secure nature, blockchain has the potential to disrupt many existing business models and transform how economic transactions and relationships are conducted.

Industries seeing blockchain innovation include supply chain, healthcare, real estate, music and entertainment, agriculture, and more. For example, blockchain is being used to track food from farm to table, secure medical records, tokenize assets, facilitate payments in the music industry, and prove ownership of digital assets.

The transparency of blockchain systems can reduce fraud and errors, while the security helps prevent tampering and hacking. Smart contracts facilitated by blockchain also allow more complex business logic and automatic execution based on predefined conditions. Overall, the technology provides the means to transform workflows, reduce costs, and enable new operating models across sectors.

As blockchain matures from its ideological beginnings, businesses across industries are recognizing its pragmatism, investing in research, and piloting use cases matched to enterprise needs. Continued adoption is expected as the technology proves its worth in delivering meaningful efficiency, trust, and accuracy.

II. Cultural Transformation Toward a Programmable Society

As we have seen, blockchain patenting activity is surging across sectors, reflecting both booming innovation and strategic corporate plays. Now we examine the cultural shifts this decentralizing technology may catalyze.

The emergence of blockchain represents a cultural shift towards decentralized approaches for organizing economic, social and political systems. As trust in centralized institutions declines, blockchain offers the means to replace intermediaries with peer-to-peer, transparent, and verifiable interactions.

Sometimes referred to as "Web3", this cultural change encompasses a transition to decentralization, openness, and user ownership enabled by blockchain and related technologies. The change promises to transform relationships between individuals, organizations, and society itself.

This shift parallels the move from mainframes to personal computing and the Internet era. Yet blockchain now takes the Internet model further by eliminating centralized platforms in favor of user-owned networks governed by consensus-based models and economic incentives.

The possibility to program governance of blockchain networks through open source software protocols allows experimentation in organizing human relationships differently. Smart contracts make the execution of rules transparent and automatic. Tokens align incentives among network participants.

Over time, ever more social and economic activity may transition to "smart contract" based models hardwired with rules decided upon collectively. This "programmable society" represents not just a technological transformation, but also a cultural one promising greater transparency, user ownership, and collaborative forms of governance.

III. Navigating the Open Source Paradox[1]

A central paradox emerges when examining intellectual property protections in the blockchain space: the clash between decentralization, open access ideals intrinsic to this technology, and the exclusive rights conferred by patents.

Blockchain architecture is fundamentally oriented as an open source public good. Cryptocurrency protocols and distributed ledgers aim to decentralize power and enable permissionless participation. The ethos emphasizes transparency, collective governance, and questioning intermediaries.

However, significant investments are accelerating commercialization. Startups continue seeking venture capital and enterprise pilot projects expecting returns. This motivates players to secure competitive advantages against rivals. Exclusive patent control over novel techniques, algorithms, and uses counter blockchain's open ethos but provides economic incentives to innovate.

While patents confer exclusivity rights that seem misaligned with blockchain's open ethos, proper policy calibration can balance incentives for innovation against equitable access priorities. Striking this balance poses complex considerations:

a. ***Interoperability vs Competitiveness***: A key advantage of blockchain networks is promoting seamless cross-platform interactions through common standards and open architectures. However, overly broad patents on essential underlying protocols risk enabling specific actors to restrict interoperability, undermining this benefit. For example, if a core encryption technique widely adopted across blockchain implementations is patented by a single company, they could theoretically leverage this IPR to compel competitors hoping to enable cross-platform data transfers into restrictive licensing arrangements. Failure to cooperate may bar alternative networks from accessing the patented technique, hampering compatibility. By allowing uncontested control over such foundational building blocks, excessive patents protections foster fragmentation as patent holders customize more closed, proprietary versions seeking competitive advantages. This jeopardizes blockchain's promise as an interoperable base layer infrastructure promoting collaboration. However, insufficient IP guarantees over pioneering inventions also pose problems. Major cryptography innovations and consensus mechanisms enabling blockchain's programmable trust infrastructure require high risk investments currently funded via anticipation of temporary monopoly profits. If patents promise no exclusivity period for commercialization before widespread replication, private sector players may pivot resources into more easily protected incremental innovations rather than ambitious pioneering projects critical for mainstream adoption. Attenuated experimentation could then delay ecosystem advancement. Appropriately bounding patents' scope and duration around fundamental blockchain techniques can sustain open access while preserving economic incentives for innovation. But overreach

1 Sung, Huang-Chih Frank (2018). When Open Source Software Encounters Patents: Blockchain as an Example to Explore the Dilemma and Solutions.

enabling even transient control over essential standards forming public good digital infrastructure damages overall progress. Finding the right contours remains imperative.

b. ***Concentration vs Decentralization***: A key promise of blockchain technology is decentralization – no single entity controls the network. Decisions occur through distributed consensus mechanisms across peer nodes. This prevents centralized points of failure distorting interactions. However, early movers in blockchain patenting have rapidly built massive portfolios. IBM, Alibaba, Mastercard, Bank of America lead filings statistics. As patent thickets controlled by handful large players emerge around foundational techniques, they accrue significant sway over ecosystem evolution. By possessing exclusive rights over platform architectures, consensus algorithms, encryption standards or other integral infrastructure, these incumbents could entrench gatekeeper roles. Competitors may face onerous licensing terms hampering their decentralized applications. Over time, the bargaining power imbalance risks undermining the democratized ethos powering this breakthrough. And yet, economic reality is that meaningful R&D investments into advancing this nascent technology require incentives like temporary monopolies over newly created techniques. The same market promise attracting venture capital and national priority status also motivates IP capture. Absent quality patents guaranteeing adequate appropriability, innovation could attenuate with lost opportunities. Reconciling this tension requires nuance, some concessions attenuating decentralization principles may be essential to catalyze progress enabling wider access downstream. Strict IP regimes however risk technological capture by self-interested actors amassing control through patents. Uraveling litigation from acts expandingly interpreting claims also looms over fast iterating developers. Calibrating policies balancing decentralization with incentive designs fostering innovation remains imperative. The blockchain community likely self-regulates through cultural norms and sanctions resisting overreach. The technology's disruptive potential hinges on preventing excessive centralization of influence over its advancement.

c. ***Secrecy vs Transparency***: A defining attribute of most blockchain implementations is radical transparency, transactions are visible to all participants and codified rules are apparent. This accountability resists concentrations of power and builds trust. However, the patent system currently enables participants to secretly secure sweeping claims over foundational techniques without peer evaluation. Only at publication 18 months after filing does the claimed scope become visible. Prior art search challenges in exponentially progressing fields also enable applicants to leverage fresh filings before wider ecosystem awareness. This confidentiality concentrates influence over platform evolution within parties first to file patents on innovations designed through collective incremental efforts. Opaqueness counters norms around transparency permeating the blockchain community's ethos.

The secrecy also prevents coordinated efforts toward interoperability standards, as participants silo proprietary extensions rather than collaborating on common architectures. It also introduces uncertainty for later inventors unsure if their work infringes on opaque pending applications. However, increased disclosure requirements around patent applications can also enable replication of novel techniques before applicants fully capitalize on invention advantages to recoup R&D expenses. Mandating transparency too early risks disincentivizing cutting-edge investments. There are no easy answers reconciling this tension between ethical norms favoring transparency and secrecy supporting profit motivations. However, mechanisms like voluntary public pre-grant peer review of patent applications could promote accountability without excessive compromise. The blockchain community may also resist overreach through cultural sanctions.

d. ***Access vs Profit***: A foundational motivation for blockchain technology was enhancing access and participation in economic systems without requiring permissions from centralized gatekeepers. It promised open infrastructure enabling value exchange through transparent decentralized consensus. However, as pioneers now seek patents over various techniques and uses, questions emerge around what constitutes fair licensing costs and terms for access. Should proprietary ownership be allowed over platforms built through cumulative open-source community contributions? How to balance upstream inventor rights versus downstream user access? On one hand, breakthrough techniques expanding blockchain applications require significant upfront investments into R&D (research and development) by specialized teams, either within startups or enterprises. Patent protections enabling recouping of expenses via temporary proprietary ownership seem ethically justified. Yet the iterative collaborative nature of much cryptography and consensus protocol design also suggests individual rewards be measured against collective contributions. There exists a risk of overreach where foundational techniques become captured behind onerous licensing restrictions, against communally developed open-access principles. Navigating this tension requires calibrating patent regulations and cultural norms around licensing conditions against blockchain's underlying access priorities. Contributors undoubtedly require upside from innovations. But restricting participation through excessive controls contradicts the democratized possibilities that excite so many about this technology's promise.

As with broader debates on intellectual property, finding equilibrium between access and profit likely involves compromise and context specific evaluation. But the blockchain community may organically self-correct extreme positions through social sanctions resisting monopoly behaviors. The technology's ethos should guide resolution around this tension at the intersection of economics and ethics. Innovation policy has struggled balancing these aspects across technologies like telecom and pharma. Blockchain's unique structural challenges complicates this further.

IV. The Hype Seems Over but Blockchain is Just Getting Started

The year 2021 marked a slowdown for the blockchain industry after the boom and bust of the 2017 crypto bubble, not to mention the dampening impact of COVID-19 across all sectors[2]. Headlines declared that blockchain failed to meet expectations as adoption slowed.

In truth, however, these assessments failed to recognize that blockchain is still in its early days. Many promising applications have matured from proof-of-concept stage to pilots and first live applications over the past years. More enterprise involvement has brought blockchain usage closer to reality across various industries.

Like other major innovations such as personal computers and the Internet, it takes time to build the infrastructure and ecosystem required for mass adoption. Use cases must be carefully matched to real industry needs rather than hype. Appropriate regulations and standards are also needed for promoting adoption.

Given steady advances on many such fronts, blockchain appears poised for much greater utilization in enabling transparent, efficient, and cost-effective economic systems. The foundational technology to build true peer-to-peer networks and redefine relationships is maturing rapidly.

Hence, blockchain still remains an epochal innovation promising to reshape business and society. While still in its early stages, rapid development is underway in solving real problems with decentralized software architectures. The hype has cooled but the impact is only beginning to crystallize.

Many major multinational institutions and governments are actively researching, building, and initiating blockchain projects recognizing the technology's immense potential. These initiatives span public, private and joint models across industries and domains.

For example, the United Nations has set up labs exploring blockchain for supply chains, digital identity, and transparency. The EU has funded pilot projects and introduced regulations supporting blockchain adoption. National governments including the US, China, UAE, Sweden and others have made blockchain a strategic priority with new policies, programs, and sandboxes to drive progress.

Among industries, trade finance, insurance, healthcare, education, retail, mobility, and more sectors are trialing blockchains to streamline data sharing, authentication, payments and contract execution. Just a few examples include IBM and Maersk's cross-border supply chain (discontinued)[3], Accenture and Microsoft working with aerospace firms[4], and other large scale enterprise projects at Societe Generale, Walmart, and so on[5].

---

2 https://www.bcg.com/publications/2019/decoding-slowdown-blockchain-startups.

3 https://cointelegraph.com/news/the-ibm-maersk-blockchain-effort-was-doomed-to-fail-from-the-start.

4 https://newsroom.accenture.com/news/2018/vast-majority-of-aerospace-and-defense-companies-plan-to-integrate-blockchain-by-2021-accenture-report-finds.

5 https://wholesale.banking.societegenerale.com/en/news-insights/all-news-insights/news-details/news/china-raising-the-bar-for-cash-management-and-trade-finance/.

Ultimately blockchain promises to unlock value, reduce risks and costs, and enable new marketplaces across domains. The next decade of adoption at scale appears imminent as global initiatives pilot applications across sectors and domains. Both public and private models show potential to rewire economic systems and relationships through blockchain's architectural shift toward decentralization.

V. Patent Trends and China's Surge in Blockchain Patent Filings

The patents describe technologies aiming to revolutionize various sectors with its decentralized and transparent nature. In the US, most of the blockchain related patents are pending for examination in front of Technology Center 2400 that provides examination for patent application including Computer Networks, Multiplex, Cable and Cryptography/Security and in front of Technology Center 3600 that provides examination for patent applications including Transportation, Electronic Commerce, Construction, Agriculture, Licensing and Review. Another prominent Technology Center is Technology Center 2100 that provides examination for patent applications including Computer Architecture Software and Information Security[6].

Some of the top industry leaders holding blockchain-related patents are provided in the following table (indicative of the number of published applications filed and pending for examination in December 2023; based on a claim keyword search):

Name	*Pending*	*Filed*
IBM	164	890
ADVANCED NEW TECHNOLOGIES CO., LTD.	11	644
NCHAIN LICENSING AG	151	276
MASTERCARD INC.	74	245
ALIBABA GROUP	0	175
ALIPAY (HANGZHOU) INFORMATION TECHNOLOGY CO. LTD.	10	158
TOYOTA MOTOR CORP.	37	152
BANK OF AMERICA CORP.	37	146
CAPITAL ONE FINANCIAL CORPORATION	29	131
STATE FARM MUTUAL AUTOMOBILE INSURANCE COMPANY	31	128
TENCENT HOLDINGS LTD.	79	125
VISA INC.	35	116
INTEL CORP.	29	94
ACCENTURE PLC	6	90
SALESFORCE.COM	15	80
SIEMENS AG	36	78
AT&T	18	76
WELLS FARGO & CO.	9	74
PAYPAL HOLDINGS, INC.	32	72

[6] Based on a word search indicative of a usage of the word blockchain in patent applications filed in the years 2014–2022.

These companies operate in a variety of industries and geographic locations, most notably the financial and technology sectors. As indicated from the above table IBM serves as an example of a company that recognizes the value of patents and IP in the blockchain industry. IBM holds an impressive number of blockchain patents, firmly establishing its position as a key player in the industry. However, this lead is a bit deceiving because, of the top individual entities filing applications in this period, Advanced New Technologies, ALIBABA and Alipay are all tied to China's Ant Group Co., Ltd., which itself is an affiliate of Alibaba Group Holding Limited. Counting these together establishes the Alibaba Group as being far and away the most active filer.

As indicated above, it seems that currently China is leading the global race in blockchain patent applications. According to the 2023 Blockchain Patent Report[7], an astounding 90% of all blockchain patent applications filed worldwide this year have Chinese origins. This torrent of patent filings aligns with China's longstanding national strategy of actively promoting blockchain adoption.

Included in the country's 13th Five-Year plan and endorsed by President Xi Jinping as an engine for innovation[8]. Blockchain development is a clear national priority. As indicated above by the the 2023 Blockchain Patent Report[9] and the table above, Chinese firms act to hold most of the world's blockchain patents. The message from Beijing is unambiguous – China aims to be the global hub for blockchain innovation.

However, China's approach towards blockchain technology diverges from common perceptions in the West. The underlying ledger system of blockchain was created to be decentralized, with no central authority controlling transactions or data. This concept of a distributed ledger seems at odds with China's political environment. According to Xu Hao, a Chinese official, blockchain in China is not about decentralization but rather "de-intermediarization", eliminating middlemen without eliminating central oversight.

While the West usually associates blockchain with libertarian ideals, China sees the technology as a tool for efficiency, cost-savings, and social good, so long as it operates on the state's terms. Chinese blockchain firms must comply with cyberspace regulations, providing authorities access to data and censoring prohibited information. This curated version of blockchain, stripped of purported anti-establishment ideals, is termed "blockchain with Chinese characteristics".

The sheer volume of patent applications, aligned with staunch national support, confirms China's intent to lead the global blockchain industry on its own terms. However, despite filing over 4,000 patents in 2019 alone, only 19% have been approved so far, highlighting the strict standards enforced by Chinese patent authorities.

---

7 https://coincub.com/ranking/blockchain-patent-report-2023/.

8 https://www.uscc.gov/sites/default/files/Research/The%2013th%20Five-Year%20Plan_Final_2.14.17_Updated%20(002).pdf.

9 https://coincub.com/ranking/blockchain-patent-report-2023/.

Despite China's dominance in blockchain patent filings, cryptocurrency-related patents are explicitly excluded. According to China's Guidelines for Patent Examination, inventions considered "detrimental to public interests" cannot be granted patents. The guidelines provide examples such as inventions that could harm people, damage property, pollute the environment, or advocate superstition. Importantly, the guidelines state that inventions with potential for abuse or unintended negative consequences can still receive patent protection. However, cryptocurrencies fall under the category of "detrimental to public interests" in China. China's Law on the People's Bank of China tasks the central bank with formulating monetary policies, preventing financial risks, and maintaining banking stability. Cryptocurrencies like Bitcoin and Litecoin are decentralized digital currencies not issued by the central bank. Their circulation is considered to violate the provision that no entity can issue monetary substitutes for the renminbi. Additionally, China's Announcement on Preventing the Financing Risks of Token Issuance bans token financing platforms from exchanging legal tender for cryptocurrencies. It also prohibits buying, selling, or acting as a central counterparty for transactions involving cryptocurrencies. In summary, while China dominates global blockchain patent filings, cryptocurrency-related patents do not align with stated goals to maintain financial stability and oversight. As such, cryptocurrency inventions are excluded from patent eligibility in China under the category of "detrimental to public interests".

VI. Cybersecurity Blockchain Patents: Protecting the Vulnerable Links in Blockchain Security

While blockchain technology holds immense promise for enhancing security and transparency, it's not immune to exploitation. Recognizing the unique cybersecurity challenges posed by blockchain, technological solutions can play a crucial role in safeguarding the integrity of this transformative technology and might yield pivotal patents[10]. Here's a closer examination of why patents hold significance in addressing key blockchain cybersecurity risks[11]:

a. *Key Theft*:

The crown jewels of blockchain security, private keys, are prime targets for attackers. Infamous incidents like the Mt. Gox hack exemplifies the potential for devastating losses. To address this, patents can incentivize and protect advancements in key management:

- **Decentralized Storage:** By distributing keys across geographically diverse nodes, patents can shield against single points of failure and increase attacker difficulty.
- **Multi-Signature Authentication:** Requiring multiple authorized parties to sign transactions fosters collective control and reduces reliance on individual custodians.

[10] https://www.insideglobaltech.com/2023/08/18/patenting-for-blockchain-and-crypto-tech/

[11] https://cipher.com/blog/how-blockchain-can-be-hacked-the-51-rule-and-more/

- **Biometric-Based Protection:** Leveraging fingerprints, iris scans, or other unique identifiers adds an extra layer of security, making unauthorized key access far more challenging.

Furthermore, patents can safeguard innovative solutions like hardware-based secure enclaves, isolated environments within dedicated chips that protect private keys even from compromised operating systems. This creates a physical barrier against digital intrusions, further bolstering key security.

This revised version condenses the information while highlighting the most impactful aspects: the real-world threat, specific patented solutions, and the unique advantage of hardware enclaves.

b. ***Code Exploitation: Guarding the Gates of Smart Contracts*****:** Blockchain's heart lies in its code, and smart contracts, the self-executing agreements that automate transactions, are prime targets for exploitation. To safeguard this cornerstone, patents can play a crucial role in promoting secure smart contract development:

- **Formal Verification:** Patents can protect rigorous mathematical techniques that ensure code correctness and absence of vulnerabilities, acting as a pre-deployment shield against potential exploits.
- **Enhanced Access Controls:** By patenting fine-grained permission systems that restrict who can modify or execute code, unauthorized access, and manipulation can be significantly curtailed.
- **Sophisticated Bug Bounty Programs:** Patent protection can encourage the creation of structured programs that incentivize ethical hackers to discover and report vulnerabilities, fostering proactive defense against code exploitation.

These patented solutions, working in tandem, can create a more secure and trustworthy foundation for smart contract interactions, fostering greater confidence in blockchain's reliability.

c. ***51% Attacks: Fortifying the Fortress against Centralization*****:** The specter of a 51% attack looms large, where a malicious entity seizes control of the blockchain by accumulating over half the network's mining power. This grants them the ability to rewrite transactions, freeze funds, and disrupt the very essence of decentralization. While security best practices remain paramount, innovative technological solutions can bolster blockchain's resilience against such centralized threats. Patents can play a key role in safeguarding these advancements:

- **Decentralized Consensus Mechanisms:** Patent protection can incentivize the development of novel consensus algorithms that enhance resistance to 51% attacks by distributing power more evenly across the network. Examples include Delegated Proof-of-Stake with dynamic validator selection and Proof-of-Activity that rewards network participation beyond mere mining.
- **Reputation-Based Systems:** Encouraging and patenting algorithms that assign trust scores to network participants based on their past behavior can serve as a powerful disincentive against malicious activities. These scores

could influence block validation, transaction fees, and access to network resources, effectively isolating and neutralizing bad actors.

By fostering the development of such innovative solutions, patents can contribute to a more robust and resilient blockchain ecosystem, one where decentralization holds firm against the temptations of centralized control.

This revised version condenses the existing information, emphasizes the threat of centralization, highlights specific solution categories (decentralized consensus and reputation-based systems), and emphasizes the role of patents in safeguarding these advancements.

d. *Physical Device Theft: Plugging the Weak Point of Hot Wallets*:
While the blockchain itself might be a fortress, the access points can be vulnerable. Hot wallets, like those on mobile devices or web browsers, offer convenience but pose a security risk if stolen. Patents can safeguard crucial innovations in this area:

- **Secure Remote Deactivation and Data Erasure:** Imagine a stolen phone instantly wiping its wallet data or becoming unusable for transactions. Patents can protect advancements in remote device management that enable swift deactivation and data erasure, minimizing potential losses.
- **Advanced Tracking and Recovery:** Losing a device doesn't have to mean losing your funds. Patents can incentivize the development of sophisticated tracking technologies and recovery procedures. Imagine locating a stolen phone or remotely disabling its wallet functionality, mitigating the damage, and potentially retrieving your precious crypto.

By promoting such innovations, patents can help bridge the gap between blockchain's robust core and the vulnerability of access points, providing an extra layer of security for crypto assets held in hot wallets.

This revised version condenses the information, emphasizes the specific vulnerability of hot wallets, and highlights the two key patent-protected solutions: secure remote deactivation and advanced tracking and recovery. It also uses evocative language to illustrate the benefits of these solutions.

Blockchain's promise hinges on its security, but even fortresses have hidden vulnerabilities. To safeguard this transformative technology, strategic use of patents can be a game-changer. By incentivizing the development of robust cybersecurity solutions specifically tailored to blockchain's unique risks, patents foster a more secure and resilient ecosystem. This not only mitigates risks but also builds trust and confidence, ultimately unlocking the full potential of blockchain for widespread adoption and transformative impact

VII. Litigation and Patent Eligibility?_
While activity has surged amongst these major players, court battles have yet to ignite around blockchain IP and Blockchain has not been a significant source of patent litigation. The current state of patentability law is a potential hurdle for blockchain litigation, and should the case proceed, it may provide an early indication of how claim construction and any validity challenges could play out for blockchain patents.

There is reason to believe more patent litigation related to blockchain is coming, but only time will tell how many more cases are filed. As more patents are issued, the chances that they will come into the hands of a party willing to assert them increase as well. Additionally, as companies dealing with this technology mature, they are making more revenue. This revenue will attract those wanting to extract value from their patents. This is certainly an area to keep an eye on for all those interested in patent litigation.

In addition to the overall low litigation levels, the complex patentability landscape presents hurdles for would-be blockchain innovators. For inventors in the rapidly evolving field of blockchain, securing patent protection presents a unique challenge: the unforgiving landscape of patent eligibility under 35 U.S.C. §101. This hurdle becomes readily apparent in cases like Rady v. Boston Consulting Group, LLC[12], where the district court deemed the patent claims ineligible. The court's rationale boiled down to three key points:

- **Abstract Idea:** The claims were deemed directed at the "abstract idea of collecting, analyzing, and storing data," falling within the Alice test's abstract idea category.
- **No Transformative Step:** Adding the blockchain infrastructure's hardware elements, considered akin to "Alice's" computer hardware, failed to provide a transformative step beyond the abstract idea.
- **No Improved Functionality:** The court found no substantial improvement in computer functionality from recording and logging data in a peer-to-peer network, nor did the plaintiff convincingly demonstrate how the claims advanced blockchain technology itself.

While patent ineligibility remains a common defense in blockchain patent litigation, accused infringers rarely seek immediate dismissal based solely on this ground. This reflects a wider trend of courts preferring to address eligibility later in proceedings, after a more complete factual record emerges, especially when the patent application alleges supporting facts.

Given the complexities of the patent eligibility landscape, crafting a robust application for your blockchain invention demands strategic precision.

VIII. Tips for Better Blockchain Patent Application Drafting

To tackle this abstract idea pitfall, tailored drafting techniques are imperative when crafting blockchain patent applications. Here are some key elements to consider fortifying a Blockchain_patent application:

1. Strike First in the Blockchain Gold Rush: Prioritize Early Filing for Maximum Protection

In the breakneck sprint of blockchain innovation, "first mover" advantage takes on a whole new meaning. Just like any invention, your groundbreaking blockchain idea needs to be demonstrably novel and non-obvious, a significant step

[12] Rady v. Bos. Consulting Grp., 1:20-cv-02285 (ALC) (S.D.N.Y. Mar. 31, 2022).

forward from existing technology. Yet, the exponential growth of the blockchain space presents a unique challenge: even a comprehensive prior art search may not guarantee your invention hasn't been preempted by another enterprising inventor.

This is where early filing becomes your most potent weapon. Patent applications remain confidential for 18 months after filing, granting you a crucial window to solidify your priority date. Think of it as staking your claim in the burgeoning blockchain gold rush – the earlier you plant your flag, the stronger your chances of securing exclusive rights to your invention.

Here's why prioritizing early filing is critical:

- **Rapidly Evolving Landscape:** With new breakthroughs emerging seemingly every day, the line between cutting-edge and "obvious" can blur in a heartbeat. Waiting can inadvertently expose your invention to unforeseen prior art, jeopardizing its patentability.
- **Global Race to Innovation:** The global blockchain community is teeming with talent, all striving to leave their mark. An early filing establishes your presence and strengthens your competitive edge, deterring potential copycats and bolstering your position in future licensing or partnership negotiations.
- **Strategic Maneuvering:** An early filing date grants you valuable leverage in potential litigation scenarios. It provides a clear timeline and strengthens your claim of independent invention, making it significantly harder for infringers to challenge your ownership.

Don't let the complexities of the blockchain space and the 18-month publication window gamble with your invention's future. By prioritizing early filing, you secure a head start in the race to patent protection, maximizing your chances of reaping the rewards of your ingenuity.

2. Subject Matter Eligibility: Navigating the Blockchain Minefield:

1. Forging a path through the patent eligibility landscape can be particularly treacherous for blockchain inventions. Several factors conspire to create this unique challenge:

a. ***Abstract Ideas and Algorithms*:** Blockchain often leverages intricate algorithms and consensus mechanisms, which may be deemed abstract concepts under patent law. This legal hurdle requires convincing the examiner that your invention transcends mere ideas and offers a concrete technical solution.

b. ***Financial Transactions and Business Methods*:** Innovations related to managing cryptocurrency transactions or optimizing supply chains might fall under the umbrella of financial transactions or business methods, which are generally excluded from patent protection. Even if your invention utilizes blockchain technology, its primary purpose can impact its eligibility.

c. ***Uncertain Legal Terrain*:** The legal landscape surrounding blockchain patents remains a shifting landscape, with inconsistent rulings and interpretations across jurisdictions. This lack of clarity poses a significant risk for inventors, potentially discouraging them from pursuing patent protection.

3. Overcoming the Abstract vs. Concrete Dichotomy: Drafting Robust Claims: Bridging the chasm between abstract ideas and concrete implementations is crucial for securing patent protection for your blockchain invention. Here's how your claim drafting can navigate this critical hurdle:

- **Focus on Technical Solutions:** Steer clear of generic concepts and emphasize the specific technical problems your invention tackles within the blockchain domain. Quantify the technical improvements (e.g., efficiency gains, enhanced security) your invention offers.
- **Delve into Technical Details:** Provide a detailed, technically-sound description of how your invention operates. Avoid generic references to financial or economic outcomes, instead demystifying the underlying blockchain mechanisms that enable your solution.
- **Frame Mathematical Concepts in Context:** Describe mathematical components through the lens of their practical application in blockchain technology. Illustrate how these tools optimize performance, bolster security, or enhance other technical aspects of your invention.
- **Embrace the Power of Infrastructure:** Leverage references to the specific operations of the blockchain infrastructure to demonstrate that your solution cannot be simply performed by human manipulation.

By embracing these strategies, you can craft claims that showcase the concrete technical essence of your invention, increasing your chances of navigating the subject matter eligibility minefield and securing the well-deserved patent protection for your groundbreaking blockchain contribution.

4. Tightening the Net: Strategies for Precise Blockchain Patent Claims: Protecting your revolutionary blockchain invention requires crafting claims that are laser-focused and impervious to unwanted broad interpretations. Here's how to achieve surgical precision in your drafting:

- **Speak the Language of the Chain:** Immerse your claims in the industry's vernacular. Utilize technical terms familiar to blockchain experts, avoiding ambiguity and open-ended phrases. This ensures precise communication with examiners and potential infringers.
- **Under the Hood, Not the Outcome:** Keep your claims firmly anchored in the technical nitty-gritty of your invention. Instead of focusing on abstract concepts or desired results, dissect the specific mechanisms and steps your blockchain magic employs. Detail the algorithms, consensus protocols, and cryptographic processes that power your innovation.
- **Layering Your Defenses:** Build a multilayered protection scheme with claims of varying scope. Nest narrower, dependent claims within broader ones, providing a safety net if broader interpretations are challenged. This creates a buffer zone around the core of your invention.

- **Exclude, Don't Include:** Don't let unwanted baggage weigh down your claims. Explicitly exclude specific prior art embodiments that fall within broader reading of the claims. This preemptively disarms potential claim rejections who might try to stretch your claims beyond their intended reach.
- **Show, Don't Just Tell:** Paint a vivid picture of your invention in the specification. Provide exhaustive technical details, diagrams, and flowcharts to illuminate its inner workings. Don't just explain what your invention does, but also how it achieves its purpose. This detailed roadmap clarifies your claims and leaves no room for misinterpretations.

By meticulously implementing these strategies, you can forge claims that are both technically precise and legally bulletproof. This ensures your valuable blockchain invention receives the robust protection it deserves, empowering you to reap the rewards of your ingenuity.

5. Casting a Wider Net: Claiming Beyond the Specific Blockchain

While specific blockchain protocols might form the foundation of your invention, remember: true patent value lies in its ability to defend against actual infringement. To maximize commercial potential, your patent claims need to act as sturdy bridges across critical technological pathways – bridges that potential competitors must traverse to effectively compete. This means broadening your focus beyond just one protocol.

Here's how to craft claims that encompass the necessary breadth:

- **Think Beyond the Specific:** Avoid pigeonholing your invention to a single blockchain technique. Instead, seek to identify and claim the core innovative elements that can be applied across multiple blockchain implementations. Highlight the novel algorithms, data structures, or processes that form the heart of your invention.
- **Embrace Generality, Not Vagueness:** While aiming for broader coverage, steer clear of vague generalities. Strike a balance between claiming functionality relevant to multiple techniques and maintaining technical precision. Precisely define key terms and technical concepts to avoid ambiguity.
- **Focus on Function, Not Form:** Emphasize the "how" of your invention, the specific mechanisms and processes that achieve its desired outcome. This functional approach allows your claims to adapt to future variations and implementations within the broader blockchain landscape.
- **Leave No Bridge Unguarded:** Identify critical "choke points" within the blockchain ecosystem – areas where competitors must necessarily pass through to replicate your invention's functionality. Focus your claims on these strategic points, creating solid barriers to entry and maximizing your competitive advantage.

Remember, the goal is not to claim every blockchain technique under the sun. Instead, strategically target the unique and inventive elements that form the backbone of your invention, ensuring your claims cover multiple pathways to success while remaining technically sound and defensible.

By adopting this broader perspective, you can craft claims that stand as formidable guardians of your valuable blockchain invention, securing your rightful rewards in the ever-evolving world of distributed ledgers.

6. Shining a Light on Infringement: Crafting Detectable Claims

While securing a patent is a significant achievement, its true power lies in your ability to enforce it against infringers. This means crafting claims that not only protect your invention's core but also make infringement visible and verifiable. In the decentralized realm of blockchain, this challenge becomes even more crucial. Here's how to navigate this critical aspect of blockchain patent drafting:

a. ***Balancing Protection and Detectability: Know When to Hold 'Em, Know When to Fold 'Em***: Not every secret is meant for the patent spotlight. For algorithms or processes deeply embedded within backend systems, where detection is inherently challenging, consider the strategic advantages of trade secret protection. This approach can safeguard your intellectual property while maintaining a veil of secrecy.

b. ***Claiming with Visibility in Mind: Focus on the Observable***: Prioritize features that manifest themselves in user-facing interactions, network communications, or other readily detectable aspects of the blockchain system. This provides tangible evidence of infringement that can be identified and pursued.

c. ***Avoid the Shadows of Decentralization***: Refrain from claiming features solely implemented by anonymous or untraceable entities within the decentralized network. While blockchain's distributed nature offers many benefits, it can also create challenges in identifying and pursuing infringers.

d. ***Think Like a Detective***: Craft claims that provide clear breadcrumbs leading to infringement. Envision the evidence you'd need to prove unauthorized use and ensure your claims encompass these detectable elements.

e. ***Look Beyond Code***: Explore claiming non-software elements of your blockchain invention, such as unique system architectures, data structures, or user interfaces. These elements can often be more easily observed and traced.

f. ***Embrace External Evidence***: Gather supporting evidence from industry publications, conferences, or online forums to demonstrate the visibility and impact of your invention's features. This can strengthen your case in infringement scenarios.

By carefully considering detectability during claim drafting, you'll create a patent that not only safeguards your innovation but also empowers you to effectively enforce your rights in the dynamic and often-opaque world of blockchain technology.

Remember: A patent is only as powerful as its ability to be enforced. By prioritizing detectable claims, you'll ensure your blockchain invention can truly shine in the marketplace, deterring infringement and securing your rightful rewards.

11.2 Conclusion

With blockchain innovation proliferating across sectors, patent filings continue their exponential climb - cementing intellectual property as a critical frontier in this technological gold rush. From global superpowers like China to tech titans like IBM and Alibaba, the race is on to stake IP claims around this paradigm-shifting technology. As blockchain transitions from hype to real-world transformation, sustained patent growth seems inevitable. However, crafting balanced policy and equitable access remain imperative as ownership concentrates among early movers. Ultimately blockchain's disruptive potential relies on avoiding excessive centralization of influence over its advancement.

References

1. Sung, Huang-Chih Frank. (2018). *When Open Source Software Encounters Patents: Blockchain as an Example to Explore the Dilemma and Solutions*. University of Illinois, Chicago.
2. https://www.bcg.com/publications/2019/decoding-slowdown-blockchain-startups.
3. https://cointelegraph.com/news/the-ibm-maersk-blockchain-effort-was-doomed-to-fail-from-the-start.
4. https://newsroom.accenture.com/news/2018/vast-majority-of-aerospace-and-defense-companies-plan-to-integrate-blockchain-by-2021-accenture-report-finds.
5. https://wholesale.banking.societegenerale.com/en/news-insights/all-news-insights/news-details/news/china-raising-the-bar-for-cash-management-and-trade-finance/.
6. Based on a word search indicative of a usage of the word blockchain in patent applications filed in the years 2014–2022.
7. https://coincub.com/ranking/blockchain-patent-report-2023/.
8. https://www.uscc.gov/sites/default/files/Research/The%2013th%20Five-Year%20Plan_Final_2.14.17_Updated%20(002).pdf.
9. https://coincub.com/ranking/blockchain-patent-report-2023/.
10. https://www.insideglobaltech.com/2023/08/18/patenting-for-blockchain-and-crypto-tech/.
11. https://cipher.com/blog/how-blockchain-can-be-hacked-the-51-rule-and-more/.
12. Rady v. Bos. Consulting Grp., 1:20-cv-02285 (ALC) (S.D.N.Y. Mar. 31, 2022).

12 Investigation of Virtual Water, Applications, and Water Footprint in the Middle East (Case Study: Iran)

Saeid Eslamian[1] and Yaser Sabzevari[2*]

Water scarcity is a critical challenge in the Middle East, making efficient water management essential for sustainable development. This chapter investigates the concepts of virtual water and water footprint, emphasizing their relevance to regional water resource planning and policy. Virtual water refers to the amount of water embedded in the production and trade of goods and services, while water footprint quantifies direct and indirect water use across economic sectors. Using Iran as a case study, the chapter analyzes patterns of water consumption in agriculture, industry, and domestic use, highlighting the country's reliance on virtual water imports and exports. In addition to conventional assessment approaches, the chapter explores the potential role of blockchain technology as a secure and transparent data management framework for tracking water consumption and virtual water flows across supply chains. Methodologies for calculating water footprint and virtual water trade are discussed alongside blockchain-enabled data validation, traceability, and smart contract applications that can enhance accuracy and accountability. The chapter further examines the implications of water footprint management for food security, trade policy, and sustainable water governance in arid and semi-arid regions. By integrating water resource theory, quantitative methodologies, regional analysis, and emerging digital technologies, this work provides a comprehensive framework for understanding and optimizing water resource utilization in the Middle East.

12.1 Introduction

Considering that the world population is increasing, it is expected to reach 10 billion people by 2055, up from 5. 7 billion today. This is happening because people are living better lives, more people are moving to cities, and there is a greater need for food. Animals and energy from plants are being used more than ever. People are also using more water each, and we're taking more water from rivers and lakes than we did in the past. On the other hand, 2.97% of the world's water is salty

[1,2] Department of Water Science and Engineering, College of Agriculture, Isfahan University of Technology, Isfahan, Iran.
* Corresponding author: yasersabzevari1511@gmail.com

DOI: 10.1201/9781003686026-12

water from seas and oceans, which we can't use because it's too salty. Freshwater reserves include only 2.8% of the volume of water reserves on the earth's surface, and 76.7% of this water is concentrated in the form of polar ice on the two poles of the earth and is not suitable for human use. 22.8% of freshwater is in the form of groundwater and 0.5% of this water is in the form of other waters. Finally, it can be said that the share of fresh water in rivers and lakes, which are the main sources that can be used by humans, is 0.003% of all water globally available on Earth. Iran has a predominantly arid climate, though certain regions experience higher levels of moisture. The country gets about 240 mm of rain each year. This is about a third of the world's average and half of what is typical for Asia. Most of the rain falls in the northwest and west parts of the country, while the eastern and southeastern areas get less than 100 millimeters of rain each year. These problems and the changing situation for water resources mean we need new methods for planning how to use water. In recent years, ideas like virtual water, water footprint, and virtual water trade have become important because of the growing population, the need to provide food for people, protecting the environment, and managing water resources wisely, both above and below the ground. When talking about managing water resources, all areas are taken into account. Virtual water is a new idea that looks at the hidden ways water is used in making products and providing services for people.

12.2 Virtual Water

Water used during different steps of making a product is called virtual water that is kept in the product (Quddosi and Davari, 2015; Nasrabadi et al., 2014; Omidi and Homai, 2015). For example, in the world, 15,400 liters of water is consumed on average to produce one kilogram of beef. This concept was first proposed in 1993 by J.A. Allen. Global trade in goods creates an international flow of virtual water. Other names mentioned for virtual water include external water or embedded water. It took almost 10 years from the time Allan proposed virtual water until it was noticed by the scientific community. The first international meeting about this topic took place in December 2002 in Delft, Netherlands. The work of this researcher received a lot of attention from the fields of economics and politics. In 2008, he received the famous Stockholm Award. Virtual water is an important concept for measuring and understanding the real amount of water used in a country. By looking at how much water is used when a country exports and imports goods, we can figure out the actual water needs of that country based on how its people consume things from the world's water supply. This index helps researchers see if a country's water resources can meet its water needs. The researchers also determine the extent to which each country depends on water supplied by exporting nations (Sedghamiz et al., 2018). The main goal of using the virtual water solution is to lower water use in places that have a shortage of water. One goal of using virtual water is to meet water needs in different areas by importing products that have a lot of water in them.

Examining the water consumption involved in cultivating crops across various regions and adapting our farming methods accordingly can promote smarter and more efficient water usage.

When figuring out the virtual water used in farming products, we also look at something called the water ecological footprint. This includes both local and outside water sources. The ecological footprint measures how much producing a product in farming affects water resources.

12.3 General Reviews

Virtual water was characterized in Allen 1997 as the water "contained" in an item, caught on not as the genuine physical sum within the item but also as the sum of water required to produce it. It could be a progressive concept since it portrays something that has not been conceptualized some time recently, water "inserted" in commercial items. Hence, virtual water moreover makes a difference in how virtual water streams between nations or locales. This thought was created in detail by Allan 2003 and Chapagain et al. (2005). It was distributed among other reports within the Esteem of Water inquiry about report arrangement supported by the UNESCO-IHE Founded for Water Instruction. This work served as a springboard for the primary distribution in an unused area, creating a ponder of virtual water streams between nations. Depending on the thought of virtual water exchange, look to measure virtual water streams related to worldwide nourishment exchange and hence recognize the concept of water impression, which is characterized as the volume of water required to deliver buyer merchandise and administrations. Oki and Kanae (2014) point out that this definition of water impression adjusted the starting understanding of virtual water preservation through virtual water exchanging, in this way causing perplexity. Their creators contend that, agreeing to Teacher Allen's introductory definition, the water requirement saved by virtual water exchange ought to be considered as the potential sum of water required to deliver items within the "bringing in" (customer) nation.

Graham et al. (2020) examined future changes in virtual water trading. By accounting for the evolution in socioeconomic and climatic conditions, we estimated the future interregional virtual water trade and the results showed that the trade of renewable water resources may triple by 2100, while the trade of non-renewable groundwater may at least double. Basins in North America and the La Plata and Nile rivers account for the largest share of virtual water exports, while most of Africa, India, and the Middle East will rely heavily on virtual water imports by the end of the century.

12.4 Types of Virtual Water

The water used to grow each farming product can be split into three types: blue water, green water, and gray water.

12.4.1 Green Water

If rainfall occurs in an area during the growing season, the effective rainfall is directly stored in the root area and used by the plant. This type of water resource to meet different plant needs is called green water. In fact, green water is a part of the rain that is stored in the soil and evaporates or transpires between two successive rains.

12.4.2 Blue Water

The resource of the water supply is precipitation. However, since this rainfall occurs in the non-cultivation season, in order to use this rainfall to meet the plant's water needs, water is collected from surface and underground sources during the cultivation season. In other words, blue water is the accessible surface and groundwater.

12.4.3 Gray Water

Gray water is the amount of water that gets dirty during making things and is not as clean as it was before. This kind of water goes into natural water sources. The amount of water needed to clean polluted water to a safe level is called virtual gray water (Ramazani et al., 2017; Mircholi et al., 2015; Montaseri et al., 2015; Nasrabadi et al., 2015; Quddosii and Davari, 2015; Nasrabadi et al., 2014; Omidi and Homai, 2015).

12.5 The Relationship Between the Concepts of Virtual Water

Since around 90% of water is used in farming, the idea of virtual water and virtual water trade is very important for agriculture. Many studies have looked at how virtual water use and virtual water trade connect in different areas, countries, and around the world (Ramazani et al., 2017). This index will show how much water is used to make a product. So, this idea covers everything and includes the water used during making and preparing products. Water footprint means how much water is used to make a product. The green water footprint is about the amount of water from rain that is actually used, while gray water is the amount of fresh water needed to clean up pollution and harmful substances produced during manufacturing. The idea of white water refers to how much irrigation water is wasted.

12.6 Virtual Water in Iran

In the sixth development plan, Iran wants to focus on the idea of virtual water. This idea has been recognized in developed countries for over 30 years. Due to the lack of attention to this concept in the country, we are exporting virtual water abroad by exporting seven watermelons to the water storage aqueduct behind the Karkheh dam. Even though we can't see this water, it is very important for the world's economy today because there is not enough water available. It takes 10 liters of water to make one sheet of A4 paper. Considering the negative balance of underground water in the country, the cultivation of water-intensive crops in the agricultural sector should be avoided. It takes 300 liters of water to grow one kilogram of watermelon. This amount of water is used in countries where farming is very effective. In the best case, it takes 500 liters of water to grow one kilogram of watermelon in Iran. According to the statistics of the Ministry of Agricultural Jihad of Iran, this country ranks fourth in the world with an annual production of more than 2 million and 200 thousand tons of watermelon. This product is grown in four seasons and more than 100 thousand tons are exported to Iraq, UAE, and Kuwait. With the export of 100 thousand tons of watermelon, more than 50 billion cubic meters of water are exported abroad every year.

Figure 12.1 shows how much each part of the water footprint adds to the production of the main crops grown with irrigation and rain. As can be seen, among the water products, four products: wheat, barley, tomato, and rapeseed have similar values of white and gray water footprint ratio to the total water footprint of the same product, i.e., around 10% respectively. In the case of fodder and grain corn, this is 50%. The share of white and gray water footprints in grain and fodder corn is 28% and 18%, respectively. These results show the management of the agronomy regarding the aforementioned crops in the study area (Ramazani Mederani et al., 2017).

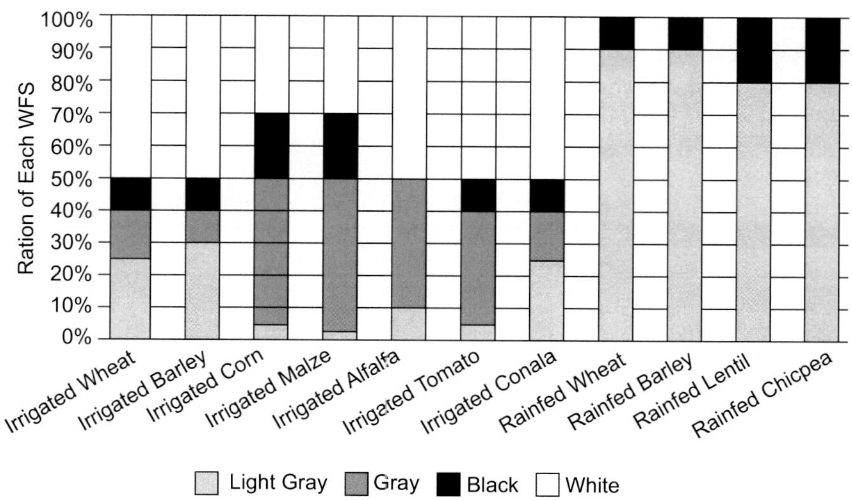

Fig. 12.1 The contribution of each component of the virtual water footprint in the main water and rain products.

In dry lands, the share of green water is higher than the share of gray water. The main reason for this can be the adaptation of the growth period of these plants to the rainfall period and the use of less nitrogen fertilizers on the surface of rainfed lands compared to irrigated lands. Due to the low yield potential in these lands, rainfed wheat and barley have 88% and 12% green and gray water footprints, respectively. In lentils and chickpeas, these values are around 79 and 21%, respectively (Ramazani et al., 2017).

The water footprint shows how much water is used from rivers, lakes, and underground sources. When a country imports virtual water, it adds to its water supply. However, when it exports virtual water, it takes water away from its own resources. This problem is called the water footprint of a country (Mircholi et al., 2015).

With a per capita water of 1,600 cubic meters, Iran is one of the dry and semi-arid countries, so according to Mr. Allen's prediction, Iran should be a net importer of virtual water. Examining export and import statistics, contrary to expectations, shows the export of virtual water in Iran (Qudousi and Davari, 2015).

Figure 12-2 shows the footprint sources of Iran's commercial products by separating green, blue, and gray water. Contrary to expectations, it can be seen that Iran has a net export of blue water and an import of green water. This means that Iran exports its valuable water resources and imports less valuable resources of green water from other countries (Qudousi and Davari, 2015).

The practical use of virtual water in the country's water resources management level faces many challenges from strategic and national to practical and local levels. Changing the policy of self-sufficiency in the agricultural sector as well as accepting the role of reliable and stable relations with neighbors and business partners in ensuring the country's food security should be clearly explained in macro policies and communicated to other levels. But the most basic problem is the lack of necessary information to calculate virtual water exchanges. To perform the calculations, the amount of exchanged products and goods as well as the virtual water of each product are required.

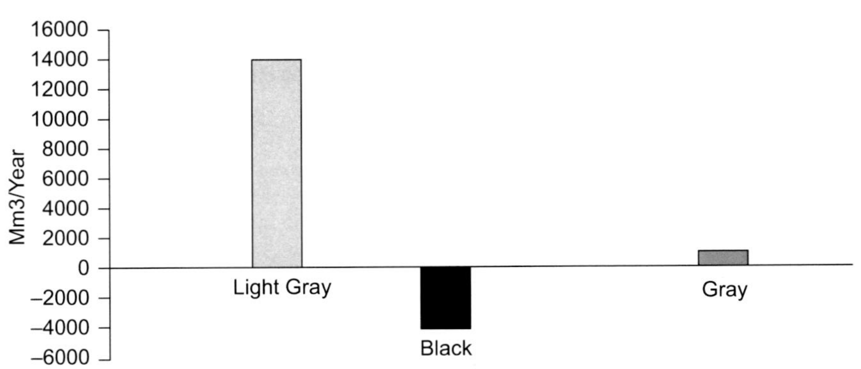

Fig. 12.2 Iran's virtual water trade by separating green, blue, and gray water.

12.7 Virtual Water Business

The trade and exchange of virtual water aims to use water more efficiently and manage its use better in different regions around the world. This is to help achieve water security in areas that are poor. Research shows that from 1995 to 1999, about 1,031 cubic kilometers of virtual water moved through the trade of some water products each year. Out of this, 695 cubic kilometers were linked to agricultural products and 336 cubic kilometers were linked to livestock trade. It comes from products. This means that 13% of the water used to make farm products is sold in the market. When we look at the trade of animals and their products, we can say that 20% of the water used to produce food from farming and livestock has been part of the water flow between countries. The virtual water business is the main way to use the idea of virtual water. Countries with not enough water can save their water by buying products that need a lot of water to make. This trade happens because some countries are better at making food than others. So, a country sells food to another country if it can make that food more easily or efficiently (Ehsani et al., 2008). Another way to use this idea is to look at virtual water trade within countries. For instance, in 1995, Egypt brought in 7. 5 million tons of grains from other countries, and growing that much grain in Egypt needs 9. 9 cubic kilometers of water. That year, Japan saved 37 cubic kilometers of its water by bringing in 27 million tons of grain from other places. If countries that sell products use less water to make them than the countries that buy them, we can save water worldwide.

12.8 Investigating Water Security in the Climate of Iran From the Point of View of the Water Footprint Index

In today's world, water is of particular importance as a socioeconomic commodity, and the growing need for water in all aspects of life has made water available to societies as a scarce commodity. This issue becomes more important in areas with arid and semi-arid climates that have limited renewable resources and require the necessity of managing water resources with a comprehensive and accurate approach. Statistics show that despite the increase in water demand, Iran's annual renewable water resources per capita will decrease to about 800 cubic meters by 1,400, which is lower than the water scarcity limit (1,000 cubic meters). And the population of Iran will reach the limit of 100 million people, and to meet the food needs of this population, we need more than 150 billion cubic meters of water.

12.9 Virtual Water Trade in Iran

For Iran, this comparison can be made between the provinces of the country, and the potential of different regions in the more economical production of agricultural products, industrial goods, processing industries, and services can be identified and investigated from the point of view of water consumption (Ehsani et al., 2008). Statistics show that between 1997-2001, by importing all kinds of agricultural products, Iran saved 37 cubic kilometers of water to produce the same products domestically. Sure, please provide the text you'd like me to simplify. It's expected that by 2025, Iran will use 47. 27 cubic kilometers of water by importing 11.36 million tons of grain. In recent years, Iran has often imported virtual water, which means it relies on food from other countries rather than producing it itself. However, the food imports are not mainly for the purpose of virtual water trade. There are new methods that can help assess Iran's strengths in growing agricultural products. By using these methods, Iran could adjust its food import and export policies. This can be done through both long-term and short-term plans that focus on national interests and food security while also using the benefits gained in other areas of the economy. Ardakanian and Sohraabi (2006), in the field of virtual water and the necessity and importance of using this type of business in Iran, pointed out that the use of virtual water and related business will have important positive effects. Among these effects, the following can be mentioned:

- Increasing access to water resources in the world;
- Reducing the pressure caused by water consumption on the domestic sources of the country.

In 2012, Mohammadi Kani Golzar investigated the amount of virtual water exchange for 32 major agricultural products from 1380 to 1388. The results of this research showed that during this period of time, Iran was a net importer of virtual water for the production of selected products and saved 13.7 billion cubic meters of water annually through these exchanges (Abedi and Tahamipour, 2016; Mohammadi Kani Golzar, 2012).

The results obtained by Abedi and Tahami (2016) indicate that wheat and almond crops have the highest amount of virtual water among selected crops and garden crops, respectively. Surveys show that most virtual water of these products is in line with their low performance. Therefore, based on the results, from the set of factors that are involved in the amount of virtual water of each product, we can mention the water requirement of the product, climatic conditions, and also the yield of the product per unit of cultivated area. Also, the results showed that almond, apple, grape, and wheat products have the lowest water productivity. On the other hand, onion, watermelon, and garlic have the highest water productivity, respectively. Increasing the production and export of products that require a lot of water, as well as importing products and goods with a low water requirement, will be a harmful policy for water resources and long-term food security in areas that are facing a water shortage crisis (Abedi and Tahamipour, 2016). Climatic and cultural conditions, place of production, management, and planning are quite effective in the volume of virtual water and it is definitely different for the same product in different regions. Virtual water trading can lead to the import of products from other regions, and in this way, the negative effects of using water resources are avoided. In this way, the involvement of virtual water trade in commercial policies will protect water resources and ensure food security. Therefore, due to the aggravation of the water shortage crisis in different regions, the issue of virtual water trade is important in the planning and policies of agricultural products trade, especially in arid and semi-arid regions. The exchange of virtual water at the internal level of countries, between countries and even continents can be used as a reservoir for improving the efficiency of water consumption at the global level, providing water for water security in water-poor areas and solving environmental limitations, with determining suitable locations for production should be used. By calculating the amount of virtual water contained in food imports in water-scarce countries, the close relationship between the availability of water and dependence on food imports is revealed; Therefore, the import of foodstuffs can be considered as one of the strong indicators in determining the level of water scarcity in countries.

According to the importance and necessity of investigating virtual water and its use in the agricultural sector, some research about this index is mentioned below.

Arabi-Yazdi et al. (2014) conducted research and found the average amount of virtual water for each ton of production in the agricultural sector of South Khorasan province to be 1312 cubic meters and also the amount of water productivity in the agricultural sector was 0.62 kg per cubic meter. They also stated that in 2008–2012, the largest amount of water imported in the form of virtual water was 7.76 million cubic meters and related to the rice crop.

Salari et al. (2014) conducted research in the Sistan and Baluchistan provinces. In this study, spatial and temporal changes in the amount of virtual water of the wheat agricultural product as one of the strategic products of Sistan Baluchistan province were investigated and evaluated. For this purpose, 12-year meteorological data, crop yield, cultivated area, and irrigation methods were used. The 12-year average of net water consumption for wheat production in the studied province was 303 million cubic meters, of which 9.2% is provided by green water. The net amount of virtual water varied between 1.59 in Khash city and 6.18 in Zabol city in terms of cubic meters per kilogram. Its average value for 12 years in the mentioned province was found to be 3.42 cubic meters per kilogram. The average virtual water import and export of the studied province was 1,043 and 43.2 million cubic meters, respectively. Also, the cities of Zahedan and Zahk were recognized as the largest importer and exporter of virtual water in the province, respectively. This research found that if we reduce water loss by 10% by making irrigation more efficient, we can improve self-sufficiency by 18.2%. This also means we would use 11.7% less water to grow wheat. Decrease the reliance on virtual water imports to the province by 3. 62%.

Abyaneh et al. (2015), during their research in the Hamedan study area, studied the amount of virtual water exchange of some major crops of the province such as potatoes and garlic (vegetables), sugar beet (industrial), alfalfa (forage) and wheat and barley (cereals) in the period from 2001 to 2010 and reviewed and evaluated it. The results of their water evaluation showed that the amount of virtual water for the vegetable and industrial group is less than 1 cubic meter of water per kilogram of produced product. In addition, both groups are included in the category of low-consumption and high-efficiency products. Due to having virtual water content higher than 2 cubic meters per kilogram, cereals are included in the category of high-consumption products. Also, the investigation and evaluation of the total virtual water content imported and exported virtual water of the studied province during the period 2001 to 2010 showed the departure of 1,538.4 million cubic meters of pure water and 2,564 million cubic meters of real water from the province.

In a study was done to find out how much virtual water is needed to grow wheat crops with irrigation and rain in different cities of Fars province during the farming years of 2003, 2008, and 2013 by Shirazi et al. (2016), virtual water was measured and calculated based on total demand, rainfall, and a 40% efficiency in irrigation. Virtual water is the amount of water required to make one kilogram of a crop.

Finally, based on virtual water values, the priority of cities for rainfed and irrigated wheat cultivation will be stated. The amount of virtual water for the cultivation of irrigated and rainfed wheat crops in different cities of Fars province during different years varies between 2-6 and 2-14 cubic meters per kilogram, respectively, which is much higher than the international standard of 1 meter. The cube per kilogram will be blue in order to cultivate the crop.

Shokohi et al. (2016) conducted research to investigate and calculate the total virtual water footprint as well as the footprint of water resources for the production of the main agricultural products of the Qazvin Plain study area during the period of 2003-2014. These parameters were estimated at 2,053 and 1,411 million cubic meters per year, respectively. In order to determine the composition of cultivation, taking into account the maintenance of the cultivated area, maintaining the amount of income in the sector related to agriculture, as well as saving water consumption, 12 scenarios were proposed. Scenarios 9 and 11 will save 23.3% and 23.8% in terms of water consumption from water resources, respectively, while maintaining income.

Aabdi and Tahamipour (2016) conducted research in the Zanjan province. The results of the research showed that the export of the studied products caused a volume equivalent to 26.53 million cubic meters of virtual water export. Also, 95% of the amount of exported virtual water was related to garden products and 5% of it was related to agricultural products. In addition to the mentioned cases, the intensity of water consumption in the sector related to farmers in Zanjan province was estimated to be 24% of the renewable water resources of the province. Therefore, it is necessary for the authorities of the agricultural sector in Zanjan province to achieve relative advantages in the production of various agricultural products in such a way that the least amount of pressure is placed on groundwater resources.

Saleh Nia and Bastani (2017) conducted research to investigate and evaluate the trade of horticultural products as well as agricultural products with an emphasis on virtual water. The results of the research showed that in 2014, Iran conserved about 9719.35 million cubic meters of water through trade in agriculture and horticulture through the import of various products to the country.

Arabi-Yazdi et al. (2010) during research conducted in Iran, stated that the agricultural product of date, followed by legumes and oilseeds, has the highest amount of virtual water, and greenhouses have the lowest amount of virtual water.

12.10 Conclusion

In recent years, declining precipitation and rapid population growth across many regions have intensified concerns regarding water scarcity, particularly in agriculture, which remains the largest consumer of freshwater resources. The amount of water required for agricultural production, total water withdrawal, and crop yields varies significantly across regions due to differences in climate, soil conditions, and farming practices. Consequently, both the virtual water content and the actual water use associated with agricultural products differ spatially, underscoring the importance of region-specific assessments of virtual water and water footprint indicators. Evaluating these indices provides valuable insights into the volume of water embedded in products throughout their entire production chain, as well as the extent of virtual water trade through the import and export of agricultural commodities. Analysis of virtual water flows highlights how water resources are effectively transferred between regions and countries via trade, offering a strategic mechanism for alleviating water stress in arid and semi-arid areas. By aligning crop production with local climatic conditions and water availability, virtual water assessments support more efficient agricultural planning and water-saving strategies. The growing global interest in virtual water stems from the substantial volumes of water embedded in food trade and its potential role in reducing pressure on limited freshwater resources. For water-scarce countries, importing water-intensive products can contribute to economic development, enhance food security, and protect domestic water reserves. However, the optimal balance between domestic production and food imports depends on national water availability, agricultural capacity, land resources, and food security policies. In this context, emerging digital technologies such as blockchain offer new opportunities to strengthen virtual water and water footprint management. Blockchain can provide a secure and transparent platform for recording water use data, validating water footprint calculations, and tracing virtual water flows across agricultural supply chains. By ensuring data integrity and improving trust among stakeholders, blockchain-based systems can support evidence-based decision-making and facilitate coordinated water governance at both national and international levels. In recent years, Iran has become a major importer of virtual water, reflecting the growing mismatch between water availability and food demand. Incorporating virtual water considerations into national water policies, supported by blockchain-enabled data transparency and traceability, can play a crucial role in addressing Iran's water scarcity challenges and advancing sustainable water and food systems at the regional and global scale.

References

1. Abedi, S. and Tahamipour, M. (2016). Measuring and analyzing the trade balance of virtual water in the agricultural sector of Zanjan province. *Journal of Economic Research and Agricultural Development of Iran, 2-47* (4), 805–814 (in Persian).

3. Allan, J.A. (1997). *'Virtual Water': A Long-term Solution for Water-short Middle Eastern Economies?* (Vol. 5145). London, UK: School of Oriental and African Studies, University of London.
4. Allan, J.A. (2003). Virtual water eliminates water wars? A case study from the Middle East.
5. Allan, T. (2011). *Virtual Water: Tackling the Threat to Our Planet's Most Precious Resource*. Bloomsbury Publishing.
6. Ardakanian, R. and Sohrabi, R. (2008). Virtual water trade, international literature, and application in Iran. In: *Proceedings of the 2nd Iran Water Resources Management Conference, Isfahan* (in Persian).
9. Chapagain, A., Hoekstra, A.Y. and Savenije, H.H. (2005). Saving water through global trade. 40 pp. https://www.academia.edu/24354649/Saving_water_through_global_trade.
10. Ehsani, M. Khalidi, H. and Barkhi, Y. (2008). Introduction to virtual water of National Irrigation and Drainage Committee of Iran, Tehran. First, No. 1 March 2012, attributes 9–18 of *Water and Sustainable Development Journal* (in Persian).
11. Graham, N.T., Hejazi, M.I., Kim, S.H., Davies, E.G., Edmonds, J.A. and Miralles-Wilhelm, F. (2020). Future changes in the trading of virtual water. *Nature Communications*, *11*(1), 3632 (in Persian).
12. Mircholi, F. Soltani Kopaei, S. & Farmarzi, M. (2016). Evaluation of virtual water exchanges and water footprint of some agricultural products in Iran. *Iranian Water Research*, *10*(1), 49–58 (in Persian).
13. Mohammadi Kani Golzar, F. (2012). *Water Management Based on Virtual Water trade in the Products of the Country*. MS Thesis, School of Economics and Agricultural Development, Tehran University (in Persian).
14. Muntaseri, M., Rasouli Majd, N., Bahman Manesh, J. and Rezaei, H. (2015). Evaluation of changes in the water footprint of crops in Urmia lake basin using the LMDI method. *Danesh Water and Soil Journal*, *25*(3/20), 210–199 (in Persian).
15. Nasrabadi, T. Arab, A. Pourasghar and Sangachin, F. (2015). Investigating the adequacy of wheat cultivation areas in Iran with water requirement and wheat yield with virtual water approach. *Environment*, *41*(3), 529–543 (in Persian).
16. Nasrabadi, T., Motaghi Far, H., Pourasghar and Sangachin, F. (2015). Life cycle evaluation in order to optimize water consumption in industrial cattle farming with the virtual water approach of Feka cattle breeding case. *Journal of Environmental Science*, *4*, 719–733 (in Persian).
17. Oki, T. and Kanae, S. (2004). Virtual water trade and world water resources. *Water Science and Technology*, *49*(7), 203–209.
18. Omidi, F. and Homai, M. (1394). Derivation of production functions for estimation of virtual water and irrigation water price Wheat. *Cereal Research*, *5*(2), 131–143 (in Persian).
19. Omidi, F., Homai, A., Pour Asghar and Sangachin, F. (2015). Derivation of production functions for estimation of virtual water and wheat irrigation water price. *Cereal Research*, *5*(2), 131–143 (in Persian).
20. Prendoni Shirazi, N. Ahmadi, S.H. Kamgar Haghighi A.O. and Sepaskhah, A.R. (2016). Temporal and spatial distribution of the amount of virtual water in the production of irrigated and rainfed wheat in Fars province. In: *The 6th National Conference on Water Resources Management of Iran with the Approach: Ecological Link with the Water Cycle for the Sustainability of the Land 13 May 2015*. University of Kurdistan (in Persian).
21. Qudousi, H. and Davari, H. (2016). Criticism of virtual water in terms of policy analysis. *Water and Sustainable Development*, *3*(1), 47–58 (in Persian).
22. Ramazani Mederani, H. Shokohi, A. & Jatboi, A. (2017). Utilizing the concept of virtual water footprint in the production of main products to overcome the water crisis in Qazvin region. *Water and Soil*, *31*(2), 422–433 (in Persian).

23. Salari, S., Karandish, F. and Darzi Naftchali, A. (2013). Spatial and temporal analysis of virtual water changes in wheat in Sistan and Baluchestan province. *Irrigation and Water Engineering Research Quarterly*, *18*, 9481 (in Persian).
24. Salehnia, N. and Bastani, M. (2017). Investigating the virtual water trading strategy of agricultural and horticultural products in Iran. *Iranian Irrigation and Drainage Journal*, *5*(11), 750–762 (in Persian).
25. Sedghamiz, A., Heidarpour, M., Nikoo, M. R. and Eslamian, S. (2018). A Game Theory Approach for Conjunctive Use Optimization Model Based on Virtual Water Concept., *Civil Engineering Journa*l, *4*(6), 1315–1325.
26. Shokohi, A., Ramazani Mederani, H., Mojtbavi, S.A. and Singh, P. (2015). Using water footprint accounting to determine the optimal crop composition in sustainable development, a case study of the plain (Qazvin). *Iran Water Resources Research*, *12*(3), 113–199 (in Persian).
27. *Velázquez, E., Madrid, C. and Beltrán, M.J. (2011). Rethinking the concepts of virtual water and water footprint in relation to the production–consumption binomial and the water–energy nexus. *Water Resources Management*, *25*, 743–761.
28. World Water Council. (2004, March). *E-Conference Synthesis: Virtual Water trade-conscious Choices*. World Water Council Publication No. 2.
29. Yazdi Arabic, A., Nik Nia, N., Majidi, N. and Emami, H. (2015). Investigating water security in arid climates from the perspective of the water footprint index, a case study of South Khorasan Province. *Iranian Journal of Irrigation and Drainage*, *8*, (4,) 746–735 (in Persian).

13 | Using Blockchain Technology for Virtual Water and Water Rights Trade

Yaser Sabzevari[1*], Saeid Eslamian[2] and Fatemeh Dadvand[3]

13.1 Introduction

Water plays a crucial role in sustaining life. It aids individuals, fosters community support, and maintains environmental stability. Agriculture uses more water than any other industry. Studies show that farming uses over 70% of the world's freshwater. With the increasing global demand for water, effective management of water resources is becoming increasingly critical. But, often not enough focus is given to managing water rights for farming. Using smart management techniques and allowing people to buy and sell water rights helps to use water resources in the best way possible. So, managing water rights in farming is very important to use water resources wisely and sustainably. Water rights come from the idea of owning property and related ideas, which means having control over water resources. The main idea is to see water as something that can be owned and bought or sold. The study of trading water rights started in developed countries like the USA and Australia. In recent years, countries like Chile and China have quickly made progress in buying and selling water rights (Delorit et al., 2019). Due to variations in national conditions, social systems, and cultures, different principles exist for the allocation of water rights. In China, for instance, the government is recognized as the owner of water resources according to the law. Water rights trading predominantly involves the rights to use, allocate, and trade water resources.

In 2016, China established a marketplace for trading water rights. This platform makes it easy to buy and sell both big water rights for regions and smaller water rights for farming. When talking about trading water rights in a region, the main goal is to use water fairly and efficiently (Ge et al., 2017). On the other hand, agricultural water rights tend to be traded more frequently and require regular updates due to the dynamic nature of agricultural practices. The trading platform serves as a mechanism to enable efficient and transparent transactions for both types of water rights, contributing to improved management and allocation of water resources in China.

[1-3] Department of Water Science and Engineering, College of Agriculture, Isfahan University of Technology, Isfahan, Iran.
* Corresponding author email: yasersabzevari1511@gmail.com

DOI: 10.1201/9781003686026-13

Blockchain combines cryptography and a distributed system to ensure trust, immutability, and security in transactions (Yli-Huumo et al., 2017; Saberi et al., 2019). Cryptographic techniques such as hash functions, asymmetric encryption, digital signatures, and digital certificates are employed to verify the integrity and authenticity of transaction participants and information. The distributed system ensures that transaction data in a multi-center network is decentralized and securely stored, making it difficult to manipulate (Predescu et al., 2021; Mahmoud et al., 2021). Moreover, when integrated with the Internet of Things (IoT) and cloud technologies, blockchain has the potential to advance environmental sustainability initiatives (Chohan, 2022; Jagtap et al., 2021). Using blockchain technology, IoT devices can safely and clearly record and share information about the environment (Eslamian and Eslamian, 2025). This helps to better track and manage resources. A block in a blockchain usually has two main parts: a block header and a block body, as shown in Figure 13.1. The block header links to the next block in the main chain, and the block body holds the real data or transaction details. Whenever a block is added to the chain, a time stamp is automatically made. This makes the blockchain system more advanced and trustworthy.

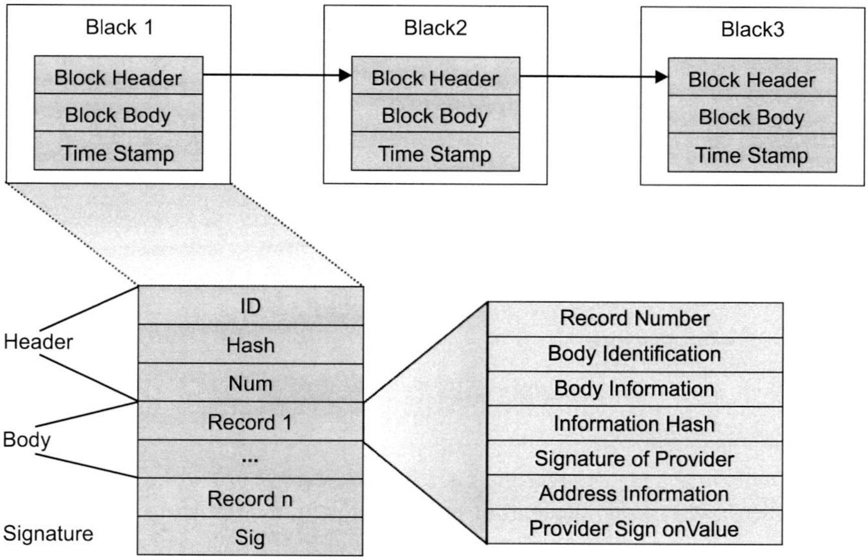

Fig. 13.1 A block structure of blockchain.

The first and most recognized application of blockchain technology, Bitcoin, was created in 2009 by a person or group operating under the name Nakamoto. Since then, blockchain technology has grown and spread quickly (Zheng et al., 2018). New types of blockchain, like Ethereum and Hyperledger Fabric, have come out and are being used in different areas. For example, Hyperledger Fabric can be used to create important games for managing city water by using ideas from mobile crowdsensing. Blockchain technology is being used a lot in saving water smartly. It helps create platforms that make water conservation easier (Thakur et al., 2021). It has brought new energy to improving water resources for farming, helping to create smart water management systems for agriculture (Torky et al., 2020). Researchers like Enescu and Bizon (Enescu et al., 2020) have proposed solutions for implementing blockchain technology in irrigation systems, showcasing its potential in this domain. Nonetheless, there is a lack of extensive research on the application of blockchain technology in trading water rights. The goal of the paper you mentioned is to use blockchain technology to improve the way people buy and sell water rights. This plan wants to make things better and bring in new ideas in this field by using the advantages of blockchain technology. Indeed, blockchain technology's verifiability and immutability make it highly valuable for enhancing security and efficiency in various transactional processes. When applied to the local energy market, blockchain can facilitate the creation of a home energy management system aimed at optimizing energy consumption and reducing electricity costs (Yahaya et al., 2020). By leveraging blockchain, transaction schemes can be made more secure, thus benefiting bilateral trade management (van Leeuwen et al., 2020). Blockchain technology is also being used to help businesses keep track of what people want. But using blockchain in managing water rights for agriculture is still quite rare. The paper you talked about wants to fix this problem by using blockchain technology to make transactions easier, reduce costs, and help grow trading in agricultural water rights. This paper aims to improve how agricultural water rights transactions are handled by using blockchain technology. This would help make things run more smoothly and support the growth of the market for agricultural water rights.

13.2 Methodology

13.2.1 *Water Production and Water Allocation Function*

Water rights trading involves the buying and selling of water rights to improve the management and utilization of water resources. In China, it's important to manage the buying and selling of water rights for farming in a smart way, so that water is used efficiently and meets everyone's needs. The effective use of water for farming in irrigation areas mostly depends on understanding how well crops use water. To use water resources wisely, we need to figure out how much water each type of crop needs for irrigation. Different crops need different amounts of water. The first way to share water equally might not be the best for using water resources. This means there could be a gap between how much water crops need for irrigation and how much water they actually get. To find out how much water each user needs and the best way to distribute it for the most benefit, we need to figure out how water affects crop production. The way to calculate how much water is produced is:

$$Y_i = f(X_i) \quad V_i = k_i \times Y_i \quad V = \sum V_i \times A_i \qquad ...(1)$$

where Y_i is the performance of product i. X_i is the irrigation quota of crop i. V_i is the output value of crop i. k_i is the coefficient of performance of product i. V is the total net value of production of the entire irrigated area, and A_i is the planted area of product i. Based on the above mathematical relationship between water and crops in an irrigation area, a suitable model will follow the principle of maximizing the overall economic benefits of the irrigation area. So we have the objective function:

$$Max\left(\sum V_i\right)\frac{d\left(\sum V_i\right)}{dX_i} = 0 \qquad ...(2)$$

and the limitation condition:

$$W = \sum (X_i \times A_i) \qquad ...(3)$$

The total amount of irrigation water is denoted as W. To ensure fairness, the objective function dictates that the marginal benefits of different agricultural products should be equal. By considering the economic irrigation quota and crop levels, it becomes possible to calculate the optimal allocation of water for each crop. This optimal water allocation may significantly differ from the initial water rights allocation, creating a greater demand for agricultural water rights trading.

Blockchain technology encompasses three main types: public chains, alliance chains, and private chains. Public chains have no restrictions on node access, allowing anyone to read data, conduct transactions, and validate them reliably. Examples of public chains include Bitcoin and Ethereum. Private chains, on the other hand, are typically utilized within a specific entity, where permissions within the chain depend on the rules established by that entity. Alliance chains are an intermediate form between public and private chains, where only authorized nodes are allowed to join. The comparison of these three features is shown in Table 13.1.

Table 13.1 Comparison of blockchain types.

Type	*Public*	*Alliance*	*Private*
Nodes	Any	Authorized Users	In an Organization
Centralization student	Decentralized	Polycentric	Centralized
Consensus mechanism	POW, POS: DPOS	Ratt	Solo
Incentive necessity	Yes	Optional	No
Openness	Complete	Partially	Not
Speed	slow	Faster	Fastest
Application	Bitcoin	Hyperledger	Ant Financial

A union chain is less open than regular chains. Still, it keeps some level of decentralization and works better and more specifically. In trading water rights for farming, there are high agreement costs among people who need irrigation water. When using blockchain technology, it should be balanced with having no single control. A series of partnerships can ensure that data stays private, which is important for businesses that use public resources. Given how developed blockchain technology is now, we have decided to use an alliance chain for the system that trades agricultural water rights.

13.3 Blockchain Technology in Trading Agricultural Water Rights

13.3.1 Blockchain-Based Trading System Architecture

The layout of the system for trading agricultural water rights is shown in Figure 13.2. It has six layers.

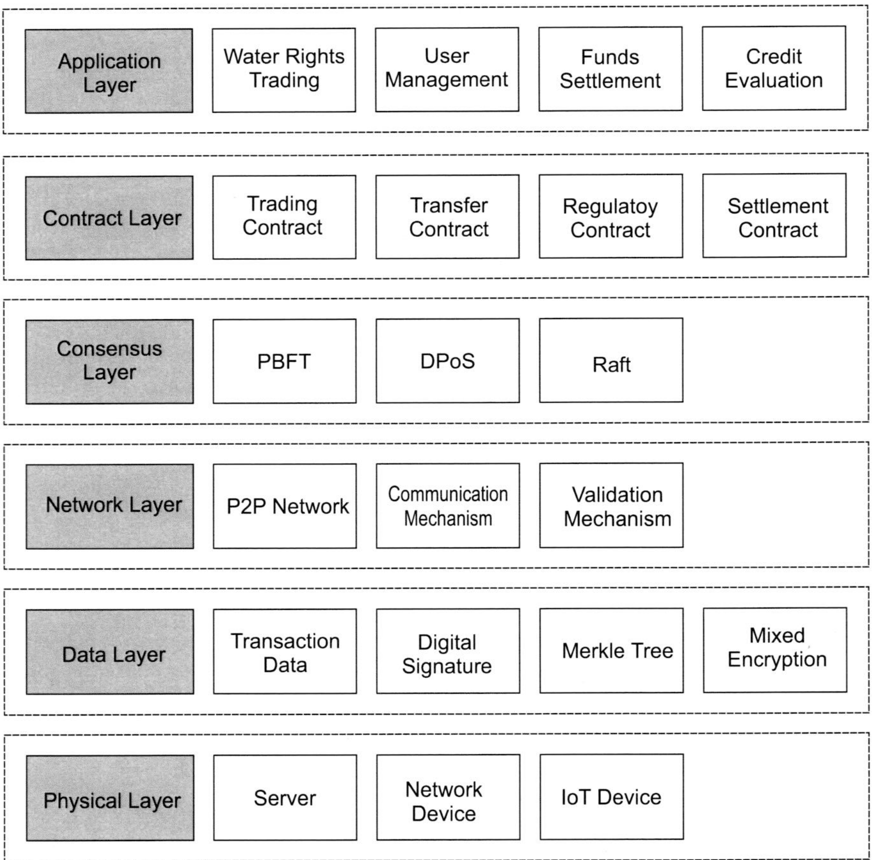

Fig. 13.2 Architecture of agricultural water rights trading system (Liu and Shang, 2022).

The water rights trading system using alliance chain technology needs a strong physical setup, which includes important hardware and software that help the system work properly. The data layer keeps information about transactions, users, and settlements in the water rights transaction system. This information is saved in blocks that are marked with time in the alliance chain. These blocks are made using a math method called hashing. The network layer uses a peer-to-peer (P2P) system with a way to check data for correctness. The nodes in the alliance chain of the transaction system talk and share information with each other through this network. The consensus layer is important for keeping data the same for all parts of the network. It uses a way for everyone to agree, called the Byzantine Fault Tolerant (PBFT) method, which works best for groups of connected chains. The contact layer has scripts, codes, rules for how things work, and smart contracts. People using the water rights trading system make deals, complete payments, and keep track of water resources with automatic contracts called smart contracts. Different kinds of contracts are used in trading water rights. These smart contracts run on code and automatically carry out actions when certain conditions are met. In the end, the application layer manages the various programs used for trading agricultural water rights in the system.

13.3.2 Designing a Trading Platform Based on Consensus Algorithms

The water rights trading platform for farming needs to use a good agreement system. Consensus algorithms are a key part of blockchain technology and have various features (see Table 13.2). Proof-of-work (PoW) means that people in a network must work hard to solve a math puzzle. When they solve it, they earn the chance to change the record of transactions. This is the most spread-out algorithm. Proof of Stake (PoS) means putting your coins aside to be chosen randomly to help confirm transactions. It is a bit centralized because the coin system decides who will be the next people to create new blocks. In PBFT, each node is aware of all the other nodes. So, even though there are fewer clusters, each point has a lot of connections. The Delegated Proof of Stake (DPoS) system is more centralized than the regular PoS because users in the network vote to choose delegates who will validate the next block. The people chosen to represent us become the main decision-makers for the next group. The Proof of Elapsed Time (PoET) system creates a random waiting time for each computer in the blockchain network. The node that has waited the least time wakes up first and gets to add a new block to the blockchain. It has no level of focus, since the next miner is picked by a random timer.

Table 13.2 Consensus algorithm components.

Consensus Algorithms	*Pow*	*Pos*	*DPOS*	*PBFT*	*POET*
Decentralization	Complete	High	Normal	Low	Complete
Efficiency	Slow	Medium	Medium	Fast	Medium
Energy consumption	Very high	Normal	Low	Very low	Low
Security	High	Relatively high	Relatively high	Medium	Relatively high

In the process of trading water rights for farming, the main parts of the system are government groups and important organizations. These nodes have the most access and power. Each node takes care of its own jobs and keeps things running smoothly. The whole system for trading water rights in farming is set up as one group through partnerships with organizations. Once a group or person gets approval, they can join the network and access important information by linking to the platform. After looking at different ways for systems to agree on things, it has been found that the PBFT (Practical Byzantine Fault Tolerance) method works best for keeping the system nodes in sync. So, the farming water rights trading system uses this method to make buying and selling easier and more trustworthy.

13.4 Virtual Water

In terms of water resources, virtual water means the amount of fresh water used to make a certain product or service. It is measured at the exact place where the product is made. (Hoekstra & Chapagain, 2007; Hoekstra et al., 2011). Furthermore, Velázquez et al. (2011) define virtual water as the quantity of water utilized in the creation and transformation of raw food into food products. On the other hand, the water footprint encompasses not only the virtual water content but also includes the water used in delivering the final product to the consumer. The water footprint shows how much water is used to make a product. This can be looked at from the perspective of the person using it or in specific places. This paper looks only at virtual water, which means the water used to make and put together products. The consideration of water footprint, which includes additional factors like product transportation and delivery, is beyond the scope of this study.

The idea of virtual water is helpful for understanding and managing the water needed to make products and provide services. It helps make sure that water is used wisely by moving it from places that have little water to places that have plenty of water. This method helps reduce water shortages in certain countries by managing water resources carefully and wisely. In this way, trading water virtually becomes very important. However, managing virtual water trading effectively and clearly does not have enough complete tools and technologies. Blockchain technology has become popular lately because it can check and keep track of complicated transactions that need to be confirmed. Blockchain is a system that keeps track of information in a way that is shared by many people in a network that doesn't have a central control. It acts as an enabler and innovation catalyst, transforming how supply chain partners collaborate. By utilizing blockchain, secure information sharing is facilitated while maintaining proper control over access privileges. It minimizes the risk of data loss and ensures a high level of data integrity. Implementing a blockchain-based tracking solution accelerates supply chain processes and holds great potential for enhancing virtual water business operations. Thus, blockchain is expected to be a suitable and effective solution in the realm of virtual water business, enabling efficient and transparent management of water resources within the supply chain.

This study aims to investigate and explore the connection between virtual water and the utilization of blockchain technology in the context of virtual water. The research methodology employed for this purpose involved conducting a systematic literature review. The review process focused on gathering relevant information from various sources, including 16 IWA Publishing journal repositories containing 153 journal publications related to virtual water, as well as five prominent journal databases, namely, IEEE Xplore, ScienceDirect, ACM Digital Library, Springer Link, and Wiley Online Library, A total of 5,026 articles that discussed the links between blockchain and water management were covered in the review. The review looked at 5,026 articles about how blockchain is connected to water management. The literature review part of the study gives a clear summary of how the research was done and what was found during it. This section looks at the main discoveries made from studying the gathered research. Then, the Research Findings section shows a suggested plan for creating an international system for trading virtual water using blockchain technology. Also, the study looks into possible credit risks that come with this system. In the end, the study wraps up by highlighting what it adds to the topics of virtual water and blockchain technology.

13.5 Literature Survey

This survey aims to gain insights into the key technical, cultural, and management factors influencing virtual water trading globally. The main question this study looks at is: "What motivates people to trade virtual water around the world." To understand this topic better, the researchers did a careful review of the existing information and studies. The main source used for this survey was the IWA Publishing collection. IWA Publishing is known for having a large collection of online information about water management and related topics. It includes top magazines about water resources from around the world. The study also wanted to look at how blockchain technology can be used in water management. To do this, we looked at different journals to find important information about how blockchain and water studies are connected. This survey uses different research methods to understand how virtual water trading works around the world. It also looks at how using blockchain technology could help improve water management.

To find important things about managing virtual water around the world, we looked at and studied existing data through a method called content analysis. This method helped us to understand the topic clearly and thoroughly (Daven-Wambolt, 1992). Specifically, our study focused on examining written texts from various sources such as journal articles, books, and web resources. Through this method, we sought to uncover valuable insights from the textual data, treating it as "silent evidence" for our investigation (Hodder, 1994).

The virtual water codebook consists of nine categories, each containing a description of the article (year, journal or conference name, title name, volume number, doi), author keywords, abstract, connection to virtual water, location, technology solution, business.

13.6 Data Collection Process

An essential phrase search technique was employed across 16 journal collections from IWA Publishing (International Water Association), as detailed in Table 13.3. We looked up the term "virtual water". We included journal articles (without pictures) and different types of articles (like research articles, review articles, and editor's picks) without looking at specific dates. There were 153 visits in total, as shown in Table 13.3. We also looked into how blockchain is being used in water and similar fields. We searched for the words "Blockchain" and "Water Management". We finished the search in July 2021, like we said earlier. Table 13.4 displays the results. We have looked at these journal databases. We also applied these filter rules to different databases.

IEEE Xplore filter options: Search for the words ("full text and info": Blockchain) and ("full text and info": water management). Time period: 1884 to 2022. Only conferences, journals, and articles that are available early are included. Books and magazines are not included in the search.

- ScienceDirect filter settings: Search term - blockchain for managing water. Types of articles: review articles, research articles, conference summaries, and case reports.
- ACM Digital Library: Filter Options: Search for the words "blockchain" and "water management". Type of content: Research papers.
- Springer Link: Search settings: Use the words "blockchain AND water". Content type: Articles, conference papers, and collections of conference papers.
- Wiley Online Library: Search for the words "blockchain for water management."
- Type of content: Magazines.

Table 13.3 Virtual water search references (Jayasri et al., 2022).

Sr. No.#	*Journal Name*	*No. of Hits*
1	AQUA	31
2	Blue-Green Systems	3
3	H_2Open Journal	5
4	Hydrology Research	5
5	Ingenieria del agua	1
6	Journal of Hydroinformatics	6
7	Journal of Water & Climate Change	9
8	Journal of Water & Health	0
9	Journal of Water, Sanitation, & Hygiene for Development	0
10	Water Policy	57
11	Water Practice & Technology	0

Contd...

Table 13.3 *Contd...*

Sr. No.#	*Journal Name*	*No. of Hits*
12	Water Quality Research Journal	0
13	Water Research	0
14	Water Reuse	0
15	Water Science & Technology	12
16	Water Supply	24
Total		153

Table 13.4 Blockchain and water management search references.

Sr. No.#	*Database Name*	*No. of Hits*
1	IEEE Xplore	1295
2	ScienceDirect	823
3	ACM Digital Library	1732
4	Springer Link	915
5	Wiley Online Library	261
Total		5026

13.7 Part 1: Demographics of the Article (Virtual Water in IWA Publications)

In our exploration of the repositories of various journals published by IWA, we discovered a total of 153 articles relating to virtual water. The distribution of these articles over the years is presented in Figure 13.3. However, it's worth noting that out of the 153 articles, 10 were identified as duplicates. Our analysis revealed that a significant majority, approximately 76%, of the articles were published between the years 2015 and 2021.

As shown in Figure 13.4, only 68 articles out of 153 articles had a close relationship with virtual water and water footprint.

As shown in Figure 13.5, China had the largest share of virtual water-related articles with 28% of articles. We noticed that 30% of the articles had no specific location links. The next major share was from Africa (11%), followed by Europe (9%), India (6%), and the Middle East (6%).

As shown in Figure 13.6, about 43% of articles linked to virtual water are in the field of agriculture. We found that few articles discuss the topic of industrial trade in a virtual water context.

13.8 Part 2: Article Demography (Blockchain for Water Management in Journal Databases)

We noticed that after 2015, IEEE Xplore issued articles about blockchain and water management. The biggest share was in 2020 with 39% of articles, then it was 29%

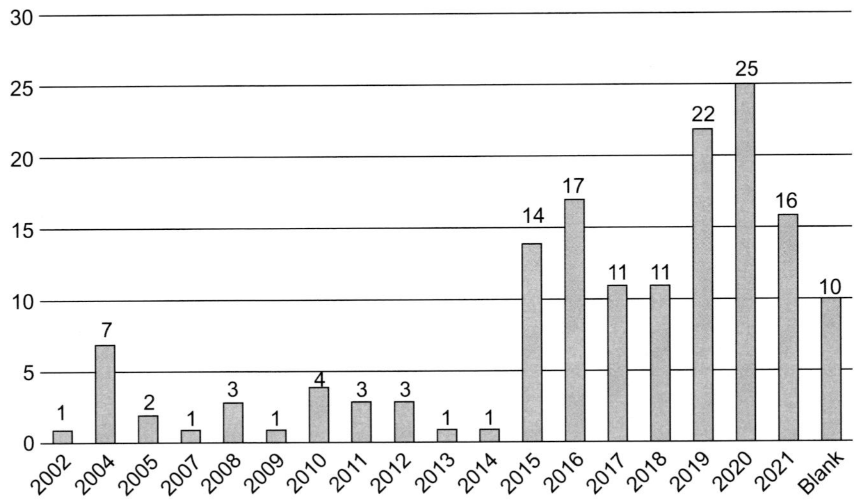

Fig. 13.3 Annual distribution of articles (Jayasri et al., 2022).

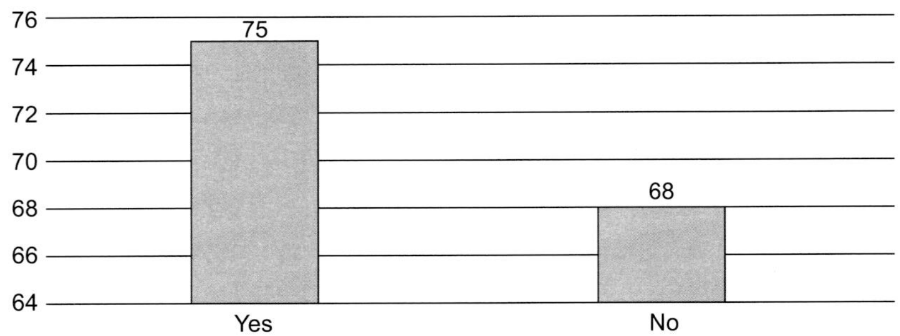

Fig. 13.4 Relevancy of articles (Jayasri et al., 2022).

in 2019 and 17% in 2021. This just shows how blockchain technology is starting to be used in the water sector (like in Figure 13.7 for IEEE). We also noticed that ScienceDirect published articles about Blockchain and Water Management after 2016. We only looked at review articles, research articles, conference summaries, and case reports. In 2021, there were more articles than in any other year, making up 44% of the total. The least number of articles were from 2022, with just one article, and from 2016, with two articles. The yearly ScienceDirect template looks a lot like the IEEE Xplore template.

We saw that the ACM Digital Library published articles about blockchain and water management starting in 2013, with the most articles published in 2019.

During our analysis, we noted that SpringerLink published numerous articles concerning blockchain and water management, particularly after 2016. The highest concentration of these publications was observed in 2021, accounting for 53% of the total articles. Similarly, Wiley Online Library reported a substantial number

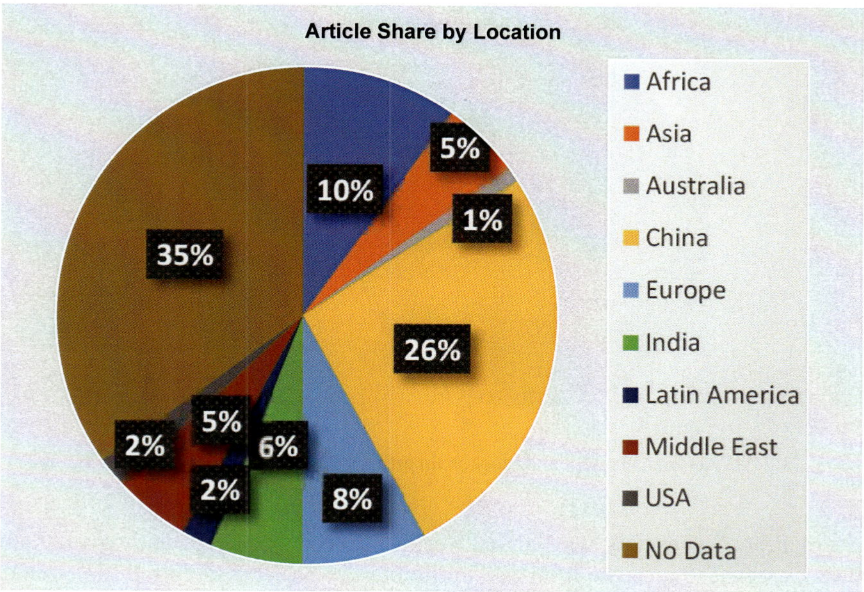

Fig. 13.5 Spatial distribution of articles (Jayasri et al., 2022).

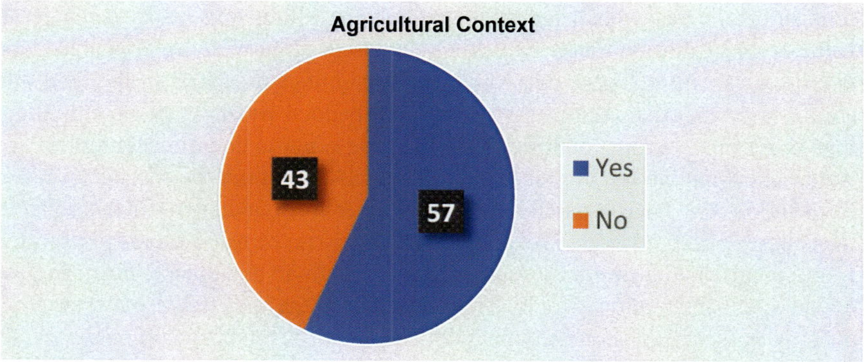

Fig. 13.6 Articles related to virtual agricultural water (Jayasri et al., 2022).

of articles, precisely 261, between the years 2016 and 2021. Among the various fields of study, the business and management domain contributed the largest share of articles, with a total of 60 publications.

13.9 Research Findings

Part 1: The Status of Virtual Water Trading

In 1999, the International Food Policy Research Institute released a study called "Effects on Food Security and Rural Development from Water Used in

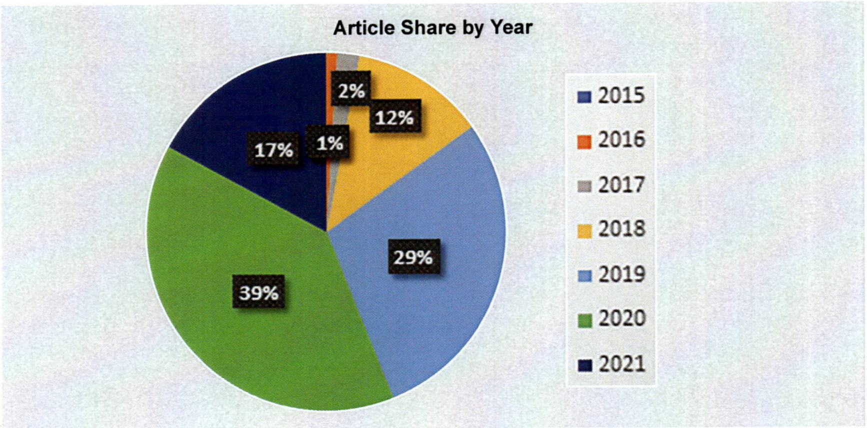

Fig. 13.7 Water management and blockchain articles in IEEE Xplore (Jayasri et al., 2022).

Agriculture." The study showed that worldwide food trade, especially grain trade, was expected to increase from 186 million tonnes in 1993 to 349 million tonnes by 2020 (Rosegrant and Ringler, 1999). The Food and Agriculture Organization of the United Nations report, known as the *World Grain Inventory in 2021*, provided further evidence supporting these forecasts. According to this report, the actual production of cereal worldwide amounted to 2,710 million tons, with a commercial value of 440.1 million tons—numbers that closely align with the estimates made two decades earlier. These data validate two significant aspects in the field: the ability to predict the value and volume of embedded water. They also highlight how water moves from countries that have little water to countries with plenty of water, like what we see in global trade. They also point out the transfer of water from nations that have a scarcity of it to those that are abundant in water, similar to the patterns observed in global trade. They emphasize the way water is transported from countries with limited resources to those with an abundance, mirroring the dynamics of international trade. They underscore the movement of water from areas with low supplies to regions with rich water resources, akin to the trends seen in global commerce.

The terms "virtual water", "embedded water", or "indirect water" mean the amount of water used when making products and providing services. In this situation, virtual water can be seen as something that changes and moves.

The Middle East region is currently grappling with a critical water shortage. To address this issue, The study indicates that countries need to join forces to more effectively manage local water resources and enhance their quality. The study suggests using virtual water trading to help save water and ensure there is enough water in the region. One way to improve farming is to bring in crops that use a lot of water to grow. This can help make smart investments in agriculture in today's quick-moving global market.

Zhang et al. (2018) worked on solving the problem of sharing water rights between farmers and factories in China. They pointed out that China doesn't have a clear system to share water use credits between farming and industry. To solve this problem, the study suggested a new pricing system that allows farmers to sell their water rights to businesses. This model operates based on water savings achieved through the adoption of conservation practices in agriculture. By implementing this approach, it is possible to facilitate the exchange of water rights between these sectors while promoting water conservation.

Huang et al. (2017) examined the distribution patterns of virtual water export and import across different regions of China. They estimated that the total amount of virtual water associated with exports in China reached 106.3 billion cubic meters. However, they discovered significant disparities in the distribution of virtual water among the regions. The eastern region of China stood out for its efficient utilization of water resources in virtual water exports.

Wu et al. (2018) highlighted the importance of local economies using virtual water as a helpful resource. They suggested using virtual water trading by buying water from areas that have a lot of it to help solve water shortages in different regions. This trading would improve how water is used and help send more water to businesses and industries. In a separate study, Wu et al. (2021) conducted research on water consumption in southwestern China. They observed that the Yunnan region had the lowest water footprint for rice production, while the Northeast region had the highest. The northeast region was identified as being at high risk concerning water scarcity. The authors advocated for the formulation of improved policies related to groundwater extraction and international trade in order to ensure food security and sustainable water usage at both regional and national levels.

Furthermore, Wang et al. (2019) regarded virtual water import as an effective approach for importing low-yield crops and promoting the development of high-efficiency agriculture, ultimately increasing water productivity. This method is considered essential for future agricultural practices and for enhancing water management.

In the study by Kumar and Jain (2011), The authors looked at how water is traded between different states in India, focusing on a main source. They said that exporting virtual water shouldn't just rely on how much water is available. It should also consider things like productivity, land use, needed changes, infrastructure, business culture, and access to credit. By thinking about these things, managing virtual water can be better and last longer.

Janjua and Hasan (2020) suggested a new way to handle bankruptcies that helps manage water resources better and share them more fairly. This method looks to solve problems with not having enough water by taking into account the importance of different users or groups involved. This method can help share water fairly and in a way that can be maintained over time.

Zeng et al. (2015) Created a two-step water trading model to share limited water among different users in a market-based way. This model boosts economic productivity and offers solutions for water shortages by using market methods to manage water resources effectively.

Menshadi et al. (2015) mentioned that managing virtual water movement should cater to economic aspirations alongside environmental necessities. They created a plan that looks at money, fairness, and the environment to help encourage sustainable development. Their suggested teamwork method helps distribute profits in the best way, promoting better and more sustainable use of virtual water.

Likewise, Sepúlveda (2015) discussed the connection between the equilibrium of exports and imports and its relevance to examining water consumption. The writer looked at how people use water around the world and in their own areas using a method called the water footprint. They highlighted how important it is to understand nature and the economy together. This study helps us understand how moving virtual water affects the environment and shows the importance of thinking about ecological factors when making decisions.

In the study by Wu et al. (2018), the authors emphasized the status of a virtual water plan in endorsing local water security and improving living standards. They observed that automation was starting to compete with agriculture for water resources and highlighted how virtual water can enhance water productivity. The authors proposed the advance of a "virtual water integrated water resource management system" to facilitate virtual water trading among regions, particularly benefiting arid regions through virtual water imports.

Reuter et al. (2018) conducted an assessment of virtual water flow throughout the supply chain of the entire Australian economy. They employed an input-output footprinting method, assigning a water footprint to each industry and allocating it to customers based on total sales. This analysis provides insights into the virtual water embedded in different industries and highlights the water usage patterns within the economy.

Youkhana and Laube (2009) emphasized the significance of sourcing food from nations that have abundant water resources. This can help places that do not have enough water. The writers stressed that it's important to have strong support from leaders, better education, and changes in finance to successfully use virtual water trading and take advantage of importing products that use a lot of water.

Salim and Abdalbaki (2019) analyzed the structure of virtual water exports in the Nile basin countries. Their study revealed that Kenya successfully saved a substantial amount of virtual water through exports, while other countries experienced losses. Social network analysis was utilized to derive these statistics, providing insights into the interdependencies and trade dynamics of virtual water among the countries in the Nile basin.

According to Lopez-Gan et al. (2012), there is a relationship between food, water, and a range of social, economic, and political matters. They emphasized how water security, food security, safety for people, and environmental safety are all connected. They pointed out that we need to look at all these issues together. The writers said that it's important to understand virtual water trade to ensure water and food security in the future. They emphasized looking at both money matters and ethical issues.

Parveen and Faisal (2004) highlighted how important virtual water trade is for making sure there is enough food. They proposed that we change the definition of food security to mean "being able to grow enough food for ourselves" instead of just being self-sufficient in general.

Deng et al. (2018) Looked at how the amount of food and water used in China changed from 1997 to 2011. They supported a "green channel" to help grow local food trade and virtual water trade, considering the economy and climate.

Hassan and Tiam (2015) used a wide economic approach to study how virtual water moves. They suggested using water more efficiently and making rainfed farming more competitive to help ensure food security.

Britella and others (2008) suggested looking into how water tax rules affect international trade. They argued that implementing a water tax could lead to greater efficiency, changes in production and consumption patterns, and alterations in international trade dynamics.

Emanuel and Clayton (2017) proposed the inclusion of virtual water trading in water policy. They argued that virtual water trading could alleviate the need for a country to possess abundant water resources.

Chen et al. (2019) investigated virtual water through the lens of water and energy. They suggested that virtual water is like hidden amounts of water that are used to make and move electricity from different places. They discovered that the eastern part of China used a lot of virtual water, consumed a lot of energy, and produced a lot of carbon emissions. This affected the environment and caused social problems.

Achampong and others (2016) suggested a new method for managing water resources which includes a strategy to separate different water uses. They talked about virtual water trading as part of Integrated Water Resources Management (IWRM). The goal is to improve how water is managed to be more cost-effective, fair to everyone, and good for the environment.

Mamo (2015) showed a model to help choose the best ways to reduce risks by taking into account group opinions and improving how decisions are made.

Wicaksono et al. (2017), was suggested to create a system that helps make decisions by showing and measuring how water, energy, and food work together. They suggested using this system to make, assess, and improve plans for sustainable development.

Doeffinger et al. (2020), suggested creating a dashboard to check the status of water security in medical tests.

Table 13.5 provides an overview of the extensive research topics covered by various authors in the field of virtual water trade.

Table 13.5 Research topics presented by different authors.

Sr. No.#	*Key Area*	*Author(s)*
1	Process of managing water resources; Evaluation of water for making decisions	Hellegers & van Halsema (2019)
2	The transfer of virtual water and improving WUE can help reduce inequality in water usage.	He et al. (2020)
3	Virtual water trade (VWT) is a potential solution for water-stressed countries.	Rogers et al. (2004); Oki & Kanae (2004); Johannessen (2004)
4	Variables connected to virtual water imports: Average income, Population, Agriculture as value-added, Irrigated area, Exports of goods and services	Ramirez-Vallejo & Rogers (2004)
5	Establishment of an international virtual water trading council	Rahman et al. (2002
6	Food Miles (the distance that food is transported from the time of its making till it reaches the consumer) and virtual water for urban management	Liebman et al. (2011)
7	Diet changes impact water footprint in a particular region; Winery water footprint calculation and wastewater calculation.	Saraiva et al. (2019)
8	The need for Virtual Water Credits and a Trading System	Ravi Shankar & Jayasri (2015)
9	Assessment of virtual water exchange between two areas; Examination of virtual water usage for developing sustainable cities; Water footprint (WF) viewed from production and consumption angles and the importation of water-intensive agricultural goods	De Miguel et al. (2010); Vanham (2011); Kang et al. (2017)
10	Redesigning existing water pricing policies using a virtual water strategy	Hristov et al. (2015)
11	A Virtual Water inflow and outflows are utilized in the creation of a multi-region computable general equilibrium (CGE) model. The writers examined how Virtual Water flows impact the GDP and employment rate	Zhao et al. (2021)
12	The method used to assess changes in Water Resources Metabolism Efficiency (WME) involves material flow analysis. Beneficial for enhancing the metabolic effectiveness of both real and digital water	Meng et al. (2020)
13	The role of virtual water in assisting with managing water stress; the significance of the water footprint (Iwf) and external water dependency (WD) within Virtual Water Trade	Huang et al. (2019); Sun et al. (2016)

Contd...

Contd...

Sr. No.#	*Key Area*	*Author(s)*
14	A water rights trading program; a pricing strategy for moving water rights from farming to industrial use in regions with water scarcity to promote the growth of future water markets	Zhang et al. (2019)

13.10 Conclusion

Effectively managing water resources relies heavily on the utilization of water rights, making it crucial to apply them wisely and sustainably. In this situation, using blockchain technology has become popular as a modern solution. This paper suggests using blockchain technology to improve the trading of water rights in agriculture, with the goal of managing water resources better. One important part of blockchain technology is that it keeps a clear and unchangeable record of transactions. By utilizing a Byzantine Fault Tolerant (PBFT) consensus mechanism, the proposed blockchain-based trading platform can strike a balance between decentralization and efficiency. On this platform, essential information such as date, water quantity, water quality, time, and trading rules can be recorded in an unalterable and transparent manner. Implementing blockchain technology in water rights trading brings several advantages. Firstly, the system becomes more reliable and automated, ensuring the availability of information. Secondly, the use of smart contracts streamlines the process, replacing manual checks with automated verification. Additionally, the trading parties can negotiate and match independently, improving efficiency. Lastly, as the water rights exchange data is securely stored and traceable on the blockchain, it enhances data security and By reducing the need for manual processing, costs are lowered, and overall efficiency is improved. By cutting down on handwork, we save money and get things done more efficiently. The blockchain platform helps to manage the buying and selling of water rights for farming better, supporting a healthy environment and economy. Designing, manufacturing, and trading water-sensitive or water-aware products plays a vital role in protecting our environment. It not only helps in preserving natural resources but also has a significant impact on operating costs, making it a business imperative. One effective approach to increase public awareness about water consumption is through labeling products based on their water usage. Digital product traceability can greatly assist in this process by providing a transparent, secure, and comprehensive experience. It enables consumers to track the origins of products, including their water footprint, throughout the supply chain. This information empowers consumers to make informed choices and encourages the purchase of water-efficient products. Integrating the concepts of virtual water usage and crop cultivation presents an innovative approach to enhancing water management. Blockchain technology is very promising for helping to connect different systems. It talks about important things like being able to track products, keeping information clear, and handling product recalls in manufacturing. Using blockchain can stop fake parts from being used in products. This makes sure that the amount of water used for each part is shown correctly and counted. Also, using virtual water management along with blockchain makes it possible for people to choose water-saving products. It also supports the use of virtual water currency along with regular money. Using virtual water management together with blockchain technology could help us use water more efficiently, support sustainable practices, and help people understand how much water is used to make products.

References

1. Acheampong, E.N., Swilling, M. and Urama, K. (2016). Developing a framework for supporting the implementation of integrated water resource management (IWRM) with a decoupling strategy. *Water Policy*, *18*(6), 1317–1333. https://doi.org/10.2166/ wp.2016.155.
6. Berrittella, M., Rehdanz, K., Roson, R. and Tol, R.S.J. (2008). The economic impact of water taxes: A computable general equilibrium analysis with an international data set. *Water Policy*, *10*(3), 259–271. https://doi.org/10.2166/wp.2008.003.
13. Chen, D., Zhang, D., Luo, Z., Webber, M. and Rogers, S. (2019). Water–energy nexus of the eastern route of China's south-tonorth water transfer project. *Water Policy*, *21*(5), 945–963. https://doi.org/10.2166/wp.2019.188.
15. Chohan, U.W. (2019). Blockchain and environmental sustainability: Case of IBM's blockchain water management. *Notes on the 21st Century* (*CBRI*).
17. De Miguel, Á., García, E. and De Buestamante, I. (2010). Estimation of the virtual water trade between two Spanish regions: Castilla-la Mancha and Murcia. *Water Supply*, *10*(5), 831–840. https://doi.org/10.2166/ws.2010.477.
18. Delorit, J.D., Parker, D.P. and Block, P.J. (2019). An agro-economic approach to framing perennial farm-scale water resources demand management for water rights markets. *Agricultural Water Management*, *218*, 68–81.
19. Deng, G., Xu, Y. and Yu, Z. (2018). Accounting and change trend analysis of food production water footprint in China. *Water Policy*, *20*(4), 758–776. https://doi.org/10.2166/wp.2018.173.
20. Doeffinger, T., Borgomeo, E., Young, W.J., Sadoff, C. and Hall, J.W. (2020). A diagnostic dashboard to evaluate country water security. *Water Policy*, *22*(5), 825–849. https://doi.org/10.2166/wp.2020.235.
22. Emmanuel, K. and Clayton, A. (2017). A strategic framework for sustainable water resource management in small island nations: The case of Barbados. *Water Policy*, *19*(4), 601–619. https://doi.org/10.2166/wp.2017.137.
23. Enescu, F.M., Bizon, N., Onu, A., Răboacă, M.S., Thounthong, P., Mazare, A.G. and Şerban, G. (2020). Implementing blockchain technology in irrigation systems that integrate photovoltaic energy generation systems. *Sustainability*, *12*(4), 1540.
24. Eslamian, S., and Eslamian, F. (2025). *Blockchain Technology in Water and Environmental Systems*. CRC Press, Taylor and Francis, USA, p. 282.
25. Ge, M., Wu, F.P. and Chen, X.P. (2017). A coupled allocation for regional initial water rights in Dalinghe Basin, China. *Sustainability*, *9*(3), 428.
28. Hassan, R. and Thiam, D.R. (2015). Implications of water policy reforms for virtual water trade between South Africa and its trade partners: Economy-wide approach. *Water Policy*, *17*(4), 649–663. https://doi.org/10.2166/wp.2014.242.
29. He, Y., Lin, Z. and Chen, X. (2020). Regional difference of water use in a significantly unbalanced developing region. *Water Policy*, *22*(6), 1182–1199. https://doi.org/10.2166/wp.2020.061.

30. Hellegers, P. and van Halsema, G. (2019). Weighing economic values against societal needs: Questioning the roles of valuing water in practice. *Water Policy*, *21*(3), 514–525. https://doi.org/10.2166/wp.2019.048.
31. Hodder, I. (1994). The interpretation of documents and material culture. In: Denzin, N.K. & Lincoln, Y.S. (Eds.), *Handbook of Qualitative Research*. Sage, Thousand Oaks, CA, pp. 673–715.
32. Hoekstra, A.Y. and Chapagain, A.K. (2007). Water footprints of nations: Water use by people as a function of their consumption pattern. *Water Resources Management*, *21*(1), 35–48. doi:10.1007/s11269-006-9039-x.
33. Hoekstra, A.Y., Chapagain, A.K., Aldaga, M. and Mekonnen, M.M. (2011). *The WF Assessment Manual: Setting the Global Standard*. Earthscan, Abingdon, UK.
35. Hristov, J., Martinovska-Stojcheska, A. and Surry, Y. (2015). Virtual water and input–output framework: An alternative method for assessing trade and water consumption in FYR Macedonia. *Water Supply*, *15*(2), 317–326. https://doi.org/10.2166/ws.2014.118.
36. Huang, H., Wang, J., Han, Y., Wang, L. and Li, X. (2019). Assessing impacts of water regulations on alleviating regional water stress with a system dynamics model. *Water Supply*, *19*(2), 635–643. https://doi.org/10.2166/ws.2018.112.
37. Huang, Y., Lei, Y. and Wu, S. (2017). Virtual water embodied in the export from various provinces of China using multi-regional input–output analysis. *Water Policy*, *19*(2), 197–215. https://doi.org/10.2166/wp.2016.002.
40. Janjua, S. and Hassan, I. (2020). Transboundary water allocation in critical scarcity conditions: A stochastic bankruptcy approach. *Journal of Water Supply: Research and Technology-Aqua*, *69*(3), 224–237. https://doi.org/10.2166/aqua.2020.014.
41. Johannessen, A. (2004). Summary and conclusions from the SIWI Seminar for Young Water Professionals Drainage basin security: Implications of virtual water trade and agricultural subsidies at regional, national, and local levels. *Water Science and Technology*, *49*(7), 215–218. PMID: 15195442.
49. Liebman, M.B., Jonasson, O.J. and Wiese, R.N. (2011). The urban stormwater farm. *Water Science and Technology*, *64*(1), 239–246. doi: 10.2166/wst.2011.182. PMID: 22053481.
50. Liu, Y. and Shang, C. (2022). Application of blockchain technology in agricultural water rights trade management. *Sustainability*, *14*(12), 7017.
51. Lopez-Gunn, E., De Stefano, L. and Ramón Llamas, M. (2012). The role of ethics in water and food security: Balancing utilitarian and intangible values. *Water Policy*, *14*(S1), 89–105. https://doi.org/10.2166/wp.2012.008.
53. Mahmoud, H.H., Wu, W. and Wang, Y. (2021). Wdschain: A toolbox for enhancing the security using blockchain technology in water distribution system. *Water*, *13*(14), 1944.
54. Mamo, T.G. (2015). Risk-based approach to manage aging urban water main infrastructure. *Journal of Water Supply: Research and Technology-Aqua*, *64*(3), 260–269. https://doi.org/10.2166/aqua.2014.052.
55. Manshadi, H.D., Niksokhan, M.H. and Ardestani, M. (2015). A quantity-quality model for inter-basin water transfer system using game theoretic and virtual water approaches. *Water Resources Management*, *29*(13), 4573–4588. https://doi.org/10.1007/ s11269-015-1076-x.
57. Meng, L., Yang, D., Ding, Z., Wang, Y. and Ma, W. (2020). Spatiotemporal variations of water resources metabolism efficiency in the Beijing-Tianjin-Hebei region, China. *Water Supply*, *20*(4), 1178–1188. https://doi.org/10.2166/ws.2020.028.
59. Nakamoto, S. (2008). *Bitcoin: A Peer-to-peer Electronic Cash System*. SCRIP Open Access.
62. Oki, T. and Kanae, S. (2004). Virtual water trade and world water resources. *Water Science and Technology*, *49*(7), 203–209. https://doi.org/10.2166/wst.2004.0456.
64. Parveen, S. & Faisal, I.M. (2004). Trading virtual water between Bangladesh and India: A politico-economic dilemma. *Water Policy*, *6*(6), 549–558.

68. Predescu, A., Arsene, D., Pahonțu, B., Mocanu, M. and Chiru, C. (2021). A serious gaming approach for crowdsensing in urban water infrastructure with blockchain support. *Applied Sciences*, *11*(4), 1449.
69. Predescu, A., Mocanu, M., Lupu, C. and Bercovici, A. (2019). A real-time architecture for collaborative IoT applications in urban water management. In: *2019 23rd International Conference on System Theory, Control, and Computing* (*ICSTCC*), pp. 839–844. doi:10.1109/ICSTCC.2019.8885509.
70. Rahman, Kadi, M.A. and Rockström, J. (2002). Workshop 7 (synthesis): Trade-offs in water for food and environmental security: Urban/agricultural trade-off. *Water Science and Technology*, *45*(8), 191–193. https://doi.org/10.2166/wst.2002. 0178.
71. Ramirez-Vallejo, J. and Rogers, P. (2004). Virtual water flows and trade liberalization. *Water Science and Technology*, *49*(7), 25–32. PMID: 15195413.
72. Ravi Shankar, S. and Jayasri, A.S.V. (2015). The need for virtual water credits and trading system. *Water Supply*, *15*(5), 933–939. https://doi.org/10.2166/ws.2015.047.
74. Reutter, B., Lant, P.A. and Lane, J.L. (2018). Direct and indirect water use within the Australian economy. *Water Policy*, *20*(6), 1227–1239. https://doi.org/10.2166/wp.2018.055.
75. Rogers, P., Nakayama, M., Lundqvist, J. and Furuyashiki, K. (2004). Workshop 7 (synthesis): Role and governance implications of virtual water trade. *Water Science and Technology*, *49*(7), 199–201. https://doi.org/10.2166/wst.2004.0455.
76. Rosegrant, M. and Ringler, C. (1999). *Impact on Food Security and Rural Development of Reallocating Water From Agriculture*. IFPRI, Washington DC.
77. Saberi, S., Kouhizadeh, M., Sarkis, J. and Shen, L. (2019). Blockchain technology and its relationships to sustainable supply chain management. *International Journal of Production Research*, *57*(7), 2117–2135.
79. Saraiva, A., Rodrigues, G., Mamede, H., Silvestre, J., Dias, I., Feliciano, M., Oliveira e Silva, P. and Oliveira, M. (2019). The impact of the winery's wastewater treatment system on the winery water footprint. *Water Science and Technology*, *80*(10), 1823–1831. https://doi.org/10.2166/wst.2019.432.
80. Selim, K.S. & Abdalbaki, S.M. (2019). On the relationship between virtual water network and crops intra-trade among Nile basin countries. *Water Policy*, *21*(3), 481–495. https://doi.org/10.2166/wp.2019.074.
81. Sepúlveda, J. (2015). Corporate social responsibility of regional institutions: Save water and money with an ecological economics perspective in a climate change context. *Journal of Water and Climate Change*, *6*(1), 104–110. https://doi.org/ 10.2166/wcc.2013.044.
83. Sun, Y., Shen, L. and Lu, C. (2016). Study on the water footprint and external water dependency of Beijing. *Water Supply*, *16*(4), 1077–1085. https://doi.org/10.2166/ws.2016.022.
84. Thakur, T., Mehra, A., Hassija, V., Chamola, V., Srinivas, R., Gupta, K.K. and Singh, A.P. (2021). Smart water conservation through a machine learning and blockchain-enabled decentralized edge computing network. *Applied Soft Computing*, *106*, 107274.
85. Torky, M. and Hassanein, A.E. (2020). Integrating blockchain and the internet of things in precision agriculture: Analysis, opportunities, and challenges. *Computers and Electronics in Agriculture*, *178*, 105476.
86. van Leeuwen, G., AlSkaif, T., Gibescu, M. and van Sark, W. (2020). An integrated blockchain-based energy management platform with bilateral trading for microgrid communities. *Applied Energy*, *263*, 114613.
87. Vanham, D. and Bidoglio, G. (2014). The water footprint of Milan. *Water Sci. Technol.*, *69*(4), 789–795. doi:10.2166/wst.2013.759. PMID: 24569278.
88. Vanham, D. (2011). How much water do we really use? A case study of the city state of Singapore. *Water Supply*, *11*(2), 219–228. https://doi.org/10.2166/ws.2011.043.

89. Velázquez, E., Madrid, C. and Beltrán, M.J. (2011). Rethinking the concepts of virtual water and water footprint in relation to the production-consumption binomial and the water-energy nexus. *Journal of Water Resource Management*, *25*, 743–761.

90. Wang, H., Liu, H., Wang, C., Bai, Y. and Fan, L. (2019). A study of industrial relative water use efficiency of Beijing: An application of data envelopment analysis. *Water Policy*, *21*(2), 326–343. https://doi.org/10.2166/wp.2019.019.

91. Wang, S., Ding, W., Li, J., Yuan, Y., Ouyang, L. and Wang, F.Y. (2019). Decentralized autonomous organizations: Concept, model, and applications. *IEEE Transactions on Computational Social Systems*, *6*(5), 870–878.

92. Wicaksono, A., Jeong, G. and Kang, D. (2017). Water, energy, and food nexus: Review of global implementation and simulation model development. *Water Policy*, *19*(3), 440–462. https://doi.org/10.2166/wp.2017.214.

93. World cereal inventories in (2021/22) expected to rise for the first time since 2017/18. http://www.fao.org/worldfoodsituation/ csdb/en/. Accessed on 02/08/2021.

94. Wu, F., Zhang, Q. and Gao, X. (2018). Does water-saving technology reduce water use in economic systems? A rebound effect in Zhangye city in the Heihe River Basin, China. *Water Policy*, *20*(2), 355–368. https://doi.org/10.2166/wp.2017.003.

96. Wu, L., Wang, M. and Avishek, K. (2021). Trans-regional rice supply paradigm reveals unsustainable water use in China. *Water Policy*, *23*(3), 783–800. https://doi.org/10.2166/wp.2021.168.

97. Yahaya, A.S., Javaid, N., Alzahrani, F.A., Rehman, A., Ullah, I., Shahid, A. and Shafiq, M. (2020). Blockchain based sustainable local energy trading considering home energy management and demurrage mechanism. *Sustainability*, *12*(8), 3385.

98. Yli-Huumo, J., Ko, D., Choi, S., Park, S. and Smolander, K. (2016). Where is current research on blockchain technology?: A systematic review. *PloS One*,–(10), e0163477.

99. Youkhana, E. and Laube, W. (2009). Virtual water trade: A realistic policy option for the countries of the Volta Basin in West Africa? *Water Policy*, *11*(5), 569–581. https://doi.org/10.2166/wp.2009.087.

101. Zeng, X. T., Li, Y.P., Huang, G.H. and Liu, J. (2015). A two-stage interval-stochastic water trading model for allocating water resources of Kaidu-Kongque River in northwestern China. *Journal of Hydroinformatics*, *17*(4), 551–569. https://doi.org/10. 2166/hydro.2015.090.

102. Zhang, W., Mao, H., Yin, H. and Guo, X. (2018). A pricing model for water rights trading between agricultural and industrial water users in China. *Journal of Water Supply: Research and Technology-Aqua*, *67*(4), 347–356. https://doi.org/10.2166/ aqua.2018.142.

103. Zhang, W., Tan, L., Yin, H. and Guo, X. (2019). Study on the price of water rights trading between agriculture and industry based on emergy theory. *Water Supply*, *19*(7), 2044–2053. https://doi.org/10.2166/ws.2019.083.

104. Zhao, J., Liu, J., Ni, H., Zhao, Y., Han, Y. and Diao, Z. (2021). Virtual water and its effect in Huaihe River basin: An analysis using a dynamic multi-region computable general equilibrium model. *Water Supply*, *21*(3), 1090–1101. https://doi.org/10.2166/ ws.2020.373.

105. Zheng, Z., Xie, S., Dai, H.N., Chen, X. and Wang, H. (2018). Blockchain challenges and opportunities: A survey. *International Journal of Web and Grid Services*, *14*(4), 352–375.

14 Application of Blockchain Technology in Climate Change

Saeid Eslamian[1] and Yaser Sabzevari[2*]

Climate change presents complex challenges requiring transparent, efficient, and verifiable mechanisms for monitoring, reporting, and managing environmental impacts. Blockchain technology, with its decentralized, immutable, and transparent ledger system, offers innovative solutions to enhance climate action and sustainability initiatives. This chapter explores the application of blockchain in addressing climate change, including carbon credit tracking, renewable energy trading, climate finance, supply chain emissions monitoring, and decentralized environmental governance. Case studies and pilot projects are examined to illustrate how blockchain enables secure, real-time data sharing, improves accountability, and fosters stakeholder collaboration in climate mitigation and adaptation strategies. The chapter also discusses challenges such as scalability, energy consumption, regulatory compliance, and integration with existing environmental frameworks. By bridging the fields of digital technology and climate action, this chapter provides insights into how blockchain can support transparent, resilient, and efficient climate solutions.

14.1 Introduction

Global warming is a contentious topic globally, driving additional climate alterations. The ecosystem will suffer harm and living organisms will perish due to these changes, which is why blockchain and other technologies are being used to assist humans. Certain climate changes resulting from the modernization of societies are detrimental to the environment and living organisms. Technologies like blockchain are actively seeking ways to avoid this occurrence. The rise of digital currencies like Bitcoin has led to increased interest in Blockchain technology. Utilizing blockchain technology to a larger extent could open up opportunities for various other advantages, such as its application for environmental protection. Blockchain has the capability to monitor various things such as supply chains and decreasing greenhouse gas emissions, allowing for the trading of green energy, and converting plastic waste into money. Several new technology initiatives and startups are emerging in this sector. Blockchain is a technology for managing information

[1,2] Department of Water Science and Engineering, College of Agriculture, Isfahan University of Technology, Isfahan, Iran.
* Corresponding author: yasersabzevari1511@gmail.com

DOI: 10.1201/9781003686026-14

which utilizes a decentralized processing structure to enhance the security of data transactions from unauthorized alterations, breaches, and manipulation. Blockchain technology has the ability to safeguard the sender and recipient of data from unauthorized identity theft. When ransomware groups request ransom payment in cryptocurrency, the significance of blockchain technology is highlighted. Supporters of digital currencies appreciate the elimination of intermediaries and their expenses in the transfer of value through blockchain technology, resulting in a decentralized system without any oversight from external sources. The idea of blockchain was first presented in 1991 by W. Scott Stornetta and Stuart Haber. The two physicists were working on a chain of secure cryptographic blocks. The implementation of blockchain made it impossible to tamper with documents. The most important achievement of blockchain was the launch of Bitcoin. In 2009, an unknown person named "Satoshi Nakamoto" published a document in which he talked about this digital currency. Bitcoin's white paper explained blockchain's ability to increase digital trust. Processing in the blockchain is so decentralized that it will not be possible to apply any control to a single entity. This formed the basis for the launch of Bitcoin in 2010. The first currency created on the blockchain is called Bitcoin. Blockchain technology has other potential applications beyond digital currencies. Basically, blockchain can be used for any process that requires high security recorded records of data (Eslamian and Eslamian, 2025). Technology has not always been the weather's best friend. From the industrial revolution that changed our relationship with the planet forever, to modern devices that see us constantly consuming energy and adding to mountains of e-waste. In the last decade, one of the biggest human problems has been climate change, which has not yet been solved.

It seems that every human step in the future puts the earth in more danger. But the fourth industrial revolution can reduce this trend. Blockchain is at the forefront of this movement. This ostensibly incorruptible public ledger enables transparent, global digital transactions worldwide, and can power beneficial climate action.

Peer-to-peer exchanges and direct product tracking reduce costs and increase trust. The technology could also be used to reward beneficial climate action. Even the United Nations supports technologies being developed to help monitor countries' progress in reducing greenhouse gas emissions. Imagine you are the food manager at a food service company and you want to source your products sustainably.

How can you believe your supply chain? Do you believe the papers that come with the item? Ought you conduct normal and expensive reviews of your provider to check in the event that they do what they say they do? Blockchain innovation offers arrangements to these natural issues. WWF (World Wide Fund for Nature) is working with blockchain frameworks designer ConsenSys and Ocean Journey to test and actualize a blockchain following instrument for the Pacific fish industry.

Currently, scanning a tuna package with a smartphone will provide a comprehensive overview of the tuna's journey from being caught to being cooked. The blockchain ledger verifies if the fish was legally caught and not in cruel conditions. Once the fish is captured on the fishing boat, it is tagged with RFID (Radio Frequency Identification) and then given QR codes at every step from processing to the end retail. In other places, the advantages of blockchain for protecting wildlife in Africa are being examined. Three key areas where blockchain proves beneficial are environmental monitoring effectiveness, cost efficiency, and fair profit distribution.

Emerging technologies and their role in aggravating the crisis of climate change and vice versa, their role in reducing the process of these changes have always been considered in the discourses related to how to deal with the global crisis of climate change. Therefore, until today, the focus of policies to deal with climate change has been on ideas and proposals using technological tools. These proposals are called "land engineering" or "climate engineering". In recent years, many countries around the world have turned to emerging technologies in the best possible implementation of the Paris Climate Change Agreement and in an effort to reduce carbon-generating activities. In this regard, in 2017, at the United Nations Conference on Climate Change, the importance of blockchain technology in dealing with climate change at the global level was emphasized.

Each modern innovation has focal points and impediments, the focal points of which move forward our well-being and life, and the drawbacks of which can cause numerous challenges. Amid the final 10 a long time until presently, the worldwide climate emergency has ended up a hot point for talk among governments; but nowadays, the sort of talk has changed and an agreement has been reached. This agreement has driven the dialog regarding climate change and how to halt it or decrease it.

In reaching this goal, there are two important historical moments: the selection of the Joined together Countries Maintainable Advancement Objectives (SDGs), whose mission could be a diagram for accomplishing distant better; a much better; a higher; a stronger; an improved", a distant better and more feasible future for all, and the Paris Ascension to combat climate alter, which is a worldwide understanding in this field. Nearly most nations of the world marked this universal ascension six a long time back in 2015.

Climate change and blockchain has gotten to be a wrangle about how to battle the worldwide climate emergency and rising advances, and their part in this. In 2017, the Joined together Countries System Tradition on Climate Alter (UNFCCC) highlighted the significance of blockchain innovation and climate change in making a difference to combat climate change universally. The UNFCCC Secretariat has particularly said these terms:

"In particular, transparency, cost-effectiveness, and productivity benefits, which in turn may lead to greater integration of stakeholders and increased production of global public goods, are currently considered to be the main potential benefits of blockchain technology."

Blockchain could be a capable device that can essentially move forward straightforwardness, responsibility and traceability of greenhouse gas outflows. This innovation makes a different company to supply more precise, dependable, standardized, and available data for most carbon emanations.

Blockchain is a platform consisting of interconnected components that form a distributed ledger and facilitate the process of recording transactions and tracking assets in a business network (Abbasi, 2017). In fact, this technology is used to create immutable and erasable data banks, and this feature makes it the heart of technologies such as cryptocurrencies and smart contracts.

Cryptocurrency, which has been one of the popular phenomena of the 21st century, is the digital currency that can be used as a means of exchange, payment, financial aid, and remittance, which is designed to bypass the limitations of the traditional banking system. These virtual currencies have become very popular as a means of exchange and based on blockchain creation, and due to their ideal decentralization and speed in exchange and high profits. This attraction has made the people of the world invest in these currencies more and more every day. The increased demand for investing in digital currencies has made the computer systems used to mine digital currencies increasingly energy-intensive and costly.

Jointly, countries' specialists accept that cryptocurrencies and blockchain innovation can play a vital part in economic improvement and progressing the state of natural checking. One of the foremost valuable perspectives of advanced monetary forms, as distant as the Joined together Countries is concerned, is its straightforwardness. This innovation is safe to control and mutilate data, so it can give a solid and straightforward history of exchanges. This issue is exceptionally imperative in regions with powerless education and tall levels of debasement.

The potential of blockchain in environmental protection has been tested in a number of projects of the United Nations and other organizations. According to the DTU UNEP partnership program (a collaboration between UNEP, the Technical University of Denmark, and the Danish Ministry of Foreign Affairs), the three main areas where blockchain can accelerate climate action are: transparency, climate finance, and clean energy markets. The European Commission also considers the use of innovations in blockchain technology to be in line with dealing with climate change due to transparency, accountability, and the ability to track greenhouse gas emissions.

Despite all these hopes, according to a study published in late 2018 by the journal *Nature Climate Change* (a journal dedicated to publishing the most important and cutting-edge research on the nature, underlying causes or effects of global climate change and its consequences on the economy Dedicated), a cryptocurrency like Bitcoin in 2017 alone had approximately sixty-nine million metric tons (sixty-nine megatons) of carbon dioxide (CO2) emissions (Thomson, 2020).

Also recently, a study at the University of Cambridge estimated that Bitcoin mining consumes 121.36 terawatt hours of energy per year, and this amount of consumption is more than the combined consumption of Facebook, Microsoft, Apple, and Google. And because of the computational process behind how blockchain works, this energy consumption isn't going down anytime soon. According to Cambridge University, only 39% of this energy comes from renewable sources and hydropower, which can have devastating effects on ecosystems and biodiversity.

For example, after China's mining ban in spring 2021, the share of renewables powering the grid has dropped from around 41.6–25.1%. Miners already had access to a significant amount of renewable energy (i.e., hydropower during the wet season in the summer months) while they were still in China, but when they were forced to move to countries like the United States and Kazakhstan, access to these resources was also lost; because these countries currently mainly provide electricity for Bitcoin miners from coal or gas, which has increased the carbon footprint of electricity consumed for Bitcoin mining.

Therefore, the main environmental concerns caused by the mining of digital currencies are: 1) the high amount of electricity required to provide energy for complex algorithms; and 2) non-renewable energy sources used for extraction. A large amount of coal-based electricity enters the Bitcoin mining network, and coal-based electricity has high carbon emissions. Additionally, it is likely that with cryptocurrency prices fluctuating, miners will continue to search for cheaper energy. The competition for earning money also encourages countries to produce dirty energy to attract more miners and boost economic activity (Thomson, 2020).

For this reason, since early 2021, concerns about Bitcoin's carbon emissions have become a controversial topic, prompting some foreign media to address the issue of Bitcoin's energy consumption and carbon emissions. And the importance of high energy consumption of digital currencies has attracted many discussions in scientific circles and among the general public.

In addition to the increased CO2 emissions associated with cryptocurrency mining activity, it is also important to consider the environmental impact of the continued consumption and disposal of e-waste. Currently, the Bitcoin mining activity alone generates, on average, approximately eleven kilotons of e-waste per year. That number is likely to rise when you add up the e-waste emitted by every cryptocurrency except Bitcoin. Each Bitcoin transaction generates approximately 90 grams of e-waste, which is basically the size of two golf balls, and while that may seem like a small amount, there are over 366,000 Bitcoin transactions per day (Thomson, 2020).

Another reason for the amazing sum of e-waste produced by cryptocurrencies is the reality that the mining rigs utilized by most diggers wear out rapidly, frequently inside fair two a long time, and cannot be successfully repurposed for anything other than Extraction utilized. As cryptocurrency programming makes it harder and harder to mine each extra coin, mineworkers are always stuck in a never-ending cycle whereby they must always overhaul their machines in order to remain competitive. This cycle gives a solid motivation for mineworkers to disregard natural concerns for financial pick up (Thomson, 2020).

Therefore, despite the belief that decentralized technologies such as blockchain technology have the potential to contribute to the realization of sustainable development goals with their advantages such as transparency and immutability, based on research, it seems that Cryptocurrency mining has a significant impact on global warming and climate change, and the extremely high energy consumption associated with this technology is one of the main obstacles to overcome.

If in the ecosystem of blockchain and digital currencies, most of the required energy used to extract cryptocurrencies is provided from renewable sources, it will move towards a cleaner and greener future. Therefore, by using clean resources, cryptocurrencies and blockchain can work in line with environmental interests, and in a society that uses only clean energy sources, this high energy demand does not lead to an increase in harmful gas emissions. However, according to the 2021 Global Crypto Asset Measurement Study, the proportion of renewable sources in total energy consumption for cryptocurrency mining is only around 39%. But this figure alone does not show the environmental impact of digital currencies. Even the share of renewable energy, which is mainly provided by hydroelectricity production, has been shown to disrupt aquatic ecosystems and harm biodiversity. Additionally, many dams with ineffective ecological planning have higher carbon emissions than fossil fuels. These findings show that the estimated amount of carbon emissions of digital currencies does not cover the full range of its effects.

A host of industry players are working on solutions to address this problem. For example, the Ethereum Foundation, the organization behind the Ethereum digital currency, is working on a new method of verifying transactions, and believes that by changing the method to a different method (called Proof of Stake, or PoS), the energy cost of each transaction can be reduced to 99.95%.

The non-profit Blockchain for Climate Foundation also announced the launch of its Bitmo platform at Cop26, which aims to help countries achieve their climate goals. Countries that have signed the Paris Agreement can implement their commitments to reduce greenhouse gas emissions by purchasing Bitmos. The organization claims that the platform provides the necessary accounting and exchange infrastructure under Article 6 of the Paris Agreement, linking the need to reduce greenhouse gas emissions with the best opportunities for climate action around the world, and is able to meet the organization's future requirements, i.e., satisfy the nations. In other words, it is claimed that this platform will help humanity to deal with the climate crisis and at the same time strengthen global cooperation.

At the same time, many players in this industry are looking to ensure that any energy consumed in this industry is completely carbon-free and from renewable sources. Therefore, by using these resources, cryptocurrencies and blockchain can work in line with environmental interests (https://www.weforum.org/agenda/2021/06/how-blockchain-and-cryptocurrencies-can-help-build-a-greener-future). And of course, as renewable options such as solar power become more efficient and thus useful for mining, Bitcoin can be a serious incentive for miners to build and develop these technologies.

14.2 Smart Contracts and Complete Transparency

With the beginning of misuse of the cryptocurrency within the world, Ethereum, a programmable layer was included to the blockchain. This layer was called shrewd contracts. This implies that a program can be propelled from the blockchain and something happens within the physical world.

For case, envision that a wastewater administration company has sensors connected to channels and a wastewater treatment plant. Already, individuals studied the sensors and entered the information into Exceed expectations, which was at that point sent to the director.

So in the event that the contamination parameter surpassed the legitimate level, the alert would sound. Whereas it could be a couple of days or weeks afterward, the supervisor will notice this issue and take action for it. Of course, it ought to be famous that you just can turn off the caution and make a fake caution so that no one will take note of the contamination occurrence.

But in a blockchain-based system that is linked to sensors, an alarm will repeatedly sound and alert the concerned person and directly issue an acceptable to the polluter. Such technology has made it impossible for anyone to commit fraud in this field. The reason for using blockchain technology in this wastewater management system is that it is cheap and flexible compared to the old method.

14.3 Impact of Supply Chain on Climate

Following the supply chain behind "green" items is incomprehensible. But blockchain-based apps like Foodtrax permit buyers to track items from source to store rack. Multinational grocery store chains as of now utilize a comparative framework, which suggests clients can filter bundling for nitty gritty data of almost the product's source and fabricating forms.

In the meantime, WWF's blockchain supply chain following venture is making a difference. Customers maintain a strategic distance from items connected to illicit fishing and human rights mishandling by following Pacific capture from transport to general stores around the world. This innovation can grow the conceivable outcomes of carbon exchanging as well as reinforce climate assurance approaches. It moreover makes a difference to decrease nursery gas emanations by calculating nursery gas outflows and dodging information control.

IBM is working with the Beijing-based Vitality Blockchain Lab to extend straightforwardness in China's carbon exchanging, but blockchain can too make carbon exchanging more available. "Collecting and confirming information without blockchain can fetch up to $40,000 (€ 33,700), putting little companies out of trade," said Matthias Gelber, chief natural officer at Singapore-based Unused Period Vitality.

14.4 Facilitating Clean Energy Trade

Blockchain innovation is additionally utilized to create renewable vitality commerce. Customers can exchange and purchase and offer renewable vitality with each other utilizing tokens or advanced resources that speak to a certain sum of energy production.

On the off chance that you have got sun-based boards on your roof, or in the event that you claim an electric vehicle that can exchange power from its battery to the framework, blockchain can offer assistance. Blockchain and climate alter and, moreover, communicate well in this choice.

14.5 Financing Climate Projects

Financing environmental projects by banks and loans can bring many challenges. The new loaning organization called DeFi or Decentralized Finance can be used to generate capital for environmental projects. DeFi projects have been around for a few years now, but 2020 has seen a huge surge in popularity.

14.6 Better Tracking and Reporting of Greenhouse Gas Emission Reduction

As discussed in previous sections, blockchain technology can ensure greater transparency about pollution and emissions and make it easier to track and report emissions reductions, including addressing data quality issues.

14.7 Problem Solver

Until recently, the development of smart contracts was hindered by the incapability of blockchains to meaningfully interrelate with real-world state information, counting the environment. But in recent years, oracles (the entities that can supply evidence about our world on the blockchain) are finally making the blockchain to fight climate change.

For illustration, these prophets nowadays thrust agrarian datasets straightforwardly onto blockchains, meaning that shrewd contract designers can use information on edit yields, soil quality, climate reports, carbon emanations, and more on the blockchain.

Nowadays, as ecological information sets, for example, weather designs are kept on the blockchain, designers are beginning to crop a wide variety of smart agreement submissions, the vital blockchain and climate change options being:

Revival of Agriculture

One of the foremost critical ways blockchain can spur individuals to battle climate change is by empowering renewable horticulture programs. These programs incorporate endeavors to empower communities around the world to diminish their carbon emissions through sustainable rural land-use practices. A part of these programs is planting trees and securing common asset lands.

14.8 Conscious Consumption

Blockchain can authorize people and alert governments of environmental issues. For example, if somebody positively initiates a reforestation smart agreement, they can receive payments in the form of carbon credits.

This credit is in the form of tax discounts and such things. Blockchain can offer energy-conscious consumers more options and discounts.

Hazard Attention with Product Insurance
Climate change is as of now changing climate designs around the world. Ranchers around the world, the larger part of whom are uninsured, are especially defenseless to these climate changes. A few of the foremost imperative of these climate changes are overwhelming downpours, dry seasons, and surges.

Despite the fact that 75% of rural ranches within the world are uninsured, conventional protection companies have so far distanced themselves to supply satisfactory protection for ranchers in developing countries. As a result, most of them have an awfully unsteady life.

Terrible climate conditions can devastate the monetary well-being of agriculturists and lead to the evacuating of families and surrender of agrarian arrival. On the off-chance that the pace of climate change proceeds to quicken, there's no question that the coming precariousness will influence all agriculturists around the world.

Nowadays, with the advancement of blockchain innovation that can screen changing climate conditions through predictions, it may be an arrangement for individuals in rustic zones managing with climate alter and common fiascos. Protection ventures such as Arbol and Etheric are cases of shrewd edit protections advertised to ranchers around the world.

By means of blockchain technology, planters can also protect their small plans. Also, through this technology, it is possible to be informed about the weather situation in a moment, and reduce the amount of damage caused to agricultural fields by timely notification of the accident.

14.9 Review Studies

UNEP (2019) assessed climate change challenges and blockchain opportunities. This study explores the intersection of climate change challenges and blockchain technology. It outlines how blockchain's transparency, immutability, and decentralization offer innovative solutions to various climate-related issues, such as carbon tracking, renewable energy integration, and supply chain management. The paper reviews current applications and case studies that demonstrate blockchain's potential to enhance data accuracy, improve climate finance mechanisms, and facilitate more efficient environmental policies. Ultimately, it argues that while blockchain presents significant opportunities, its implementation must be carefully managed to address scalability, regulatory, and technical challenges.

Truby et al. (2022) investigates the environmental and health impacts of blockchain technologies, particularly focusing on Bitcoin mining and NFTs (non-fungible tokens). It critically examines the substantial carbon footprint associated with these technologies and their contribution to climate change and public health issues. The article proposes various policy interventions aimed at mitigating these impacts, such as regulatory frameworks to limit energy consumption, incentivizing greener mining practices, and promoting sustainable alternatives. It emphasizes the need for a balanced approach that addresses both technological advancements and environmental sustainability to achieve net-zero goals.

Pham et al. (2024) explores how blockchain technology can bolster the resilience of smallholder farmers in the face of climate change. It synthesizes recent research on the application of blockchain to agricultural practices, emphasizing its potential to improve supply chain transparency, enhance access to financing, and enable precise climate risk management. The article discusses several use cases, including blockchain's role in providing real-time weather data, verifying sustainable practices, and facilitating direct market access for smallholder farmers. It also examines challenges such as technology adoption barriers, infrastructure limitations, and the need for supportive policies. The review concludes that while blockchain offers promising tools to address climate-related challenges, successful integration requires targeted efforts to overcome existing obstacles and ensure equitable access for smallholder farmers.

Oqbi et al. (2024) examines how blockchain technology can enhance the efficiency and transparency of CCUS (carbon capture, utilization, and storage) supply chains. It highlights the potential of blockchain to improve tracking and verification of carbon capture processes, streamline data management, and ensure the integrity of emissions reductions claims. The review discusses various blockchain applications, such as smart contracts for monitoring CCUS operations, and decentralized ledgers for recording carbon credits. It also addresses challenges including integration with existing infrastructure and scalability. The article concludes that blockchain holds significant promise for advancing CCUS technologies and contributing to climate change mitigation efforts, but successful implementation will require overcoming technical and regulatory hurdles.

Padmapriya and Sujatha (2023) explores how blockchain technology influences urban environments and climate change mitigation efforts. It assesses both positive and negative impacts, including blockchain's potential to enhance transparency in environmental monitoring, optimize energy consumption, and facilitate more efficient waste management systems. The article also examines challenges such as the high energy consumption associated with certain blockchain applications and potential unintended environmental consequences. It highlights case studies of blockchain initiatives aimed at urban sustainability, discussing their effectiveness and scalability. The review concludes that while blockchain offers innovative solutions for improving urban environmental management, its deployment must be carefully managed to mitigate its own environmental footprint and achieve meaningful climate change outcomes.

Harris (2018) presents insights from leading experts on the intersection of blockchain technology and climate change. The article captures a series of interviews or discussions with thought leaders and innovators in both fields, highlighting their perspectives on how blockchain can address climate challenges. Key themes include the role of blockchain in enhancing transparency in carbon markets, improving data accuracy for climate science, and supporting sustainable practices across various industries. The article delves into specific applications such as blockchain's use in tracking carbon credits, managing renewable energy certificates, and enabling decentralized climate initiatives. It also addresses potential limitations and critiques, such as concerns about blockchain's energy consumption and its effectiveness in driving large-scale climate action. Overall, the review emphasizes the potential of blockchain technology to contribute to climate change mitigation while acknowledging the need for careful implementation and continued dialogue among stakeholders to maximize its benefits and address its drawbacks.

Mzoughi et al. (2024) explores the relationship between blockchain markets and green finance, focusing on how blockchain can influence environmental investments and outcomes. It examines how blockchain technologies are being integrated into green finance markets to enhance transparency, efficiency, and accountability in environmental investments. The article reviews case studies where blockchain has been used to track green bonds, manage carbon credits, and verify sustainable investment projects. It also discusses the environmental impacts of blockchain itself, including energy consumption concerns related to certain consensus mechanisms. The review highlights the dual nature of blockchain's influence—its potential to drive positive environmental change through improved financial mechanisms and the need to address its own environmental footprint to achieve sustainable outcomes.

Popkova et al. (2023) investigates how blockchain technology supports the transition to clean energy and promotes sustainability. It explores applications of blockchain in managing renewable energy resources, facilitating decentralized energy trading, and enhancing transparency in environmental reporting. The article discusses how blockchain can optimize energy grids, track renewable energy production, and verify sustainability claims. It also addresses challenges such as blockchain's energy consumption and integration with existing energy infrastructure. The review concludes that while blockchain offers promising tools for accelerating the clean energy transition, careful consideration of its environmental impact and strategic implementation are crucial for achieving long-term sustainability goals.

Truby et al. (2022) analyzes the environmental and health impacts associated with blockchain technologies, specifically focusing on NFTs and Bitcoin. It evaluates the substantial carbon footprint and energy consumption of these blockchain applications and their broader implications for climate change and public health. The article discusses various policy interventions aimed at mitigating these impacts, such as regulating energy-intensive blockchain practices, promoting the use of renewable energy, and setting industry standards for carbon offsets. It also explores potential strategies for achieving net-zero emissions in the blockchain sector. The review highlights the urgency of implementing effective policies to balance technological innovation with environmental and health considerations.

Hull et al. (2021) critically examines the role of blockchain technology in the realm of international climate governance. It explores both the optimistic claims and potential drawbacks associated with using blockchain to address climate change. The article includes:

Promises of Blockchain in Climate Governance: The article discusses how blockchain is touted as a tool for enhancing transparency and accountability in climate policies, tracking carbon emissions, and verifying climate finance flows. It highlights specific blockchain initiatives aimed at improving data integrity and fostering more effective international climate agreements.

Perils and Criticisms: The review also addresses criticisms related to blockchain's energy consumption, scalability issues, and the risk of creating new forms of inequality or inefficiency. It examines the challenges of integrating blockchain solutions into existing climate governance frameworks and the potential for these technologies to exacerbate existing problems rather than solve them.

Case Studies and Discourses: The article provides case studies and examples of blockchain projects within international climate politics, analyzing their impact and effectiveness. It discusses the discourse surrounding these technologies, including both advocacy and skepticism within the climate policy community.

Future Directions: The review concludes with reflections on the future of blockchain in climate governance, emphasizing the need for careful consideration of its role and impact. It calls for a balanced approach that leverages blockchain's strengths while addressing its weaknesses through thoughtful policy design and implementation.

Overall, the article provides a nuanced perspective on the integration of blockchain into climate governance, stressing the importance of critical evaluation and adaptive strategies in harnessing its potential benefits while mitigating its risks.

Luo et al. (2024) explores the use of blockchain technology to enhance the management of carbon emissions throughout the lifecycle of buildings. It focuses on a blockchain-based system designed to improve the monitoring, reporting, and verification (MRV) of carbon footprints in the construction and operation of buildings. Key aspects of the article:

Blockchain-Enabled MRV System: The article details how blockchain technology can be applied to create a traceable and immutable record of carbon emissions associated with building materials, construction processes, and building operations. This system aims to enhance the accuracy and transparency of carbon reporting and verification.

Application in Hong Kong: The review highlights a specific case study in Hong Kong, where this blockchain-based information management system has been implemented. It discusses the system's design, features, and the outcomes of its application in the local context, including any challenges encountered and benefits realized.

Advantages of Blockchain: The article emphasizes the advantages of using blockchain, such as the ability to create tamper-proof records, ensure data integrity, and facilitate real-time access to carbon data. It also covers how the technology supports compliance with sustainability regulations and standards.

Challenges and Limitations: The review addresses potential challenges, including technical limitations, the need for stakeholder buy-in, and the integration of blockchain with existing building management systems. It also discusses the scalability of the system and its adaptability to different regulatory environments.

Future Prospects: The article concludes with reflections on the future of blockchain in building life cycle management. It suggests potential improvements and expansions of the system, including broader adoption and integration with other digital technologies to further enhance sustainability efforts in the building sector.

Overall, the review provides a comprehensive analysis of how blockchain can transform carbon monitoring and management in the building industry, with practical insights from a real-world application in Hong Kong.

Dorfleitner et al. (2021) provides an in-depth analysis of how blockchain technology is being utilized to address climate protection efforts worldwide. The article aims to empirically investigate various blockchain applications and their effectiveness in promoting environmental sustainability and combating climate change. Here's a summary of the key elements covered in the study:

Overview of Blockchain Applications: The article begins with a general overview of blockchain technology and its potential applications in climate protection. It explains the fundamental features of blockchain—such as decentralization, immutability, and transparency—and how these features can be leveraged to support climate-related initiatives.

Empirical Evidence: The core of the review involves a global empirical investigation into different blockchain projects and initiatives aimed at climate protection. This includes analyzing data from various case studies and real-world implementations, such as:

Carbon Credit Trading: Blockchain-based platforms for trading carbon credits that enhance transparency and traceability in carbon markets.

Supply Chain Tracking: Systems that use blockchain to track and verify the sustainability of supply chains, including the sourcing and environmental impact of raw materials.

Renewable Energy Certification: Projects that use blockchain to certify and validate renewable energy production and consumption, ensuring that energy claims are accurate and verifiable.

Benefits and Challenges: The article discusses the benefits of using blockchain for climate protection, such as improved transparency, reduced fraud, and enhanced accountability. It also addresses challenges like high energy consumption, technical complexity, and the need for widespread adoption and integration with existing systems.

Future Directions: The review concludes with recommendations for future research and development in blockchain applications for climate protection. It suggests areas for improvement, such as increasing blockchain's scalability, reducing its environmental impact, and fostering collaboration between stakeholders to maximize the technology's potential.

Overall, the article provides a comprehensive empirical assessment of how blockchain technology is being used to support climate protection efforts globally, highlighting both the opportunities and obstacles associated with its implementation.

14.10 Conclusion

It is worth noting that cryptocurrencies are still in the early stages of development, and there are still many technical and political challenges to overcome. It seems necessary that countries and international organizations cooperate in this field and use the best practices and learn from each other. One of the authors of UNEP's recent foresight report on blockchain applications says: "The UN should continue to explore the blockchain space. The more it is tested, the more information is gained about this technology. This will improve our knowledge at the UN level about blockchain, our understanding of the environmental and social implications of mining operations, and increase our chances of dealing with the problems this technology may create in the future. If this technology is to really have a positive impact on climate crises, more technical research is needed, as well as more international dialogue involving experts, scientists, and policymakers".

Regarding the connection between blockchain technology and climate change, the most vital question that arises is whether the use of blockchain has specific aids to reduce climate change or not? According to this study, the answer to this question is yes. There are numerous ways blockchain can help climate change mitigation efforts.

Furthermore, these approaches are used in many countries today. As Alexandre Gellert, head of the UN Climate Change Mitigation Programme, says: "As countries, regions, cities, and businesses work to rapidly implement the Paris Climate Change Agreement, they must use all the innovative and advanced technologies available. Blockchain can help engage more stakeholders and create trust and new solutions to combat climate change".

The most important application of blockchain technology and climate change is:

- Improve carbon emissions
- Facilitating clean energy trade
- Increasing financial aid for environmental projects
- Better tracking and reporting of greenhouse gas emission reduction

References

1. Dorfleitner, G., Muck, F. and Scheckenbach, I. (2021). Blockchain applications for climate protection: A global empirical investigation. *Renewable and Sustainable Energy Reviews*, *149*, 111378.
2. Eslamian, S. and Eslamian, F. (2025). *Blockchain Technology in Water and Environmental Systems*. CRC Press, Taylor and Francis, USA, p. 282.
3. Harris, A. (2018). A conversation with masterminds in blockchain and climate change. In: *Transforming Climate Finance and Green Investment with Blockchains*. Academic Press, pp. 15–22.
4. Hull, J., Gupta, A. and Kloppenburg, S. (2021). Interrogating the promises and perils of climate cryptogovernance: Blockchain discourses in international climate politics. *Earth System Governance*, *9*, 100117.
5. Javad Abbasi (2017). *Blockchain: Getting to Know the Basic Concepts*. Institute of Books of Mehraban Nashr, p. 17.
6. Luo, Y., Shen, J., Liang, H., Sun, L. and Dong, L. (2024). Supporting building life cycle carbon monitoring, reporting and verification: A traceable and immutable blockchain-empowered information management system and application in Hong Kong. *Resources, Conservation, and Recycling*, *208*, 107736.
7. Mzoughi, H., Amar, A. B., Guesmi, K. and Benkraiem, R. (2024). Blockchain markets, green finance investments, and environmental impacts. *Research in International Business and Finance*, *69*, 102249.
8. Oqbi, M.Y. and Al-Mohannadi, D.M. (2024). Supporting Carbon Capture, Utilization, and Storage Supply Chains (CCUS) with Blockchain Technology: A Sustainable Solution for Climate Change Mitigation. In: *Computer Aided Chemical Engineering* (Vol. 53). Elsevier, pp. 3163–3168.
9. Padmapriya, V. and Sujatha, D.N. (2023). Impact of blockchain technology on the urban environment and climate change. In: *Green Blockchain Technology for Sustainable Smart Cities*. Elsevier, pp. 97–117.
10. Pham, H.C., Truong, H.Q., Binh, A.D.T. and Yap, T.L. (2024). Enhancing Agriculture Resilience of Smallholder Farming to Climate Change through Blockchain Technology Adoption. MDPI.
11. Popkova, E.G., Bogoviz, A.V., Lobova, S.V., Vovchenko, N.G. and Sergi, B.S. (2023). Blockchain, sustainability, and clean energy transition. *Global Transitions*, *5*, 64–78.
12. Thomson, J.C. (2020). Tragedy of the energy commons: How government regulation can help mitigate the environmental and public health consequences of cryptocurrency mining. *Seattle J. Tech. Env't & Innovation L.*, *11*, 77.
13. Truby, J., Brown, R.D., Dahdal, A. and Ibrahim, I. (2022). Blockchain, climate damage, and death: Policy interventions to reduce the carbon emissions, mortality, and net-zero implications of non-fungible tokens and Bitcoin. *Energy Research and Social Science*, *88*, 102499.
14. UNEP (2019). *Climate Change Challenges and Blockchain Opportunities*. Retrieved 5 September, 2020.

15 | A Conclusion to Block Chain Technique and Future Works

Yaser Sabzevari[1*] and Saeid Eslamian[2]

Blockchain technology has emerged as a transformative tool for decentralized, transparent, and secure digital systems across multiple sectors. This concluding chapter synthesizes the key concepts, methodologies, and applications discussed in preceding chapters, highlighting blockchain's core features, decentralization, immutability, consensus mechanisms, and smart contract programmability. The chapter reflects on the current state of blockchain adoption, successes, and challenges, including scalability, interoperability, energy consumption, and regulatory considerations. Furthermore, it outlines potential directions for future research and development, such as hybrid consensus models, integration with artificial intelligence and Internet of Things (IoT), post-quantum cryptography, privacy-preserving techniques, and innovative governance frameworks. By connecting lessons learned with emerging trends, this chapter provides a forward-looking perspective on how blockchain can continue to evolve and impact technology, business, and society.

15.1 Introduction

Today, computerized changes have become an important part of improving frameworks and procedures. Emerging technologies such as blockchain, big data, artificial intelligence (AI), the Internet of Things (IoT), and machine learning (ML) are rapidly replacing outdated systems, propelling societal advancement at an accelerated pace (Abu-Amara et al., 2022). A blockchain is a system of digital records that are connected and keep changing over time, using agreed-upon rules (Angara and Saripalle, 2022). Each block has a special code that links it to the last transaction, a time stamp, and details about the exchange. Blockchain is a new and flexible technology that can be used in different ways (Satilmsioglu et al., 2022). The benefit of blockchain technology is that it helps make better and new organized programs and turns resources into digital form. Blockchain, also known as a distributed ledger, is an important new technology that helps stop mistakes in organizations (Mia et al., 2022). It allows businesses and groups to manage their processes more efficiently. So, in water and farming services, blockchain can help with support, checking things, and making rules using computers. Blockchain can

[1,2] Department of Water Science and Engineering, College of Agriculture, Isfahan University of Technology, Isfahan, Iran.
* Corresponding author: yasersabzevari1511@gmail.com

DOI: 10.1201/9781003686026-15

also be effective in creating a confidential environment by representing reality. The underground layer of water, which people can take water from, has been used too much by industries and farmers for over ten years. This has caused water levels to drop significantly. If this keeps happening, the water level will drop soon and will be very low, like a breakwater. Now, we can track water levels using Blockchain and the IoT. Blockchain platforms aim to help track water levels quickly, thereby enabling farmers to use water resources more wisely (Eslamian and Eslamian, 2025).

15.2 Methodology

Blockchain technology will use IoT sensors, which send water extraction information to satellites (Sriyono, 2020). Finally, all this information is stored in real-time in the blockchain ledger and smart blockchain compatible with contacts. The blockchain platform will be available for use by all water consumers such as farmers and investors. Since the blockchain will have a precise tracking system, this system is also used to issue groundwater shares (Yasuno et al., 2022). These shares can be bought and traded by regional activists. For example, TFT has established the Groundwater Sustainability Agency in the North Delta in the United States. This agency brings together smaller organizations for the sustainable use of groundwater. This collaborative effort seeks to establish an equitable system for monitoring water levels with ease. Because the system will operate in real-time, the information will be accurate and clear.

Digital technologies such as blockchain provide simpler solutions to complex problems. Therefore, this technology initially assesses the complexity of the system. A research team is studying how transactions happen in New South Wales step by step. They are looking closely at water trading rules to see if blockchain technology can cover everything needed. Address deficiencies while ensuring compliance with all applicable laws and policies.

On the other hand, throughout the last three years, RVP has initiated research initiatives aimed at assisting power companies in discovering technological solutions for sustainable water resource management. These projects, such as The Future of Urban Water and Digital Tide, use different methods, from creating maps to making storyboards. So, the information collected in this system will be very helpful for people who build water systems or rely on predicting future water levels for their jobs. Also, this system can improve public understanding of water flow location.

15.3 Conclusion

15.3.1 Blockchain Technology and Helping Farmers Deal with Drought

Currently, water is considered a vital asset in all sectors, especially the agricultural sector. About 70% of water is used for farming, 20% is used for industry, and 10%

is used for drinking. To help achieve economic goals, it's important to look at the areas that rely heavily on water.

This straightforward issue requires a thorough and innovative approach that can effectively serve both public and private sectors, particularly in underserved regions. Using blockchain technology to help with problems in farming, like handling droughts, managing water distribution, reducing risks from not having enough water, and keeping track of water management systems, is a fairly new idea.

Blockchain can help organizations active in the water sector to effectively monitor the settlement of financial statements and debt settlements, achieve different drought patterns, and also to pay special attention to maintaining the income of the water resources sector during severe drought. Since farmers may not be prepared to deal with special and sudden weather conditions such as storms, droughts, and severe floods, the necessity of using new technologies such as blockchain is felt more than ever. In this regard, insurance for farmers against climate change using blockchain technology is an example of a humanitarian effort to support farmers using technology. For example, the Crypto Climate Lemonade Coalition is one of a multitude of startups and private organizations that are looking to use blockchain to tackle social, economic, and climate challenges around the world.

Lemonade Foundation in Africa is planning several different platforms, including Avalanche, Chainlink, and DAOstack, using blockchain technology to help attract investment pools through which investors can provide funds to insure farmers. These are several cases demonstrating the application of blockchain technology for managing drought challenges in countries like the United States and Australia.

a) California

IBM in California is exploring how to use blockchain in an unconventional way. This company has planned a project in which it will help to quickly track the water level by integrating the two technologies of blockchain and the IoT.TFT is a private organization that works to protect freshwater ecosystems. This group is working with IBM Research and SweetSense Inc. to use blockchain technology to measure and monitor groundwater levels in California accurately.

b) Colorado

Currently, the state of Colorado has passed a bill that paves the way for the application of blockchain technology for water management. Also, in this bill, other emerging technologies such as remote sensors, satellite systems, and drones are studied. The use of these technologies improves the monitoring and management of surface and groundwater. The bill comes at a time when the United States is facing one of the worst droughts in its history. Western states fared the worst, making 2020 the driest or second driest year since records began in 1895.

c) Australia

Australia is known around the world for being very good at managing water resources. Australia plays an important role in managing water resources for cities, industries, and farms through water trading markets and blockchain technology.

This work talks about how blockchain technology can be used to manage flood risks and looks at different studies about using this technology in the water industry. We looked at 104 academic journals and 37 non-academic studies about hydrology and water resources.

From 2017 through October 15, 2022, the volume of research publications related to blockchain applications in water management experienced overall growth. There was a drop in 2020, but after that, the numbers started to go up again. This study looked at how current uses of blockchain technology could grow with more advanced techniques, and it also explored new ways to apply it in water management. The findings from the review suggest that water-related applications using blockchain technology are still not fully developed. There is a lack of research on how to use quick data in water management projects and hydrological tasks. With more research and development to fill in the gaps, blockchain technology has a lot of promise and could be used in lasting ways in hydroinformatics. The survey showed that many investments and projects are already using blockchain technology. They aim to create a reliable environment for easy processes and to improve the creation and distribution of economic resources while working together with important partners. In the group of interested parties.

Important lessons can be drawn from this summary. Blockchain, also known as a distributed ledger, is a new technology that makes it easier to store and manage valuable resources in a network. Still, there are some risks that could make it hard to put the system in place. This might stop it from being more popular with farmers, ranchers, and water supply systems. In the near future, we will see if legal and private actions can solve these issues to make the use of distributed ledger technology safe, strong, and easy to understand.

From this point, different actions can be taken. The way money is managed for current activities needs to be looked at again. The results of these financial checks are important for speeding up the use of blockchain technology in farming and water systems.

Even with these benefits, it's clear that there are still serious problems and issues that we need to think about more deeply. For example, to make it easier to use and manage, governments and policymakers should support using blockchain technology for water management. This will help create a more open and responsible system. At the regional level, it is important to work together across different sectors to help promote the use of blockchain technology. It's really important for governments and policymakers to invest more money in research and development (R&D). This includes preparing and testing systems, installing them, encouraging the use of water management blockchain, and showing that this idea is very important.

Designing, producing, and marketing water-conscious products contributes to environmental protection. Using water efficiently is very important for saving money and is now necessary for running a business. Labeling products based on how much water they use helps people become more aware. Digital product tracking makes the process clear, safe, and valuable. So far, there hasn't been a significant link between the flow of virtual water around the world and how crops are grown. This is a new way of managing water. Blockchain technology appears to be the most promising in this area. It covers different topics like tracking products, being clear with information, and handling product recalls in making things. Using virtual water management with blockchain technology helps prevent fake parts from being used in products. This method helps people willingly choose water-saving products and suggests using virtual water money along with regular money.

Evaluations might not fully recognize the yearly water shortage around the world because water availability changes with the seasons and how water use varies among different areas. However, these studies do not explain how climate change will affect water supply in different areas. The uneven distribution of water resources becomes worse due to climate change, and the impact of climate change can vary based on location. Traditional methods focus on water problems at a national or regional level, but they don't consider how water resources and the need for water in different areas are connected through river systems. So, a good way to manage water resources for different water issues could be to use a distributed network approach. Using blockchain technology to keep a shared and secure record of public water transactions could be helpful. It can build trust and make sure the data is accurate and safe. However, some people believe there are problems with how effective it is and also raise ethical concerns. Remote water use sensors can gather a lot of data that is regularly updated, but the information is not easy for the general public to understand or access.

This study shows that many projects and investments are already using blockchain technology. They aim to create a reliable and simple environment for developing economic opportunities and working together with key partners in the store system.

We can learn important things from this summary. Blockchain is a new and exciting technology that makes it easier to keep track of and manage valuable resources in a network. But there are some risks that might make it hard to put the system in place. This might make it less popular with farmers, ranchers, and water supply systems. Soon, we will see if and how legal and private actions can solve these problems and make distributed ledger technology a safe and reliable way to ensure health and respect. In summary, think about how blockchain can work together with new technologies like AI, ML, RFID (radio frequency identification), the IoT, and NFC. It's really exciting to focus on using more robots in water supply systems, which will greatly improve how they work together.

There are several actions we can take, such as creating blockchains that do not rely on biological system evaluations to encourage eco-conscious living, inspiring novel solutions to meet our targets, and guaranteeing the ongoing maintenance of networks. Water is like planning a good system to use blockchain effectively. We need to take another look at how money is being managed in our current activities. The results from these reviews should help us quickly use blockchain technology more in farming and water systems.

Presently, the nation is experiencing critical difficulties with its water resources. These include droughts, getting drinking water to cities and villages, and providing water for industry and farming. Not having enough water and experiencing dry conditions can cause serious problems. These issues include the sinking of land, damage to lakes and wetlands, and too much groundwater being taken out. Blockchain technology is a very useful way to quickly manage water resources and deal with droughts. By using blockchain technology, we can build trust and clear communication, allowing us to create a strong and affordable system for managing important groundwater resources anywhere around the world. In conclusion, we can say that using new technologies in our environment will help us find useful ways to manage problems caused by climate change. If we don't come up with new ways to manage droughts and this problem keeps happening, the water situation can become more than just an environmental issue. It could lead to various serious problems, especially conflicts between different ethnic groups and regions.

15.4 Future Work

In this study, it was found that the application of blockchain technology has been effective in flood risk management and water resources. However, this technology is almost new in this sector, which requires more and more detailed work for the future.

One of the major challenges facing countries in the coming years is the water issue. Blockchain can play an important role in monitoring the quality of water or its proportional distribution in different geographical areas within and outside borders. Through the combination of the Internet of Things and sensors with the blockchain, it is possible to monitor the quality and quantity of water, and all consumers are assured of the correctness and accuracy of the data, and automatic warnings can be given to consumers in certain situations. On the other hand, saving water consumption and fulfilling the obligations related to the consumption and maintenance of water supply network equipment can be monitored by everyone and continuous warnings can be given to the wrongdoers. By using the blockchain, it is possible to issue digital bills and receive funds at the same time and manage it with high accuracy and reliability. This solution is being investigated for implementation in the city of Flint, United States, after problems in the supply of clean water.

To grasp the functionality and benefits of blockchain technology, we require additional studies and practical examples drawn from real-world applications. The study, which has more practical uses, will motivate more water agencies to try out blockchain technology in their current systems. We need to look into whether using blockchain for digital transformation is a good idea financially and economically.

This involves improving the blockchain system and connecting it better with real water systems. Also, it's important to research ways to improve smart contracts related to water studies and develop systems for blockchain information sharing. We need flexible designs and blockchain systems that can easily be used by different groups and for various purposes. This is important for making the most of water-related uses, like trading water rights or managing city water supplies. Water companies or authorities might want to use the good things about blockchain technology without needing to understand the details of how cryptography works.

Private and public blockchain networks have their good and bad sides for a water system, but no research has looked into this comparison. Public blockchain networks are more transparent, but they also have higher fees for transactions. This topic should be talked about, especially when it comes to water systems. In the same way, different consensus algorithms have different qualities, but there aren't many studies comparing them with a focus on hydrology.

Managing water that flows across different regions is very important. Here, people involved may not trust each other, so it's essential to have good data and to agree on a way to manage the water together. Studying water systems that cross borders is important and has a lot of potential. This is because blockchain can offer benefits like being responsible, clear, and unchangeable. Virtual water trading and water footprinting, which are parts of hydrology, could also be made digital using blockchain technology. The trade of goods between countries involves the movement of virtual water. The product supply chain can be managed by adding water footprint and virtual water trade data based on blockchain.

Lastly, it's important to mention non-fungible tokens, abbreviated as NFTs, when discussing blockchain applications. NFT is a kind of digital money. In this definition, any asset can be worth money. NFTs can include things like artwork, videos, tweets, websites, pictures, and stories made on social media. NFTs cannot be swapped for another one because each NFT is one-of-a-kind and different from all the others. This feature is called "non-fungibility". Public blockchains make these tokens rare and one-of-a-kind, while smart contracts help make sure they can't be copied or repeated. In managing water resources, each product's water usage or water quality can be saved as a digital certificate called an NFT. In the same way, you can save things like investments, water bills, or tax payments as NFTs. A smart contract sets rules for these, ensuring that it can't be changed who pays, when they pay, and how much they pay.

As a suggestion for future studies, it is possible to suggest the use of blockchain against the phenomenon of climate change. Today, blockchain and climate change are closely related to each other, and blockchain is the latest solution to combat climate change. Every new technology has its good and bad sides. The good parts help make our lives better, while the bad parts can create problems. In the past 10 years, the climate crisis has been a big topic for governments to talk about. Now, the way people discuss it has changed, and everyone mostly agrees on what needs to be done. This agreement has shifted the conversation about climate change to figuring out how to stop it or make it less serious. The conversation is now focused on how to tackle the worldwide climate problem and how new technologies can help with this effort. In 2017, the United Nations Framework Convention on Climate Change (UNFCCC) stressed how blockchain technology can help fight climate change around the world.

The UNFCCC Secretariat has pointed out some important terms: transparency (being open and clear), cost-effectiveness (getting good results for the money spent), and productivity benefits (increased output). These benefits could help involve more people and improve the production of shared global goods. Blockchain is being looked at. Blockchain is a strong tool that can really help make tracking greenhouse gas emissions clearer, more reliable, and easier to follow. This technology helps companies give clearer, trustworthy, consistent, and easy-to-get information about carbon emissions. When people began using Ethereum, the second biggest cryptocurrency, they added a special layer to the blockchain that allows programming. This part was named smart contracts. This means a program can start on the blockchain and then something occurs in the real world.

Blockchain technology is also used to grow renewable energy businesses. Customers can trade, buy, and sell renewable energy with each other using tokens or digital assets that represent a certain amount of energy production.

Financing environmental projects by banks and loans can bring many challenges. The new lending system called DeFi or Decentralized Finance can be used to generate capital for environmental projects. DeFi projects have been around for a few years now, but 2020 has seen a surge in popularity.

Blockchain technology can ensure greater transparency about pollution and emissions and make it easier to track and report emission reductions, including addressing data quality issues.

One of the best ways blockchain can encourage people to help the environment is by supporting farming that uses renewable methods. These programs aim to help communities worldwide lower their carbon footprint by using eco-friendly farming and land management methods. These programs include planting trees and taking care of natural areas.

Climate change is changing weather patterns all over the world. Farmers everywhere, most of whom do not have insurance, are especially at risk from these changes in the climate. Some of the biggest climate changes are heavy rain, dry spells, and floods. Even though 75% of farms around the world don't have insurance, regular insurance companies have not been able to offer enough coverage for farmers in developing countries. Because of this, many of them have a life that is not very stable. Bad weather can hurt farmers' money and cause families to lose their homes and leave their farms. If climate change speeds up, it will surely cause problems for all farmers everywhere. Today, thanks to blockchain technology that can track changing weather through Oracle, there is a way to help people in rural areas facing climate change and natural disasters. Insurance projects like Arbol and Etherisc provide smart crop insurance for farmers everywhere. Farmers can use blockchain technology to get insurance for their small fields. Also, through this technology, it is possible to be informed about the weather situation in a moment and reduce the amount of damage caused to agricultural fields by timely notification of the accident.

References

1. Abu-Amara, F., Alrammal, M., Al Hammadi, H., Alhameli, S., Mohamed, I., Alaydaroos, M. and Alnuaimi, Z., (2022). A blockchain solution for water and electricity management. In: *Materials Today: Proceedings* [online]. [Viewed 21 November 2022]. Available from: doi: 10.1016/j.matpr.2022.05.106.
2. Angara, J.S. and Saripalle, R.S. (2022). Towards a virtual water currency for industrial products using blockchain technology. *Water Policy*, *24*(6), 923–941.
3. Eslamian, S. and Eslamian, F. (2025). *Blockchain Technology in Water and Environmental Systems*. CRC Press, Taylor and Francis, USA, p. 282.
4. Mia, M.U., Rahman, M., Elbeltagi, A., Abdullah-Al-Mahbub, M., Sharma, G., Islam, H.T. and Washakh, R.M.A. (2022). Sustainable flood risk assessment using deep learning-based algorithms with blockchain technology. *Geocarto International*, 1–29.
5. Satilmsioglu, T.K., Sermet, Y. and Demir, I. (2022). *Blockchain Applications and Opportunities for Water Resources and Hydrology: A Systematic Review*. ResearchGate.
6. Sriyono, E. (2020). Digitizing water management: Toward the innovative use of blockchain technologies to address sustainability. *Cogent Engineering*, *7*(1), 1769366.
7. Yasuno, T., Ishii, A., Amakata, M. and Fujii, J. (2020). Smart Dam: Upstream Sensing, Hydro-Blockchain, and Flood Feature Extractions for Dam Inflow Prediction. In: *Advances in Information and Communication: Proceedings of the 2020 Future of Information and Communication Conference* (*FICC*). (Vol. 1, pp. 139–158). Springer International Publishing.

16 Blockchain Technology for Secure Data Trust: Enhancing Access Control, Provenance, and Auditability

Bharat Ramdas Pawar[1], B. Sathyasri[2] and G. Sasikala[3*]

Blockchain technology is a revolutionary development in technology that emphasizes the use of a distributed ledger system to track asset ownership. Blockchain technology (BCT) is a real curse for threat actors or data hackers, exploiters, manipulators, hackers, and in-out attackers. But users find it difficult to comprehend the technologies it provides. From years we all know blockchain is the technology architecture used for Bitcoin cryptography. The distributed ledger feature and blockchain's security can be used to address present financial and non-financial company issues. This generation of data and its analysis are essential to the Internet of Things functioning, so they must be safeguarded for the duration of the device's life cycle. While data transcription the main barrier is the trust in the channel from source to destination, many threat actors are ready to exploit the channel for their interest. CIA triad fulfillment is the challenge while data sharing by using data into channels or recusing it. According to the GDPR guideline, data must be saved during encryption channel travel. By using encryption techniques and a special type of security algorithm, blockchain technology is the kind of buzzword technology for developers or business consultants to migrate to this technology. By using extra superfine algorithms like consensus and Merkle tree algorithms blockchain technology is used as primitive solution driven technology. Regarding complete business solutions by using the parties involved to maintain consensus on the ledger, blockchain holds immense potential to overcome the obstacles associated with traditional centralized unchangeable programmes and provide a distributed and transparent administration capability. Blockchain has the potential to be faster, more flexible, reusable, and powerful thanks to programmable coding components called smart contracts. They are comparable to objects in a programming context. With the recent advancements in smart contract and NFT development on blockchain platforms, blockchain-based applications in several

[1] Department of Electronics & Computer Engineering, CSMSS Chh Shahu College of Engineering, Chh Sambhaji Nagar, Maharashtra, India.
[2] Department of Electronics and Communication Engineering, Vel Tech. Rangarajan Dr. Sagunthala R&D Institute of Science and Technology, Chennai, India.
[3] Department of Electronics and Communication Engineering, Vel Tech Rangarajan Dr. Sagunthala R&D Institute of Science and Technology, Chennai, India.
* Corresponding author: gsasikala@veltech.edu.in

DOI: 10.1201/9781003686026-16

industries such as digital identification, supply chain, legal challenges, insurance, and medical health care are now possible. This chapter investigates how a GDPR-compliant architecture for data trust could be provided by the distributed blockchain. As you examine this chapter and all of its applications, The first step is to examine smart contracts, which are a crucial part of blockchain technology that help create decentralized data trust. In response to these issues, decentralized applications have been developed and studied to handle issues with access control and data sharing are used in many different contexts, including commercial operations, physical access control systems, and healthcare data sharing. A general-purpose program based on an attribute-based access control paradigm is recommended in light of this breakthrough. Long term, this application can offer not only the trusted auditability needed for access control and data sharing systems, but also a foundation for data trust. The system offers a level of openness that might be beneficial to resource owners (data controllers) and access requesters (data users) in addition to audits. The proposed techniques have been validated using independent digital libraries as a use case. Additionally, a comprehensive performance study of the system's deployment is offered. Several databases and consensus techniques were used to compare the performance results, which showed the system's low latency and great throughput. A blockchain-based end-to-end data trust mechanism wraps off this chapter. The proposed system promotes data dependability by assessing input datasets, effectively managing access restrictions, and offering data provenance and activity monitoring. A method for determining the trust value by considering the reliability of the input data sets is offered. The trust value is used to adaptively define the number of transaction validators. By guaranteeing the accuracy and reliability of the data at the point of origin and the clear and safe use of the data at the conclusion, this research offers solutions for data owners and users. The system under presentation efficiently and with low latency handles a large number of transactions, according to a detailed experimental examination.

16.1 Introduction

"Blockchain Technology" is a word that many people use to refer to a wide range of topics. They could be discussing Bitcoin, cryptocurrencies, digital tokens, Ethereum Blockchain, smart contracts, or any combination of these at any given time. The truth is that many people misuse the term "blockchain", which has led to confusion among many people regarding both the phrase "blockchain technology" and all of the associated terminology and terminologies. However, the distributed ledger technology that underpins all of these applications unites them. Instead of being saved on a centralized server, transactions with distributed ledger technology are duplicated and stored among several computers connected to the network (Eslamian & Eslammian, 2025).

In the present era, when trust is a crucial consideration when designing any new technology, blockchain technology leads the cryptographic approach to technology when considering encryption from decades ago. The main focus of technological advancement is on data that needs to be created, transferred, and simply and securely delivered to the end user in a way that preserves all of its original form and eliminates lag and loss.

This chapter's primary goal is to introduce all of the essential elements of blockchain technology, as using it in any sector requires having these elements in place. The whole transaction execution process, as well as the block creation technique and structure, have all been thoroughly addressed in this chapter. This chapter goes into detail on the many uses of blockchain technology that are addressed, as its use is becoming more and more widespread. Furthermore, there are issues or challenges associated with applying blockchain technology in other sectors, which can be very beneficial to researchers, professors, and business professionals.

A distributed database powered by blockchain technology keeps an unchangeable public ledger of every transaction. Blockchain enables all transactions to be recorded with a time stamp. Every node in the network is in charge of upkeep and ongoing transaction verification. Digital tokens are created using blockchain technology for digital data, including transactions or documents. You may think of these digital tokens as the files' digital fingerprints. These fingerprints are stored in what are known as "block" groups. Subsequently, the separate blocks are joined to create a chain of blocks, with each new block containing a digital token from the previous block. It becomes challenging to update an older block in the chain without simultaneously changing the blocks that came before it. The primary goal of blockchain technology is to eliminate the need for middlemen in the registration, confirmation, and transfer of all types of agreements and properties (Bheemaiah, n.d).

The most widely used and well-known technology is inextricably linked to bitcoin. It is also one of the most talked about phenomena in the recent past because it makes it easier for multibillion dollar transactions on the international market to occur without interference from the government. As a result, there are many regulatory issues involving national governments as well as other financial institutions. But the concept of blockchain technology has emerged as a major topic recently, is completely uncontroversial, and has found effective applications in both the financial and non-financial sectors. Marc Andreessen, the doyen of Silicon Valley's capitalists, claims that "the Blockchain's distributed consensus process is the most significant invention in Internet history". This economy runs on the digital system, and some of nature's most reliable authorities depend on it. Any person conducting business must have confidence in some sort of system. Examples of such systems include email services that deliver information, Facebook posts that share information with the public, and bank transactions that provide confirmation that funds are being transferred to a recipient anywhere in the world (Nakamoto, 2018; Namasudra et al., 2021). We can examine the fact that we live in a digital age where security and privacy concerns must be handled by a third party. But the truth is that it is also possible to operate or hack this third source. This is where blockchain technology enters the picture. Currently, the mechanism for transactions between two people or two businesses is mostly centralized and managed by a third party. A third party is always involved when we make a digital payment in order for the transaction to be completed. In addition, a credit card company or bank may levy additional fees. A comparable tendency is also observed in other domains, such as software, gaming, and music. Thanks to the introduction of Blockchain technology, this problem has been fixed. This technology's primary goal is to establish a decentralized environment where transactions and data are handled without the need for a third party (Frame and White, 2004). With the use of distributed consensus, this technology has the ability to track every transaction made in the past and present, which can be confirmed in the future. This is accomplished even without compromising the privacy of complex and digital assets belonging to third parties. Anonymity and distributed consensus are two of blockchain technology's key characteristics.

With the aid of distributed database solutions, blockchain makes it possible to continuously increase the number of data entries that are contributed by the various nodes that make up the network. Every completed transaction is entered onto a public ledger. The decentralized approach offered by blockchain technology eliminates the need for third-party mediation. All of the nodes involved in the Blockchain exchange information about every transaction that is completed. Because of this feature, blockchain technology offers greater transparency than centralized transactions mediated by a third party. Transactions are safer for other nodes since every node in the Blockchain is anonymous. Bitcoin was the first use of blockchain technology to be introduced. Bitcoin is the digital payment network that facilitates the exchange of products in a decentralized environment (Frame and White, 2004).

Even while this technology appears to be ideal for using cryptocurrencies to execute transactions, it still has certain technical limitations and difficulties that need to be thoroughly examined and handled. Nodes in the Blockchain must remain secret in order to fend off attacks, preserve a high degree of security, and facilitate transactions (Lerner, 2006). Furthermore, with the Blockchain, processing power is needed for transaction confirmation. In this situation, it becomes crucial to comprehend the subjects that various writers have studied as well as the subjects and areas that require additional attention and are posing risks and difficulties for future research. We have employed methodical mapping to comprehend the research methodology and identify the many works pertaining to the Blockchain sector in order to investigate and comprehend these problems.

Figure 16.1 shows the block structure of the Blockchain.

16.2 Blockchain Technology Inside

It's said that Bitcoin combines already-existing technologies to generate new uses. Implementing safeguards against data fabrication and duplicate payments, along with a means to keep the system safe from malicious user attacks, are essential for the smooth operation of a decentralized electronic money system. The following is

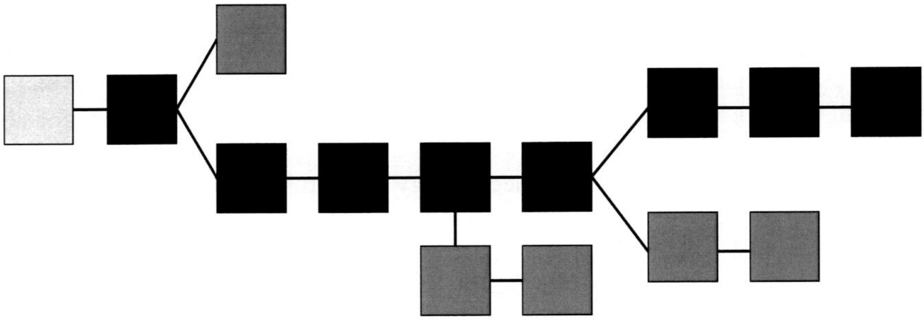

Fig. 16.1 Structure of Blockchain.

a list of major blockchain technologies that enable the proper operation of apps like Bitcoin: P2P, hash, public-key cryptography and digital signatures, Proof of Work (PoW0, and Proof of Stake (PoS). In (Tran, 2020), DLT, which is unchangeable by nature, is the foundation of blockchain technology. In a blockchain network, when a user wants to perform a transaction, they record their request in the ledger, which is accessible to all other users in the network, openly. A node is added to the network once every user has confirmed that it is authentic and has verified the required transaction in that node. It is nearly impossible for more than 51% of the network's nodes to be hacked in order for someone to change or hack the transaction. Every technology underlying blockchain technology is covered in this section.

16.2.1 Block Structure

There are two components to a block: the block body and the block header. As was previously indicated, every block creates a chain when it is connected to its parent or preceding block. The parent block is absent from the initial block, often known as the genesis block. In this case, the header of each block contains the hash of that block (Figure 16.2).

16.2.2 Peer-to-Peer Network

The roles of the two parties in a client-server network are predetermined since the client requests data from the server and is granted access to it. The server is responsible for maintaining and providing data. On the other hand, in peer-to-peer (P2P) networks, every node involved stores data independently and forms a self-contained network in which nodes can exchange requests and information. The corresponding nodes' role (server or client) is flexible in such a network. P2P networks must take data transfer and search strategies into account. Nodes can be located using search methods, and data can be transmitted between nodes (direct or relayed) using data transmission methods (Figure 16.3).

Block Version	02000000
Parent Block Hash	B6ff0b2b1680a2862a30 ca44d346d9e910d334b eb48ca0000000000000 00
Merkle Tree Root	9d10aa52ee94936ca93 85695f04ede270dda208 10decd12bc9b048aaa3 1417
Timestamp	24d95a54
nBits	30c31b18
Nonce	Fe9f0864

Transaction Counter

TX 1 | TX 2 | TX n

Fig. 16.2 Block structure.

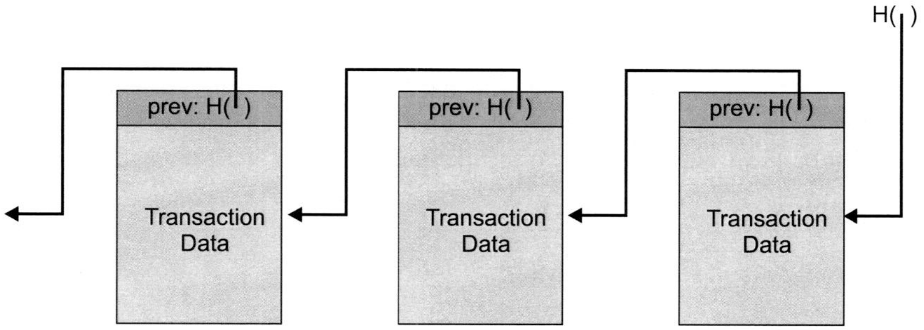

Fig. 16.3 Blockchain nodes.

16.2.3 Hash

Hashing and hash functions are heavily used in blockchain technology. A hash function is a mathematical operation that transforms an input into an output. To create a hash, the original data given as input is altered. Collision resistance, or the difficulty of reconstructing the input data from the hash value alone, is the main characteristic of the hash function. This procedure generates the same hash value from the same data, hence any modification to the original data will result in a completely different hash value. This technique is used to detect data manipulation by taking advantage of such traits.

16.2.4 Public Key Cryptography Digital Signature

A form of cryptography known as public-key cryptography uses separate keys for encryption and decryption. By splitting the key into two versions—a public key that is accessible to all users and a private key that is kept for personal use—the issue of key distribution was resolved. Symmetric-key cryptography requires a number of security precautions to ensure that the key is only exchanged with the appropriate counterparty for encryption and decryption. Public-key cryptography, however, can only be used to securely transfer and receive data if the recipient has already created a set of private and public keys and given the sender the public key. As long as the recipient appropriately handles his or her private key, safety can be preserved even while other people utilize the public key. Blockchain verifies transaction authentication by an asymmetric cryptographic technique. A digital signature, which is similar to public-key cryptography, verifies the legitimacy of data transmitted over a network using two keys. A digital signature created by encrypting the file's hash value with the sender's private key is delivered to the recipient along with the pertinent file. Using the same hash function as the sender, the recipient independently calculates the file's hash value. It then compares this value to the hash value obtained by decrypting the sender's digital signature using the sender's public key. Asymmetric cryptography is used by Blockchain to validate transaction authentication. A digital signature is a method of using two keys to prove the authenticity of data sent over a network, much to public-key cryptography. A digital signature is produced and sent to the receiver with the pertinent file by encrypting the hash value of the file using the sender's private key. The recipient uses the same hash function as the sender to independently produce the file's hash value, which it then compares to the hash value obtained by decrypting the sender's digital signature using the sender's public key.

16.2.5 Consensus Mechanism

Ensuring that all peer-to-peer network users reach the consensus protocol is a significant challenge when using blockchain technology with public ledgers (Williams, 2016). In order to ensure that transactions are validated in the correct order and that network members abide by its rules, a consensus protocol is employed. It is also used to prevent problems like the double spending problem, ensure that the information within a block is accurate, and ensure that the nodes, or miners, receive appropriate reward.

There must be some sort of vote among a known group of peers in order for algorithms for reaching consensus with arbitrary flaws. There are two primary methods that are in use: Byzantine Fault Tolerance (BFT) and the "Nakamoto consensus". In the first method, a leader is chosen by a "lottery" of sorts, after which they put forward a block that can be added to a series of blocks that have already been committed to. The second method relies on BFT algorithms and reaches the required consensus through several rounds of explicit voting. The most popular consensus methods are introduced in this part, including PoW and PoS, which are mostly tested with cryptocurrencies.

16.2.6 Proof of Work (PoW)

Anyone can take on the PoW mining challenge. The miners fight with each other to add the next block. A specific reward is given to the miner who solves the puzzle the quickest. Actually, the winner of the race is usually the node with higher computational efficiency. The PoW algorithm is applied to Bitcoin.

The general term for a system that assigns a simple but challenging work that can be easily verified as done to prove someone's innocence is PoW. One way to ensure authenticity is to give each member of the network one vote to decide which transaction belongs in the next block. Which set of transactions should be included is determined by the total number of votes. This type of consensus-process is susceptible to Sybil assaults, in which a single person pretends to be numerous in order to gain more clout in the network. The Bitcoin inventor Nakamoto found a solution to this influence problem by making voting more expensive. The quantity of processing power possessed by each individual determines their level of influence, Wagner, A. (2014), higher processing power results in higher energy requirements and higher hardware prices. The proof-of-work consensus protocol is conceptualized as follows. The PoW consensus algorithm used by bitcoins allows the network to compile all of the transactions that have been made during a given time frame into a block. The nodes' job is to validate the transactions and record them in a blockchain, hashing the data to keep it safe from hackers. Nodes have financial incentives to continue hashing and mining; the more blocks generated, the more bitcoins are awarded to the nodes. Neighboring nodes receive blocks that are created by a node. The nearby nodes independently confirm that the regulations have been observed and that the data within the block is accurate. It is advised to wait at least six blocks on a bitcoin network to ensure that the transaction is final. A blockchain fork occurs when many nodes attempt to work on the same transaction at the same time in competition to be the first to produce a block The node that creates the block with the longest blockchain first wins and is paid. Although PoW requires a lot of hardware and energy, which increases mining prices, it has been empirically shown to be reliable and safe.

Among the drawbacks of a PoW consensus method are the high energy expenditures associated with generating a single block and the possibility of a 51% attack. According to Courteous, "the proof-of-work process is going to self-destruct". The mining industry is becoming more specialized and smaller, and large organizations possessing superior resources are able to outmaneuver an individual miner. A 51% attack is becoming more likely as a result of the system's increased centralization to a few numbers of powerful companies due to this mining specialization.

16.2.7 Proof of Stake (PoS)

The blockchain community adopted PoS, a new consensus mechanism, to lower the possibility of a 51% attack and to use less energy. Rather than demonstrating that it has completed a computationally challenging activity, as required by the proof-of-work protocol, a node could instead demonstrate that it possesses a specific quantity of coins. In a PoS system, a new block is created using coins rather than processing power, and the node with the most coins has the most sway. Game theory states that the majority owner will not attack a network that is secure and stable because it serves their interests. In the event of an attack, the digital currency's value will only drop and it will become unstable. Forking is one problem with the PoS protocol. Without incurring any additional computational costs, multiple nodes may begin mining on a transaction at the same time. Comparing a PoW protocol to a blockchain increases the risk of malicious nodes forking the blockchain. This raises the possibility of greedy nodes mining on all forks in order to avoid missing out on block rewards, as well as double spending assaults. The problem of double spending attacks can be resolved by employing check-point blocks, which prevent blocks prior to a check-point from being changed. The arguments put out by Manning and the bitcoin community are unfounded since a PoS technology is still susceptible to a 51% attack. He claims that by acquiring 50% of a PoW cryptocurrency network's money base for free, a miner can take over the network (Swan, 2015).

16.2.8 Three Main Categories of BCT

16.2.8.1 Public Blockchain

A distributed, decentralized ledger that is accessible to anybody wishing to participate is called a public blockchain. Without the need for a central authority, it enables the transparent and impenetrable recording of transactions over a network of computers (nodes).

16.2.8.2 Permissioned or Private Blockchain

A private blockchain is a type of blockchain where network access is restricted to a chosen subset of users. Private blockchains are usually centralized and permissioned, in contrast to public blockchains, which are decentralized and accessible to everybody. This implies that users must have authorization in order to connect to the network, which is frequently run by a single organization or a group of organizations.

16.2.8.3 Federated or Consortium Blockchain

A consortium blockchain refers to a type of blockchain network that is not controlled by a single organization, but rather is maintained and administered by several. Consortium blockchains achieve a balance by incorporating several organizations, in contrast to private blockchains, where one entity has exclusive control, and public blockchains, like Bitcoin and Ethereum, where anybody can participate and contribute to the network.

16.3 Compute, Build, and Deploy Blockchain

At first, it was all about Bitcoin, but as the movement spread, a lot more cryptocurrencies entered the market. Some of them succeeded in making money, while other cryptocurrencies trailed behind. But before long, blockchain technology realized its true potential and expanded into a wide range of unexpected fields. Healthcare, enterprise software development, banking, insurance, and other financial sectors are among the domains where blockchain technology is presently radically changing the existing frameworks of technology.

16.3.1 Banking and Payments

The future of payment and financial systems is now blockchain. Cryptocurrencies that resemble Bitcoin have the ability to govern payment systems independently of national boundaries. Remittances based on bitcoin are exemplified by BRA.

16.3.2 Cyber Security

Cryptography is used in blockchain to safeguard and verify data. This will prevent any unapproved modifications and system hacks. By doing so, intermediaries are eliminated from the system, preventing any unauthorized alterations.

16.3.3 Supply Chain

Better accountability, transparency, and feedback mechanisms along the supply chain are just a few of the ways that the blockchain may transform the supply chain. Complete product tracking is possible with blockchain-based supply chain management. IoT sensors enable the blockchain to record every action and state of a product. A supply chain management system built on the blockchain is called Block Verify and Provenance.

16.3.4 Online Data Storage

A single point of failure can occur with OneDrive, Google Drive, and other centralized systems that hold data. Blockchain technology increases the security and dependability of distributed data storage. One such encrypted cloud storage platform is called Storj.

16.3.5 Networking and IoT

A decentralized network of IoT devices can be established by utilizing blockchain technology in networking and IoT. As a result, managing the IoT devices no longer requires a central location.

16.3.6 Insurance

The foundation of the global insurance market is trusting management. The new paradigm in trust management is blockchain. Blockchain relies on participant mistrust to maintain confidence.

16.3.7 Government

Blockchain technology will reduce red tape and bureaucratic barriers while increasing operational efficiency and transparency in government processes.

16.3.8 Crowdfunding

It is a well-liked means of financing new ventures and initiatives. Smart contracts and online reputation systems are used in blockchain-based crowdfunding platforms to establish trust, doing away with the need for a middleman who would otherwise charge exorbitant fees for this service. It is possible for new enterprises to issue their own tokens, which may then be traded for goods, services, or money.

16.3.9 Multimedia and Entertainment

Blockchain is now present in the entertainment industry, where excessive third-party influence is a problem. The middleman will no longer be present in this scenario thanks to the blockchain's implementation. One of the entertainment industries where blockchain is already being implemented is online music. For instance, Ujo and Mycelia's music.

16.3.10 Real Estate

This is another crucial area where the use of blockchain technology will bring about significant change. There are numerous ownership and transfer problems in the current real estate system. The complete real estate system with shared ledgers can be controlled by the blockchain implementation in this industry.

For instance, the state government of Andhra Pradesh in India has begun using blockchain technology to handle land registration in its entirety.

The list includes a lot more other locations such as energy management, voting, healthcare, forecasting, and transportation. Along with blockchain came industry-specific blockchain development frameworks, DApp and digital asset management software, blockchain management software, and other blockchain-related products and services. And as it expands, a ton more tools are added. Blockchain creation and management are now easier than ever thanks to all these frameworks and tools. Thus, it's now simpler than ever to build and implement blockchain-applied solutions.

16.3.11 Features of Blockchain

These days, blockchain is utilized in a wide range of applications across numerous industries and use cases thanks to its distinctive features. Below is a discussion of some of the key characteristics of blockchain.

16.3.11.1 Enhanced Security

Increased security is possible with blockchain technology since all data is stored in a decentralized network. Since each node's hash is linked to the one before it, changes to one node's hash impact the hashes of all nodes. In this scenario, data cannot be changed. As was previously said, there are various types of blockchain networks. However, because permissioned blockchain enables specific users to add new blocks to the network, it is usually utilized in industries.

16.3.11.2 Decentralized Network

Decentralization is one of the main characteristics of a blockchain network. It suggests that a blockchain network lacks a central authority. In this case, the network is decentralized, with all system control being exercised by a group of peer nodes. Users on a blockchain network are assigned specific positions. Because a blockchain network doesn't require a central authority, people can utilize it directly through the internet and keep their data there.

16.3.11.3 Immutable

A blockchain network's data is viewed as an endless transaction log. A block cannot be removed or changed once it has been uploaded to the network. In this instance, each blockchain node maintains a copy of the web ledger and verifies each transaction's authenticity before allowing it to join the network. A transaction is added to the public ledger if it is validated by the majority of nodes in a blockchain network. This guarantees a blockchain network's dependability.

16.3.11.4 Transparent

Data control over users is intended to be transparent with blockchain technology. Every peer node in this scenario has access to data blocks, which they can use anyway they see fit. This encourages openness, which dismantles boundaries across systems, organizations, and individuals.

16.3.11.5 Distributed Ledger

Distributed ledgers store, share, and synchronize transactions in their individual electronic ledgers using independent computers, or nodes. Unlike a standard ledger, this one does not store data centrally. A public ledger is typically used to make all transaction data, including the identities of the participants, available to the public. Federated and private blockchain networks function substantially differently. Nonetheless, a lot of users and administrators might find a lot of ledger facts in these networks. This is primarily because everyone who uses the blockchain network contributes to maintaining the ledger.

16.3.11.6 Consensus

The foundation of a blockchain network's architecture is its consensus algorithms. The network uses these consensus algorithms to help it decide what to do in each situation. A consensus method is necessary to validate a specific transaction that enables the seamless operation of the entire blockchain network, even with millions of nodes connected. As a result, every choice the network makes is a means—directly or indirectly—of accomplishing its objectives. One of the main advantages of DLT is its consensus method.

16.3.11.7 Faster Settlement

Conventional financial processes move slowly; a transaction may take days to complete. The quicker settlement method supported by blockchain technology is in contrast to these conventional approaches. As a result, users can finish transactions quickly, keeping the system from becoming sluggish overall. Even if there are situations when the network cannot handle a high volume of users or quick settlement, blockchain is now used for quick transactions and P2P money transfers.

16.3.11.8 Chronological Data

Blockchain technology is a network of blocks, as was previously explained. Here, each block contains transaction-related data and uses a cryptographic hash to connect to the block before it. The block that comes after with the same hash value is linked to the block that came before it. By joining together, these networked bricks create a chronological chain of data that documents ownership transfers and establishes provenance.

16.3.11.9 Smart Contracts

One of the key elements of a blockchain network is a smart contract. These programmers, scripts, or logics are used in a blockchain network for terms, conditions, or network-wide automated chores. Environments with permissioned blockchains are ideal for smart contract operation. It expedites the completion of any settlement. Many of the operations that take place in a blockchain network can be automated by smart contracts.

16.4 Risks for Adoption

New techniques for managing dispersed and decentralized systems are needed as cutting-edge technologies like data business models (Trivedi et al., 2022), on cloud technologies like Azure, AWS or GCP and the IoT. Since all of these technologies operate online, there are more and more criminal users and attackers, which leads to a constant rise in security concerns. Furthermore, due to the growing volume of data and susceptible gadgets on the market, it is critical that verified and trustworthy services are enforced today. Blockchain is a cutting-edge and exciting technology. As we know before, blockchain-based technology can be used to tackle a wide range of problems or applications. This covers both financial and non-financial uses, such as remittance and investment banking, as well as notary services. Most of them are extreme innovations. Adoption is fraught with peril, much like dramatic inventions. Shift in behavior: Although resistance to change exists, change is a constant. In the non-tangible world that BlockChain presents, users must become used to the notion that their electronic transactions are full, safe, and secure. The roles of the contemporary middlemen, like Visa or Mastercard (for credit cards), will likewise change. We expect them to invest as well as migrate to a blockchain-based system. They will continue to provide customer relationship-related services.

Scaling: One of the issues with the newly developed BlockChain services at the moment is scaling. Imagine that this is your first time executing a BlockChain transaction. You have to download and validate each and every BlockChains that are currently in existence before you can finish your first transaction. Considering the exponential increase in the quantity of blocks, it can take several hours or even longer.

16.4.1 Bootstrapping

To convert the current contracts, business papers, or frameworks to the new BlockChain based approach, a significant set of migration procedures needs to be finished. For example, present records maintained by counties or escrow services on real estate ownerships or liens need to be converted to the appropriate BlockChain format. This could include expenses and time. The fact that these agencies are trusted by their clients may help adoption in the United States. Strong opposition to adoption is expected in economies like China that have more stringent rules.

16.4.2 Sutardja Center for Entrepreneurship and Technology Technical Report

The bad guys can utilize this for fraudulent operations like money trafficking because it is easy to transfer assets and because BlockChain transactions are pseudonymous. However, if there are enough regulations and technical assistance, law enforcement agencies will be able to monitor them and file charges.

Quantum computing: The foundation of blockchain technology is the logical impossibility of a single party gaming the system due to insufficient processing power. However, if quantum computers are available, it would be feasible to crack cryptographic keys quickly enough using merely brute force. The system will collapse as a result. The counterargument is that keys ought to become stronger in order to make them harder to crack.

16.5 Conclusion

In conclusion, we always use a blockchain approach to build microservices with different versions, design the artifact, run on different platforms, compute, and deploy successful microservice applications in a best suited and adoptive framework with all reliability and comprehensiveness, whether on premises or off. To comply with the GDPR regulations, this method uses a number of novel adaptive encryption technologies, including consensus algorithms and Merkle trees. The core technology of Bitcoin is blockchain. The immutable and fast adopting distributed ledger feature, security algorithms, and protocols of Blockchain technology make it an appealing solution for contemporary financial and non-financial company issues worldwide. Due to the great amount of interest in blockchain-based commercial applications, many start-ups are developing them. Adoption certainly faces several difficulties, as was previously indicated. PayPal, Visa, and other e-payment corporations are investing in the research of integrating blockchain technology into their current business strategies. In fact, several of them are searching the blockchain space for fresh company ideas.

Given how quickly blockchain technology is being developed and used as a lead technology, it will soon become a commonplace piece of technology. The primary obstacle to the acceptance and deployment of smart contracts and related technologies is the significant challenges that must be resolved in the near future. Very few blockchain frameworks are ideal for resolving issues in the real world. Technology is more portable and easier to adopt when it is deployed via an integrated approach thanks to consensus algorithms. Continuous integration with digital signatures, NFT, and smart contracts gives us a wealth of options for end-to-end user deployments. The Blockchain approach provides a thorough evaluation of innovative and comprehensive business ideas, leading to the discovery and fulfillment of several technological problem solutions.

References

1. Bheemaiah, K. (no date). *Block Chain 2.0: The Renaissance of Money*. https://www.wired.com/insights/2015/01/block-chain-2-0/. Accessed 07 June 2021.
2. Eslamian, S. and Eslamian, F. 2025. *Blockchain Technology in Water and Environmental Systems*. CRC Press, Taylor and Francis, USA, p. 282.
3. Frame, W. and White, L. (2004). Empirical studies of financial innovation: Lots of talk, little action? *J. Econ. Literature*, *42*(1), 116–144. Available: http://www.jstor.org/stable/3217038.
4. Lerner, J. (2006). The new financial thing: The origins of financial innovations. *J. Financial Econ.*, *79*(2), 223–255. Available: http://search.proquest.com.ezproxy.lib.usf.edu/Docview/231721046?accountid=14745.

5. Nakamoto, S. (2018). Bitcoin: A peer-to-peer electronic cash system. https://www.ussc.gov/sites/default/files/pdf/training/annual-national-training-seminar/2018/Emerging_Tech_Bitcoin_Crypto.pdf. Accessed 13 Apr. 2022.
6. Namasudra, S., Deka, G.C., Johri, P., Hosseinpour, M., and Gandomi, A.H. (2021). The revolution of blockchain: State-of-the-art and research challenges. *Arch. Comput. Meth. Eng.*, *28*, 1497–1515.
7. Swan, M. (2015). *Blockchain: Blueprint for a New Economy*. Beijing, China: O'Reilly Media, Inc. (1st Edn.)
8. Tran, S. (2020). IBM and Oracle collaborate on interoperability work for their blockchains to communicate with each other. https://blockchain.news/news/ibm-oracle-collaborate-interoperability-work-block chains-to-communicate. Accessed 31 Aug. 2021.
9. Trivedi, M.H., Priyadarshi, G.V., Lalwani, D. and Eslamian, S. (2022). Risk Assessment Applications: Exposure, Safety, and Security, In: Furze, J.N., Eslamian, S., Raafat, S.M., and Swing, K. (Eds.), *Earth Systems Protection and Sustainability* (Vol. 2). Springer, Cham, Switzerland, pp. 215–240.
10. Wagner, A. (2014). Ensuring network scalability: How to fight blockchain bloat. https://bitcoinmagazine.com/technical/how-to-ensure-network-scalibility-fighting-blockchain-bloat-1415304056. Accessed 29 May 2021.
11. Williams, A. (2016). *IBM to Open First Blockchain Innovation Centre in Singapore, to Create Applications and Grow New Markets in Finance and Trade. The Straits Times* Singapore Press Holdings Ltd.,Co., Singapore.

17 Blockchain Unveiled: Evolution, Challenges, and Transformative Applications across Sectors

Prateek Pandey[1*]

In the last five years, the term "Blockchain" has transcended niche discussions within tech circles to become a mainstream topic, featuring prominently in general awareness magazines, newspaper articles, and even TEDx talks. However, contrary to popular perception, Blockchain is not an entirely new invention. Instead, it represents a culmination of existing technologies, skilfully woven together to create a powerful paradigm capable of revolutionizing established systems. This chapter aims to delve into the intricacies of Blockchain technology, unraveling its underpinnings and shedding light on the conditions and technologies that paved the way for its current transformative capabilities. Rather than a singular breakthrough, Blockchain emerges as a mosaic, crafted through the adept mix-and-match of components that have been available to us over the years. In this chapter, various types of Blockchain technologies, currently in practice, have been scrutinized by us. Each type brings its own set of advantages and challenges, offering a diverse landscape for innovation and implementation. By dissecting the pros and cons of these approaches, we gain a nuanced understanding of the intricate tapestry that forms the backbone of Blockchain. Beyond the theoretical realm, this chapter goes further to examine real-world applications of Blockchain. From finance to healthcare, supply chain management to identity verification, the transformative footprint of Blockchain is palpable. As we traverse through these varied use cases, a comprehensive narrative unfolds, showcasing the adaptability and versatility that Blockchain injects into diverse sectors.

17.1 Introduction

In the earlier stages of computing, standalone systems operated in isolation and lacked the capability to connect with each other through a network. This compelled individuals to establish computing facilities at different geographical locations, characterizing an early form of distributed computing. With the advent of networking technologies, centralized systems gained prominence. These systems, epitomized by a singular server catering to multiple clients, became the cornerstone

[1] Jaypee University of Engineering and Technology, Madhya Pradesh, India.
* Corresponding author: prateek.pandey@juet.ac.in

DOI: 10.1201/9781003686026-17

of architectures like the Internet's HTTP model. However, this centralized approach introduced vulnerabilities; the entire system was at risk if the central entity suffered a breakdown or compromise.

In response to the limitations of centralized systems, the pendulum swung back towards distributed architectures. The resurgence of distributed systems marked a strategic shift from concentrating data on a single system to distributing it across multiple nodes. This distributed model offered a fail-safe mechanism, mitigating risks associated with single points of failure or security breaches.

The transition from early standalone systems, where computing occurred in isolated environments, to the subsequent shift towards centralized architectures with the advent of networking technologies, followed by a return to distributed systems, reflects the dynamic nature of computing. This evolution is fueled by the pursuit of efficiency, reliability, and security in tandem with technological advancements and evolving user needs.

As we delve into the theory of blockchain, it's essential to appreciate this historical context. Blockchain, with its decentralized and distributed ledger technology, represents a paradigm shift in how we approach data integrity, security, and trust. In the subsequent subsections we will explore the foundational principles and mechanisms that underpin the fascinating world of blockchain.

17.1.1 Distributed vs. Decentralized Systems

A distributed system emphasizes the idea of spreading tasks across multiple nodes, while a decentralized system focuses on distributing decision-making, authority, or control. Decentralized systems are a subset of distributed systems, emphasizing a lack of central authority.

Distributed System Example: A cloud computing environment where different servers work together to process and deliver services.

Decentralized System Example: File sharing networks like BitTorrent offers a client to download a required file in pieces from multiple other clients. Thus, the authority to serve a request doesn't lie with one system, rather it is decentralized.

Relevance with Blockchain

The inherent decentralization and distribution of blockchain technology contribute to a range of prominent characteristics that make it a resilient, trustworthy, and scalable technology. Here are some key attributes associated with the decentralized and distributed nature of blockchain:

a) Immutability and Tamper Resistance

Characteristics: The distributed ledger in blockchain ensures that once data is recorded in a block and added to the chain, it becomes extremely difficult to alter. Immutability is a direct result of the decentralized consensus mechanisms that require agreement across a network of nodes.

Impact: This characteristic enhances trust in the integrity of data, making blockchain a reliable and tamper-resistant technology.

b) Decentralized Consensus Mechanisms

Characteristics: Blockchain relies on consensus mechanisms, such as Proof of Work (PoW) or Proof of Stake (PoS), to validate and agree on the state of the ledger. These mechanisms distribute decision-making across nodes, avoiding the need for a central authority.

Impact: Decentralized consensus ensures that no single entity has control over the entire network, fostering trust and resilience against attacks or manipulation.

c) Enhanced Security through Decentralization

Characteristics: Decentralization reduces the vulnerability of a blockchain network to single points of failure. Each node in the network holds a copy of the entire ledger, making it robust against attacks attempting to compromise a single node.

Impact: This distributed security model contributes to the overall resilience and trustworthiness of the blockchain.

d) Censorship Resistance

Characteristics: Since blockchain is decentralized, it is resistant to censorship attempts. No central authority can control or restrict access to the network, providing individuals with autonomy over their transactions and data.

Impact: Censorship resistance ensures that blockchain remains an open and inclusive technology, accessible to participants without centralized interference.

e) Redundancy and Fault Tolerance

Characteristics: The distributed nature of blockchain ensures redundancy, with multiple copies of the ledger stored across nodes. This redundancy enhances fault tolerance, as the network can continue to operate even if some nodes fail.

Impact: Blockchain remains resilient in the face of network disruptions, contributing to its reliability and continuous functionality.

f) Scalability and Parallel Processing

Characteristics: Blockchain's distributed architecture allows for parallel processing of transactions across multiple nodes. This contributes to scalability, enabling the network to handle a larger number of transactions simultaneously.

Impact: Scalability is essential for accommodating the growing demands of various applications built on blockchain technology.

g) Peer-to-Peer Network Model

Characteristics: Blockchain operates on a peer-to-peer network model, where nodes can directly interact with each other. This eliminates the need for intermediaries and facilitates direct, efficient communication.

Impact: The peer-to-peer model aligns with the decentralized philosophy, promoting a more democratic and transparent exchange of information.

17.1.2 Linked List Data Structure

A linked list is a fundamental data structure used in computer science and programming. It consists of a sequence of elements, where each element points to the next one in the sequence. The elements in a linked list are called nodes. Here are some key concepts related to linked lists:

a) ***Node***: Each element in a linked list is a node. A node contains two main parts: data and a reference (or link) to the next node in the sequence.
b) ***Head***: The head is the first node in the linked list. It serves as the starting point for traversing the list.
c) ***Tail***: The tail is the last node in the linked list. Its next reference is typically null, indicating the end of the list.
d) ***Singly Linked List***: In a singly linked list, each node points to the next node in the sequence, forming a unidirectional chain.
e) ***Doubly Linked List***: In a doubly linked list, each node has references to both the next and the previous nodes, creating a bidirectional chain.

Relevance with blockchain: The concept of a linked list shares some conceptual similarities with certain aspects of blockchain design (Readers are advised to compare Figure 17.2 with Figure 17.3 for understanding the linkage between the two data structures):

a) ***Block Structure***: A blockchain can be perceived as a chain of blocks, where each block contains a list of transactions. The linkage between blocks is established through the use of cryptographic hash functions. While not a traditional linked list, this structure implies a sequential connection between blocks.

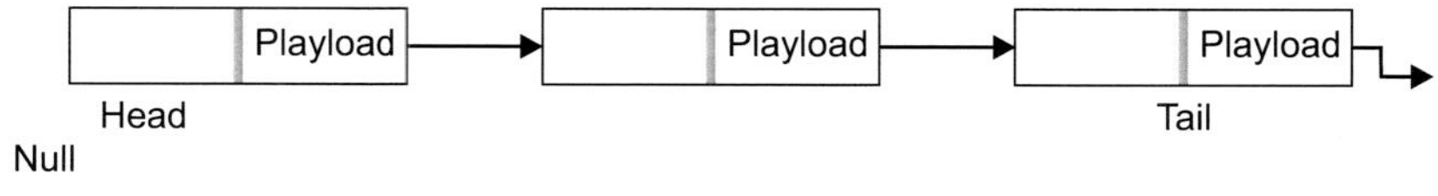

Fig. 17,1 Singly Linked List.

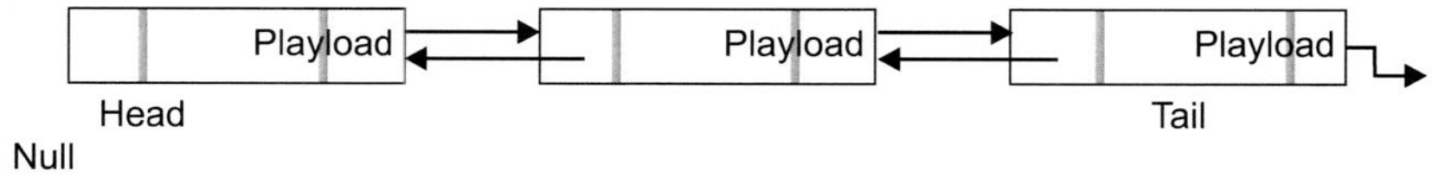

Fig. 17.2 Doubly Linked List.

b) ***Chaining of Blocks***: The chaining mechanism in blockchain ensures that each block is cryptographically linked to the previous one. This creates a continuous and unbroken chain of blocks, similar to the linkage in a linked list. The hash of the previous block is included in the current block's header, ensuring the integrity of the entire chain.

17.1.3 Cryptography and Hashing

Cryptography is the practice of techniques for securing communication and data from adversaries or malicious actors. It encompasses various methods for encoding and decoding information, ensuring confidentiality, integrity, and authentication in the process.

Here are key aspects and concepts related to cryptography:

- **a)** ***Encryption:*** The process of converting plaintext (readable data) into ciphertext (unreadable data) using an algorithm and a key.
- **b)** ***Decryption:*** The reverse process of converting ciphertext back to plaintext using the same algorithm and a corresponding key.
- **c)** ***Types of Cryptography:***
  - **i. Symmetric Cryptography:** Uses a single key for both encryption and decryption. Both communicating parties must share the same secret key.
  - **ii. Asymmetric Cryptography:** Uses pairs of public and private keys. The public key is shared openly, while the private key is kept secret.

Hashing: Hashing is a fundamental concept in computer science and cryptography, but it's important to note that hashing itself is not encryption. Hashing is a one-way function that takes an input (or message) and produces a fixed-size string of characters, which is typically a hash value. A hash function is designed to be fast and deterministic, meaning the same input will always produce the same hash value, and it should be computationally infeasible to reverse the process (find the original input from the hash value).

Relevance with Blockchain

a) ***Data Integrity*****:** Blockchain relies heavily on cryptographic hash functions to maintain data integrity. Each block in a blockchain contains a hash of the previous block, forming a chain. If any data in a block is tampered with, it would change the hash, and since this hash is referenced in subsequent blocks, it would be immediately apparent that something is amiss. This makes it extremely difficult for any single entity to alter the contents of a block without detection.

b) ***Authentication*****:** Cryptographic digital signatures are used to verify the authenticity of transactions. Each participant in a blockchain network has a pair of cryptographic keys: a public key (known to everyone) and a private key (known only to the owner). Transactions are signed with the private key, and others can verify the signature using the public key. This ensures that transactions are authentic and have been approved by the rightful owner.

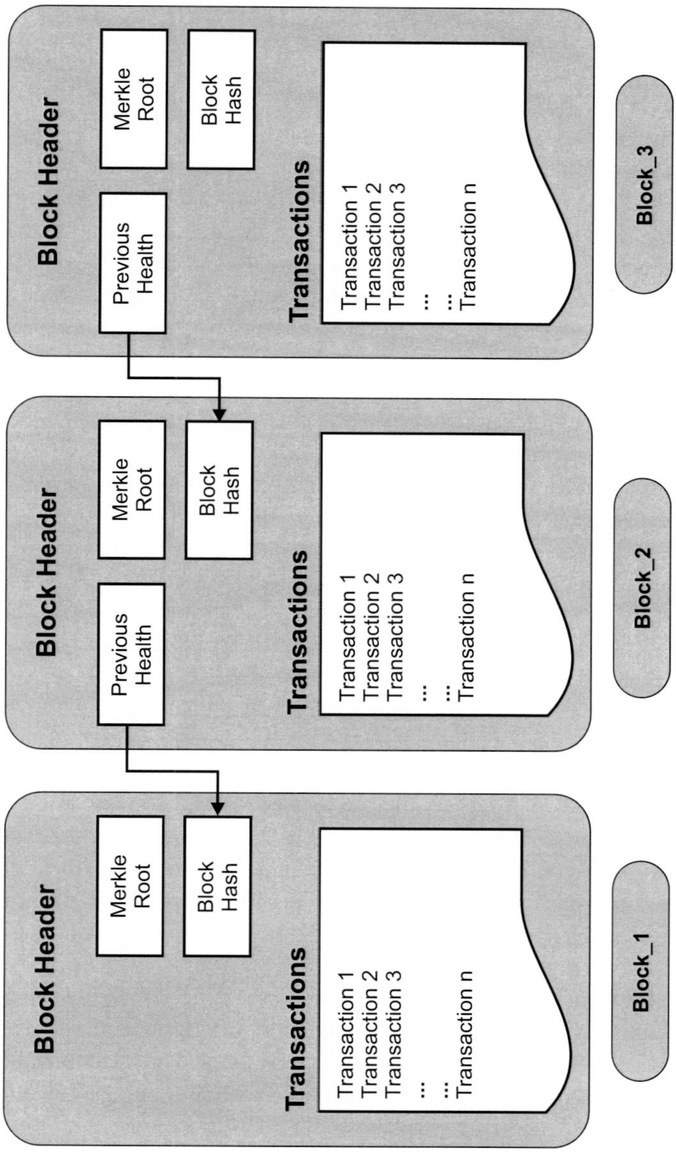

Fig. 17.3 Blocks with their composition forming a Blockchain (Pandey & Litoriya, 2020e).

c) ***Consensus Mechanism*****:** Cryptographic puzzles, such as those used in PoW, are employed to achieve consensus in blockchain networks. These mechanisms prevent malicious actors from easily manipulating the blockchain by requiring participants to solve complex mathematical problems or prove ownership of a certain amount of cryptocurrency.

17.2 Literature Review

Knowing the history and the underpinnings of the blockchain propels us to look at what has been done till now by blockchain for the industry and society. In this section, a marathon review of the works done in different walks of business and society using blockchain and how blockchain itself evolves over time to meet the industry requirements is presented (Eslamian and Eslamian, 2025).

Blockchain technology stands out as a pivotal and disruptive force globally. Initially designed to facilitate secure cryptocurrency transactions, particularly with Bitcoin, it has since transcended its financial origins and found applications in diverse sectors. Notably, the integration of smart contracts has bolstered its capabilities (Androulaki et al., 2018; Ying et al., 2020; Zheng et al., 2017), making it a versatile tool in various industries.

The health and life care sector emerges as a promising arena for the adoption of blockchain technology. The innovation holds the potential to revolutionize healthcare and life science services by prioritizing the patient within the medical ecosystem. This shift promises advancements in security, fortification, interoperability, and traceability of health-related information. While blockchain is recognized as a potential game-changer in healthcare, the path to its adoption is not straightforward (Pandey & Litoriya, 2020e).

The transformative impact of blockchain in healthcare extends to the realm of electronic health management (EHM), offering a more secure, decentralized, and distributed approach to electronic health records (EHR). It is crucial to note that while blockchain is a valuable tool in healthcare, it is not a panacea. The rapid evolution of this technology provides an opportunity for investment, experimentation, venture, and proof-of-concept testing in the healthcare industry. In essence, the passage underscores the profound potential of blockchain technology in reshaping healthcare while acknowledging the complexities and ongoing evolution in its adoption.

Within the healthcare domain, there are numerous activities and use cases that can benefit from efficient and secure management through the implementation of blockchain technology. These include but are not limited to medical record-keeping, relief efforts, tracking clinical trials, facilitating patient-doctor interactions, and detecting counterfeit medicines (Casino et al., 2019).

The momentum of blockchain technology adoption is steadily increasing, with organizations swiftly embracing this distinctive technology for managing EHRs and securing medical transactions. According to a 2016 report by IBM, 16% of healthcare providers were identified as trailblazers, indicating their readiness to scale up blockchain implementation in 2023. The projection was that the majority of healthcare organizations would become mass adopters by the year 2026 (Mettler, 2016). Various reports and articles have succinctly outlined the applications of blockchain technology in the healthcare domain (Ben Fekih & Lahami, 2020; Leeming et al., 2019; Yoon, 2019). Moreover, discussions on blockchain fundamentals, along with its current and future use cases in the fields of healthcare and EHR, have been explored in depth (Zhang et al., 2016).

The deepening connection between the healthcare domain and blockchain technology has been thoroughly investigated and acknowledged. However, there has been limited attention given to the expanding influence of blockchains in the health insurance sector. Health insurance companies routinely grapple with the issue of fraudulent claims. Smart contracts, if utilized, could play a pivotal role in determining the validity of alleged claims by assessing the accuracy of information related to health and treatment submitted to a health insurer. The work done by (Pandey & Litoriya, 2020b) throws some issues and possible remedies using blockchain technology.

In the realm of the energy sector, blockchain technology emerges as a transformative force, offering innovative solutions across various domains. This chapter explores five pivotal areas where blockchain stands poised to revolutionize energy practices:

(a) ***Smart Metering:*** Blockchain introduces a secure and transparent framework for recording and validating smart meter data (Roshandel and Eslamian, 2023), ensuring the integrity and accuracy of energy consumption information. Automated billing processes and tamper-proof invoicing through smart contracts streamline payment procedures, reducing disputes (Moroni et al., 2019).

(b) ***Energy Communities:*** Decentralized, peer-to-peer energy trading within local communities becomes a reality with blockchain. This technology facilitates efficient, localized energy markets, where participants can seamlessly buy and sell excess energy. Additionally, the creation of decentralized energy grids fosters autonomy within communities (Andreadou et al., 2019).

(c) ***Flexibility Services:*** Implementing smart contracts for demand response and grid balancing, blockchain enables dynamic adjustments to energy usage based on real-time market conditions. The decentralization of grid control enhances resilience, mitigating the risks associated with single points of failure (Gungor et al., 2011).

(d) ***Certification of Origin:*** Blockchain's immutability is leveraged to certify the origin of energy sources, particularly crucial for verifying claims related to renewable energy. Beyond provenance tracking, blockchain aids in emissions verification, supporting carbon credits and emission reduction initiatives (Galli et al., 2011).

(e) ***Electro-Mobility:*** In the realm of electric vehicles (EVs), blockchain streamlines the management of charging stations, handling transactions, authentication, and data sharing among stakeholders. Additionally, it enhances transparency in the supply chain for EV components, ensuring authenticity and supporting sustainable sourcing practices (Andreadou et al., 2018).

This exploration showcases the potential of blockchain to usher in a new era of efficiency, transparency, and resilience within the energy sector. These applications align with the overarching goal of establishing a decentralized and sustainable energy ecosystem, marking a significant chapter in the ongoing evolution of energy technologies (Andreadou et al., 2016).

A ground-breaking approach to revolutionize information exchange in construction projects has also been explored using a Blockchain-based Building Information Modeling (BIM) data provenance model (Celik et al., 2023). Recent advances in BIM have provided new avenues for process and data integration, necessitating mechanisms to define information priority, provenance, and suitability. The integration of Blockchain technology offers a secure and transparent framework for recording metadata associated with BIM objects, including issuing discipline, object version, and responsibility/liability. Through real-world testing in a bridge construction scenario and a comprehensive cost analysis supported by an Ethereum public test network, the model demonstrates its ability to recognize levels of competence and significantly improve the BIM implementation process. The result is increased stakeholder confidence in data sharing, reduced costs, and enhanced risk management in construction projects. The inclusion of a user-friendly front-end web page further facilitates interaction with smart contracts, making the solution not only robust but also accessible to industry practitioners (Mohammadi & Taylor, 2017).

Blockchain technology holds immense potential in transforming and optimizing various aspects of smart city development. In a smart city context, blockchain can contribute to enhancing security, efficiency, and transparency in several key areas:

(a) ***Secure Transactions*:** Blockchain's inherent security features can fortify financial transactions within a smart city. Whether it's for smart payments, utility billing, or financial transactions related to city services, the decentralized and tamper-resistant nature of blockchain helps prevent fraud and ensures the integrity of transactions (Albino et al., 2015).

(b) ***Identity Management*:** Implementing blockchain for identity verification and management can enhance the security and privacy of citizen data. This is particularly crucial for accessing smart city services, managing personal information securely, and ensuring that only authorized entities have access to sensitive data (Sriyono, 2020).

(c) ***Supply Chain Management*:** Blockchain can be employed to trace and authenticate the origin and delivery of goods within a smart city. This is especially relevant for supply chains related to public services, emergency response, and infrastructure development, providing a transparent and immutable ledger of the entire supply chain (Ren et al., 2019).

(d) ***Smart Contracts for Governance*:** Smart contracts, self-executing contracts with the terms of the agreement directly written into code, can automate and streamline various governance processes in a smart city. This includes contract management, regulatory compliance, and transparent execution of city policies, reducing bureaucracy and enhancing efficiency (Wong et al., 2020).

(e) ***Energy Management***: Blockchain can optimize energy distribution and consumption in a smart city. Through a decentralized energy grid, residents and businesses can efficiently trade excess energy, and blockchain can enable transparent and automated energy transactions, leading to a more sustainable and resilient energy infrastructure (Kim & Huh, 2018).

(f) ***Data Security and Privacy***: The decentralized and cryptographic features of blockchain enhance data security and privacy. In a smart city, where vast amounts of data are collected from various sensors and devices, ensuring the confidentiality and integrity of this data is crucial. Blockchain can provide a secure and transparent framework for managing and sharing data (Fernandez-Carames & Fraga-Lamas, 2019).

(g) ***Citizen Engagement***: Blockchain can facilitate secure and transparent voting systems, enabling citizens to participate in decision-making processes in a trusted and tamper-proof manner. This fosters greater civic engagement and trust in the democratic processes within a smart city.

Blockchain technology can play a transformative role in electronic voting (e-voting) systems by addressing key challenges related to security, transparency, and trust. Here are several ways in which blockchain can be applied to enhance e-voting:

(a) ***Immutable Record***: Blockchain's core feature of immutability ensures that once a vote is recorded, it cannot be altered or tampered with. Each vote is securely stored in a block, creating a transparent and unchangeable ledger of all transactions (Alvi et al., 2022).

(b) ***Transparency and Audibility***: Blockchain provides a transparent and publicly accessible ledger of all transactions. This transparency allows voters, election officials, and other stakeholders to independently verify the integrity of the voting process, thereby enhancing trust in the electoral system (Abayomi-Zannu et al., 2020).

(c) ***Decentralization***: The decentralized nature of blockchain makes it resistant to single points of failure or manipulation. By distributing the voting records across a network of nodes, the risk of a centralized attack or manipulation is significantly reduced, enhancing the overall security of the e-voting system (Cui et al., 2018).

(d) ***Cryptography for Security***: Blockchain employs cryptographic techniques to secure transactions. In e-voting, this ensures the confidentiality and integrity of each vote. Advanced cryptographic methods can also enable voters to verify that their vote is included in the final tally without compromising the secrecy of their individual choices (Arumugam et al., 2018).

(e) ***Smart Contracts for Automation***: Smart contracts, self-executing code on the blockchain, can automate various aspects of the voting process. This includes verifying voter eligibility, ensuring the accuracy of votes, and automatically tallying results. Smart contracts can reduce the potential for human errors and manipulation (Caro et al., 2018).

(f) ***Voter Authentication*****:** Blockchain can be used to implement robust voter authentication systems. Each voter can have a unique cryptographic key, and only those with the correct credentials can participate in the voting process. This helps prevent unauthorized access and ensures the integrity of the voter pool (Reinecke & Prinsloo, 2017).

(g) ***Post-Election Audits*****:** After the election, blockchain allows for efficient and transparent post-election audits. Election officials, as well as external auditors, can independently verify the results by examining the blockchain, contributing to the overall credibility of the electoral process (Hassini, 2008).

(h) ***Accessibility and Convenience*****:** Blockchain-based e-voting systems can improve accessibility for voters, particularly those who may face challenges in traditional voting methods. Blockchain can enable secure and convenient voting from remote locations, contributing to increased voter turnout (Hassini, 2008).

Blockchain technology is increasingly being leveraged in provenance applications to enhance transparency, traceability, and authenticity across various industries. Provenance refers to the origin, authenticity, and history of a product or asset, and blockchain's decentralized and immutable nature makes it particularly well-suited for addressing provenance challenges. Here are key ways in which blockchain is utilized in provenance (Sigwart et al., 2020):

(a) ***Supply Chain Traceability*****:** Blockchain enables end-to-end traceability in supply chains by recording every transaction or movement of goods on a tamper-resistant ledger. This is particularly crucial in industries such as food and pharmaceuticals, where consumers and regulators demand transparency regarding the origin and journey of products (Dutta et al., 2020).

(b) ***Authentication of Origin*****:** Blockchain provides a secure and unalterable record of the origin of products. This is valuable for industries dealing with high-value goods, art, luxury items, and even agricultural products, as it assures consumers of the authenticity and quality of the products they purchase (Pandey & Litoriya, 2020a).

(c) ***Anti-Counterfeiting Measures*****:** By recording each step in the production and distribution process, blockchain helps combat counterfeiting. Consumers can verify the authenticity of products by accessing the blockchain to confirm that the product they are buying is genuine (Pandey & Litoriya, 2020e).

(d) ***Diamond and Jewelry Industry*****:** Blockchain is used to track the provenance of diamonds and other precious gems. This ensures that these items are ethically sourced, addressing concerns related to conflict diamonds and promoting fair trade practices (Thakker et al., 2021).

(e) ***Intellectual Property Protection*****:** Blockchain can be employed to timestamp and verify the ownership and creation date of digital content, including art, music, and literature. This provides creators with a secure and verifiable proof of their intellectual property rights (Luo, 2022).

(f) ***Carbon Footprint Tracking***: In industries focused on sustainability, blockchain is used to trace the environmental impact of products. This allows consumers to make informed choices based on the carbon footprint of items they purchase (Kohli et al., 2023).

(g) ***Document Verification***: Blockchain can be used to verify the authenticity of documents, certificates, and licenses. This has applications in academic credentials, professional certifications, and legal documents, ensuring the integrity and provenance of critical records (S.S. Latha et al., 2022).

(h) ***Wine and Spirits Industry***: Blockchain is employed to verify the authenticity and origin of wines and spirits. This is crucial for both connoisseurs seeking assurance about the quality and origin of the product and producers aiming to protect their brand reputation (Popović et al., 2021).

By leveraging blockchain in provenance applications, industries can instill trust among consumers, reduce fraud, and adhere to regulatory requirements. The decentralized and secure nature of blockchain ensures that the provenance data remains tamper-proof, providing a robust solution to challenges related to traceability and authenticity.

In concluding this literature survey, it is evident that the exploration of various scholarly works has provided valuable insights into the multifaceted landscape of blockchain technology. The synthesis of existing research has illuminated key trends, challenges, and innovative solutions within the field, contributing to a comprehensive understanding of the current state of knowledge.

The collective body of literature reviewed underscores the significance of blockchain use cases in a myriad of applications. Additionally, it has identified critical gaps and areas where further investigation is warranted to advance the understanding of blockchain technology in impacting human life with little impact on the environment due to its high energy requirements. The diverse perspectives presented in the literature highlight the interdisciplinary nature of blockchain technology and emphasize the need for collaborative efforts among researchers, practitioners, and policymakers.

As we move forward, it is imperative to build upon the foundations laid by previous research, embracing emerging technologies, methodologies, and theoretical frameworks. The synthesis of this literature survey serves not only as a testament to the progress made in blockchain technology but also as a call to action for continued exploration and innovation.

Ultimately, this literature survey aims to serve as a valuable resource for researchers, educators, and industry professionals alike, fostering a deeper appreciation for the complexities and opportunities inherent in blockchain technology. By acknowledging the achievements of the past and embracing the challenges of the future, we can collectively contribute to the dynamic evolution of knowledge in blockchain applications and pave the way for advancements that hold the potential to shape our understanding and practices in meaningful ways.

17.3 Method and Classification

A block comprises a structure encompassing a header and transactions. The block header incorporates a timestamp (indicating the block's creation time), the hash of the previous block, the hash of the current block, and the Merkle root. The current block's hash is calculated using the hash of the preceding block and the Merkle root of the transactions (Satoshi, 2008). Consequently, each new block is appended to the previous block, forming a sequential chain of blocks. Figure 17.3 illustrates that the block hash of block_1, referred to as the genesis block, acts as a key for computing the hash of block_2. Similarly, the block hash of block_2 is an input for block_3, and so forth. This arrangement ensures the tamper-proof nature of transactions. If an adversary manipulates a transaction to their advantage, altering the Merkle root representing the hash of all candidate transactions and consequently the hash for that specific block, any subsequent block linked to it would be affected. Such disharmony triggers rejection by the network, illustrating the collaborative role of the entire network in upholding security and guaranteeing transaction integrity.

Upon grasping the foundational aspects of Blockchain, delving into its operational methodology becomes more accessible. Figure 2 illustrates a general structure of the Blockchain, disseminated across various nodes within the distributed network. This strategic dissemination ensures fault tolerance, as the network remains functional even if one or more nodes experience downtime. Furthermore, the collective state of the network is contingent upon the majority consensus among the nodes. In the event of compromise on one or more nodes, the system defaults to the state held by the majority, adhering to a democratic principle. The efficacy of this democratic approach lies in the inherent difficulty of hacking a single node, making it a formidable task (Chang et al., 2020). The impracticality of compromising 51% of the nodes in the network further fortifies the security of the system. Thus, the larger the Blockchain network, the more impregnable its security posture becomes.

Three fundamental types of blockchains exist: public, private, and hybrid. Public and private blockchains share similarities in being distributed networks secured by consensus protocols and cryptography, and both exhibit immutability. However, they differ in terms of network participation, consensus protocol execution, and ledger maintenance (Laura Gargolinski Jaeger, 2018). A prominent example of a public blockchain is the Bitcoin blockchain, where anyone worldwide can read and write. In contrast, a private blockchain permits only trustworthy and recognized participants to engage in reading or writing activities. The hybrid blockchain combines the privacy advantages of a permissioned private blockchain with the security and transparency benefits of a public blockchain. This implementation offers flexibility to business associates, allowing them to choose which data to keep private and which to make public and transparent (Upadhyay, 2020).

Beyond permission considerations, another crucial parameter distinguishing blockchains is the consensus mechanism. Public blockchains achieve consensus through time-consuming and computationally intensive algorithms, such as PoW. PoW serves as a deterrent for potential attackers, enhancing network security (Parikshit Hooda, 2018). Given the significant computational power costs associated with PoW, the industry is exploring alternative consensus approaches like PoS.In contrast, private blockchains are discerning in granting participation permissions and employ a mechanism known as selective endorsement. In this method, predetermined endorsers are entrusted with the responsibility of validating or invalidating a transaction. Performance-wise, private blockchains exhibit higher transaction throughput than their public counterparts. Public blockchains, relying on PoW, require more time and computational power to approve a transaction, contributing to the performance differential between the two types of blockchains.

17.4 Implementation

In the exploration of blockchain theory, the selection of an appropriate platform is pivotal to align with the specific requirements of a given application. For applications necessitating the use of public blockchains, "Ethereum" emerges as a popular and versatile choice. Known for its decentralized nature and support for smart contracts, Ethereum is often preferred for decentralized applications (DApps) where transparency and a vast network of nodes are paramount.

Bitcoin, another prominent example in the area of public blockchains, serves as a foundational platform, particularly for applications seeking a secure and decentralized digital currency or a public ledger. Its use case primarily revolves around cryptocurrency transactions, showcasing the adaptability of public blockchains to various applications.

On the other hand, for those who prioritize building private blockchain-based applications, "Hyperledger Fabric" stands out as a robust implementation platform. Tailored for enterprises, Hyperledger Fabric offers a permissioned and private blockchain environment, allowing for greater control over access, privacy, and scalability. Developers can leverage multiple programming languages, such as Go and JavaScript, making it a versatile choice for tailored enterprise solutions (Dubey et al., 2020).

When navigating the theory of blockchain, considerations extend beyond the public-private spectrum. Factors such as the support for smart contracts, the consensus mechanism employed (e.g., PoW, PoS), and scalability play pivotal roles in determining the suitability of a blockchain platform. In this theoretical exploration, understanding the nuances of Ethereum, Bitcoin, and Hyperledger Fabric aids in making informed decisions based on the unique requirements and objectives of a given blockchain-based application (Leible et al., 2019).

The definition of a smart contract is warranted at this juncture, which at the implementation level are computer codes executed upon the fulfilment of some conditions. Table 17.1 shows a smart contract written for marriage contracts in Hindu community. The brief explanation of the smart contract conditions is also provided here:

(a) ***Deployment Authorization***
Condition: Only the state-authorized executive, the registrar, can deploy this contract.
Explanation: The smart contract can only be deployed by a specific entity, in this case, the registrar. This ensures that only authorized individuals have the capability to initiate the contract.

(b) ***Payment Requirement***
Condition: The parties (husband and wife) entering into the contract must pay a certain price in ethers.
Explanation: Both the husband and wife need to fulfill a payment requirement in ethers to participate in the contract. This condition could represent a financial commitment or a symbolic gesture within the context of the Hindu marriage.

(c) ***Selection of Party for Fund Release***
Condition: The registrar picks one party (husband or wife) in whose name the funds in the account will be released.
Explanation: The registrar has the authority to choose one of the parties, either the husband or the wife. The funds in the account will be released only in the name of the selected party.

Table 17.1 Smart Contract for Smart Marriage.

pragma solidity 0.4.17;
contract Marriage {
address public registrar;
address [] public parties;
function Marriage () public {
registrar = msg.sender;
}
function enter () public payable {
require (msg.value >= 1 ether);
parties.push(msg.sender);
}
	function getParties() public view returns(address[]) {
	return parties
	}
	function pickParty() public restricted {
	uint index=0; //assuming female first
	parties[index]. transfer(this.balance);
	parties = new address [] (0);
	}
	
	modifier restricted () {
	require (msg.sender == registrar);
	_;
	}
	}

Figure 17.4. Typically entails a permissioned blockchain based architecture given by (Pandey & Litoriya, 2020c). The different architecture components are explained briefly with their purpose and roles.

(a) ***Clients***
Purpose: Clients are nodes that interact with the blockchain by submitting transaction requests.
Role: They do not maintain a replica of the blockchain. Their primary purpose is to initiate transactions and interact with the network.

(b) ***Endorsers***

Purpose: Endorsers receive the transaction requests from clients.

Role: They simulate the execution of the transaction to check if it meets the endorsement policies. If satisfied, they endorse the transaction. Endorsers do not maintain a full copy of the blockchain but store the replica temporarily during the transaction validation process.

(c) ***Ordering Service***

Purpose: Collects endorsed transactions, orders them into blocks, and sends the blocks to committers and endorsers.

Role: It ensures a consistent order of transactions across the network. The ordering service is a critical component for achieving consensus on the order of transactions.

(d) ***Committers***

Purpose: Receive ordered blocks from the ordering service.

Role: They commit the blocks to the blockchain. Committers maintain a replica of the blockchain and ensure that the transactions are appended in the agreed-upon order. They form the part of the network that maintains the ledger.

The architecture is decentralized because the blockchain is replicated across multiple nodes, specifically endorsers and committers (Casino et al., 2019). This replication ensures redundancy and fault tolerance, contributing to the overall robustness of the system.

Note that this architecture is designed to ensure that transactions are endorsed by the required parties before being included in the blockchain, providing a level of trust and security (Leeming et al., 2019).

The separation of endorsement and ordering helps in achieving a modular and scalable architecture, where different components can be scaled independently based on the network's needs (FICCI & PwC, 2020).

This architecture is in line with the principles of Hyperledger Fabric, a permissioned blockchain framework suitable for enterprise use cases.

17.5 Use Cases

In the exploration of the theory of blockchain, this chapter delves into the practical applications of blockchain technology across diverse domains. The focus is on five distinct use cases—drug supply chain, health insurance, land record management, courier services, and immigration records. An underlying premise of this discussion is the recognition that the choice of blockchain type is paramount and must be tailored to the specific needs of each use case.

Use Case 1: Drug Supply Chain

The drug supply chain grapples with vulnerabilities such as counterfeit medicines and the illicit reselling of prescribed drugs. Here, the authors propose a blockchain-based solution that incorporates QR Codes on drug sachets. This innovative

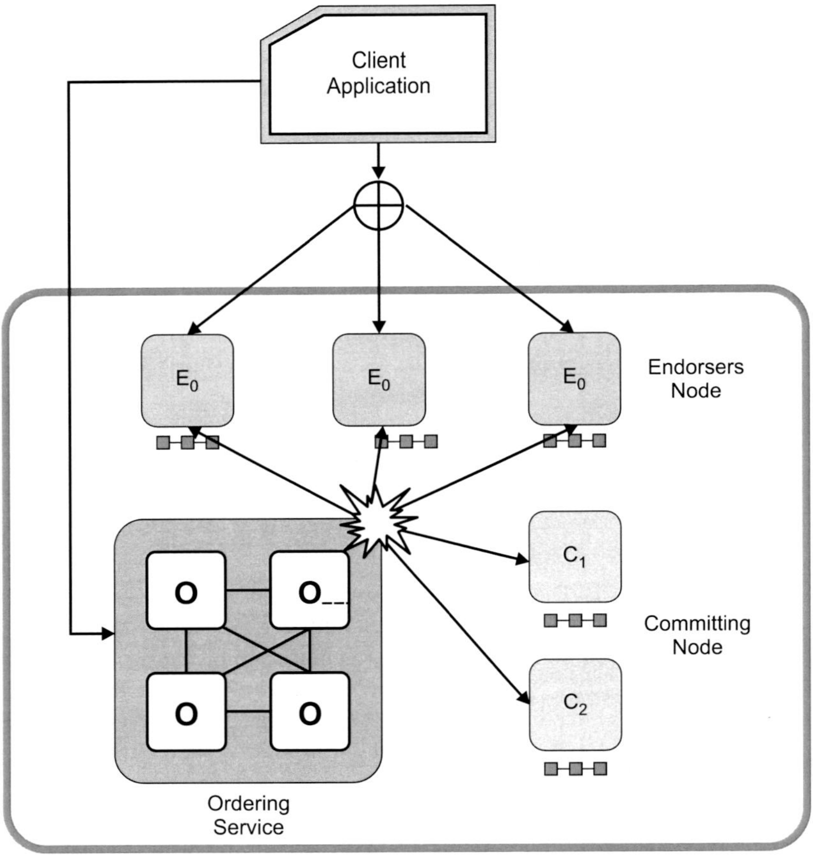

Fig. 17.4 Blockchain architecture based on Hyperledger Fabric.
Note: Typically entails a permissioned blockchain based architecture given by (Pandey & Litoriya, 2020c). The different architecture components are explained briefly with their purpose and roles.

approach ensures the traceability of drug provenance, enabling customers to authenticate their purchases through authorized mobile apps. Furthermore, the blockchain ledger records each drug sale against a prescription, mitigating the risk of unauthorized resale (Pandey & Litoriya, 2020e).

Use Case 2: Health Insurance Schemes
The health insurance industry in India faces substantial annual losses due to fraudulent claims. The chapter explores the application of blockchain technology in creating a nationwide ledger that records critical health-related events. This comprehensive blockchain not only curtails fraudulent claims by documenting hospitalization and pre-existing conditions but also enhances the quality of

treatment by providing treating physicians with instant access to patients' historical health data (Pandey & Litoriya, 2020b).

Use Case 3: Land Record Management
In the realm of land record management, particularly pertinent to India, the authors advocate for blockchain adoption to secure records from tampering. Beyond preventing forged records and property disputes, blockchain ensures transparency and reliability in land transactions by providing an immutable and shared source of truth (Soner et al., 2021).

Use Case 4: Courier Services
Fraudulent activities in courier services, especially within the burgeoning online business landscape, pose challenges to customer trust. The proposed blockchain solution involves recording consignment handovers and utilizing X-Ray imagery for enhanced security. This not only deters forgery but also restores trust in e-commerce activities by safeguarding the integrity of the delivery process.

Use Case 5: Immigration Records
The chapter concludes with a timely exploration of the potential application of blockchain in recording immigration data, especially relevant during a pandemic such as COVID-19. A blockchain-based system could effectively trace the travel history of individuals, offering a reliable means to verify visitors' travel trails and implement measures to contain the spread of infectious diseases (Pandey & Litoriya, 2021).

The authors acknowledge that centralized systems could address these challenges but assert the superiority of decentralized blockchain systems. The inherent vulnerability of centralized systems to attacks, combined with the imperative of establishing trust among stakeholders, underscores the suitability of a decentralized blockchain approach for these use cases. This chapter contributes to the theoretical foundation of blockchain by examining its real-world applications in diverse sectors, emphasizing its transformative potential.

17.6 Future Trajectories of Blockchain Technology in the Theoretical Context

Within the discourse on the theory of blockchain, an examination of the prospective trajectories of this transformative technology becomes imperative. This chapter scrutinizes anticipated future trends and developments, considering their potential impact on the theoretical underpinnings of blockchain within diverse contexts:

a) ***Integration with Established Systems:*** Within the theoretical framework, an imminent integration of blockchain with prevailing systems across industries is envisioned. This amalgamation is poised to introduce heightened transparency, security, and operational efficiency, catalyzing a paradigm shift in processes ranging from supply chain dynamics to financial transactions (Srivastava et al., 2021).

b) ***Orchestrating Interoperability:*** The theoretical landscape anticipates a concerted effort towards achieving interoperability among disparate blockchains. Initiatives and protocols facilitating seamless communication and collaboration are positioned to play a pivotal role, promoting a harmonious coexistence of blockchain networks.

c) ***Scaling Operational Efficiency:*** The theoretical narrative foresees decisive advancements in addressing the scalability challenge confronting numerous blockchain networks. Innovations, including sharding and layer-two scaling solutions, are projected to redefine the operational capabilities of blockchain, expanding its applicability.

d) ***Sustainability Imperatives:*** Within the theoretical discourse, environmental considerations cast a discerning gaze on certain blockchain implementations. A theoretical shift towards sustainability is postulated, with potential transitions to eco-friendly consensus mechanisms, such as PoS, and the integration of energy-efficient technologies.

e) ***Emergence of Central Bank Digital Currencies (CBDCs):*** A pivotal chapter in the theoretical landscape unfolds with the emergence of Central Bank Digital Currencies (CBDCs). Theoretical discussions explore the transformative potential of nations developing their digital currencies, redefining the financial landscape by leveraging the foundational principles of blockchain.

f) ***Symbiosis of Blockchain and the Internet of Things (IoT):*** Harmony between blockchain and the IoT is envisioned within the theoretical realm. The symbiotic relationship is expected to enhance the security and reliability of data generated by IoT devices, with theoretical applications spanning supply chain optimization to healthcare innovation.

g) ***Crescendo of Decentralized Finance (DeFi):*** The theoretical landscape resonates with the crescendo of decentralized finance (DeFi). The maturation and expansion of the DeFi space are poised to potentially reshape traditional financial paradigms, presenting a decentralized alternative to conventional intermediaries.

h) ***Privacy-Centric Transactions:*** Within the theoretical discourse, privacy and security enhancements take a central role. Theoretical innovations, such as zero-knowledge proofs and privacy-centric coins, are posited to fortify the confidentiality of transactions, addressing concerns and instilling trust.

i) ***Blockchain's Impact on Healthcare:*** The healthcare sector emerges prominently within the theoretical narrative, exploring blockchain's potential to securely manage health data. Theoretical considerations envision streamlined patient record-keeping, resilient medical supply chains, and optimized clinical trials, ushering in an era of efficiency and trust (Pandey & Litoriya, 2020d).

j) ***Regulatory Frameworks:*** As the theoretical framework matures, regulatory considerations take prominence. Clearer theoretical regulations are envisioned to guide the trajectory of blockchain, fostering increased institutional adoption and mainstream acceptance, and delineating the parameters within which blockchain innovations unfold.

In these theoretical deliberations, we embark on a speculative exploration, envisioning a future where the theoretical foundations of blockchain converge with practical applications, reshaping industries and redefining paradigms. The theory of blockchain, characterized by its dynamic nature, invites scholarly anticipation and contemplation of the transformative journey that lies ahead.

17.7 Conclusion

The chapter bridges the gap between theory and practice, unveiling the practical significance of concepts like distributed systems, cryptography, and linked list structures in the context of Blockchain. It exemplifies how Ethereum, Bitcoin, and Hyperledger Fabric serve as living testaments to theoretical principles, and how real-world use cases, from drug supply chains to healthcare systems, attest to Blockchain's transformative potential.

Looking ahead, the speculative trajectories provide a glimpse into a future where Blockchain seamlessly integrates with established systems, achieves interoperability, scales operational efficiency, and leaves a significant mark on diverse sectors. This chapter, therefore, is not just an endpoint but a launchpad for further exploration into the ever-evolving odyssey of Blockchain. As we stand on the cusp of technological redefinition, Blockchain beckons us to chart the unexplored territories, inviting scholars, practitioners, and enthusiasts to join in this unfolding journey of innovation, trust, and decentralized possibilities.

References

1. Abayomi-Zannu, T.P., Odun-Ayo, I., Tatama, B.F. and Misra, S. (2020). Implementing a Mobile Voting System Utilizing Blockchain Technology and Two-Factor Authentication in Nigeria. *Journal of Physics, Conference Series*, 857–872. https://doi.org/10.1007/978-981-15-3369-3_63.
2. Albino, V., Berardi, U. and Dangelico, R.M. (2015). Smart Cities: Definitions, Dimensions, Performance, and Initiatives. *Journal of Urban Technology*, *22*(1), 3–21. https://doi.org/10.1080/10630732.2014.942092.
3. Alvi, S.T., Uddin, M.N., Islam, L. and Ahamed, S. (2022). DVTChain: A blockchain-based decentralized mechanism to ensure the security of digital voting system. *Journal of King Saud University - Computer and Information Sciences*, *34*(9), 6855–6871. https://doi.org/10.1016/j.jksuci.2022.06.014.
4. Andreadou, N., Guardiola, M. and Fulli, G. (2016). Telecommunication Technologies for Smart Grid Projects with Focus on Smart Metering Applications. *Energies*, *9*(5), 375. https://doi.org/10.3390/en9050375.
5. Andreadou, N., Kotsakis, E. and Masera, M. (2018). Smart Meter Traffic in a Real LV Distribution Network. *Energies*, *11*(5), 1156. https://doi.org/10.3390/en11051156.
6. Andreadou, N., Lucas, A., Tarantola, S. and Poursanidis, I. (2019). Design of Experiments in the Methodology for Interoperability Testing: Evaluating AMI Message Exchange. *Applied Sciences*, *9*(6), 1221. https://doi.org/10.3390/app9061221.

7. Androulaki, E., Barger, A., Bortnikov, V., Cachin, C., Christidis, K., De Caro, A., Enyeart, D., Ferris, C., Laventman, G., Manevich, Y., Muralidharan, S., Murthy, C., Nguyen, B., Sethi, M., Singh, G., Smith, K., Sorniotti, A., Stathakopoulou, C., Vukolić, M., … Yellick, J. (2018). Hyperledger fabric. In: *Proceedings of the Thirteenth EuroSys Conference*, 1–15. https://doi.org/10.1145/3190508.3190538.
8. Arumugam, S.S., Umashankar, V., Narendra, N. C., Badrinath, R., Mujumdar, A.P., Holler, J. and Hernandez, A. (2018). IOT Enabled Smart Logistics Using Smart Contracts. In: *2018 8th International Conference on Logistics, Informatics and Service Sciences* (*LISS*), 1–6. https://doi.org/10.1109/LISS.2018.8593220.
9. Ben Fekih, R. and Lahami, M. (2020). *Application of Blockchain Technology in Healthcare: A Comprehensive Study*. Springer Nature, pp. 268–276. https://doi.org/10.1007/978-3-030-51517-1_23.
10. Caro, M.P., Ali, M.S., Vecchio, M. and Giaffreda, R. (2018).. Blockchain-based traceability in Agri-Food supply chain management: A practical implementation. In: *2018 IoT Vertical and Topical Summit on Agriculture - Tuscany* (*IOT Tuscany*), 1–4. https://doi.org/10.1109/IOT-TUSCANY.2018.8373021.
11. Casino, F., Dasaklis, T.K. and Patsakis, C. (2019). A systematic literature review of blockchain-based applications: Current status, classification and open issues. *Telematics and Informatics*, *36*, 55–81. https://doi.org/10.1016/j.tele.2018.11.006.
12. Celik, Y., Petri, I. and Barati, M. (2023). Blockchain supported BIM data provenance for construction projects. *Computers in Industry*, *144*, 103768. https://doi.org/10.1016/j.compind.2022.103768.
13. Chang, Y., Iakovou, E. and Shi, W. (2020). Blockchain in global supply chains and cross border trade: A critical synthesis of the state-of-the-art, challenges, and opportunities. *International Journal of Production Research*, *58*(7), 2082–2099. https://doi.org/10.1080/00207543.2019.1651946.
14. Cui, L., Xie, G., Qu, Y., Gao, L. and Yang, Y. (2018). Security and Privacy in Smart Cities: Challenges and Opportunities. *IEEE Access*, *6*, 46134–46145. https://doi.org/10.1109/ACCESS.2018.2853985.
15. Dubey, R., Gunasekaran, A., Bryde, D.J., Dwivedi, Y.K. and Papadopoulos, T. (2020). Blockchain technology for enhancing swift-trust, collaboration, and resilience within a humanitarian supply chain setting. *International Journal of Production Research*, *58*(11), 3381–3398. https://doi.org/10.1080/00207543.2020.1722860.
16. Dutta, P., Choi, T.-M., Somani, S. and Butala, R. (2020). Blockchain technology in supply chain operations: Applications, challenges, and research opportunities. *Transportation Research Part E: Logistics and Transportation Review*, *142*, 102067. https://doi.org/10.1016/j.tre.2020.102067.
17. Eslamian, S., and Eslamian, F. (2025). *Blockchain Technology in Water and Environmental Systems*. CRC Press, Taylor and Francis, USA, 282 pp. Fernandez-Carames, T.M. and Fraga-Lamas, P. (2019). A Review on the Application of Blockchain to the Next Generation of Cybersecure Industry 4.0 Smart Factories. *IEEE Access*, *7*, 45201–45218. https://doi.org/10.1109/ACCESS.2019.2908780.
18. FICCI and PwC. (2020). *Revamping India's Health Insurance Sector with Blockchain and Smart Contracts*. https://www.pwc.in/assets/pdfs/healthcare/revamping-indias-health-insurance-sector-with-blockchain-and-smart-contracts.pdf.
19. Galli, S., Scaglione, A. and Wang, Z. (2011). For the Grid and Through the Grid: The Role of Power Line Communications in the Smart Grid. In: *Proceedings of the IEEE*, *99*(6), 998–1027. https://doi.org/10.1109/JPROC.2011.2109670.
20. Gungor, V.C., Sahin, D., Kocak, T., Ergut, S., Buccella, C., Cecati, C. and Hancke, G.P. (2011). Smart Grid Technologies: Communication Technologies and Standards. *IEEE Transactions on Industrial Informatics*, *7*(4), 529–539. https://doi.org/10.1109/TII.2011.2166794.

21. Hassini, E. (2008). Supply Chain Optimization: Current Practices and Overview of Emerging Research Opportunities. *INFOR: Information Systems and Operational Research*, *46*(2), 93–96. https://doi.org/10.3138/infor.46.2.93.
22. Kim, S.-K. and Huh, J.-H. (2018). A Study on the Improvement of Smart Grid Security Performance and Blockchain Smart Grid Perspective. *Energies*, *11*(8), 1973. https://doi.org/10.3390/en11081973.
23. Kohli, V., Chakravarty, S., Chamola, V., Sangwan, K.S. and Zeadally, S. (2023). An analysis of energy consumption and carbon footprints of cryptocurrencies and possible solutions. *Digital Communications and Networks*, *9*(1), 79–89. https://doi.org/10.1016/j.dcan.2022.06.017.
24. Latha, S.S., N, Priya. and Shettar, A. (2022). Block chain Based Framework for Document Verification. In: *2022 2nd International Conference on Artificial Intelligence and Signal Processing* (*AISP*), 1–5. https://doi.org/10.1109/AISP53593.2022.9760651.
25. Laura Gargolinski Jaeger. (2018). *Public versus Private: What to Know before Getting Started with Blockchain*. IBM. https://www.ibm.com/blogs/blockchain/2018/10/public-versus-private-what-to-know-before-getting-started-with-blockchain/.
26. Leeming, G., Cunningham, J. and Ainsworth, J. (2019). A Ledger of Me: Personalizing Healthcare Using Blockchain Technology. *Frontiers in Medicine*, *6*. https://doi.org/10.3389/fmed.2019.00171.
27. Leible, S., Schlager, S., Schubotz, M. and Gipp, B. (2019). A Review on Blockchain Technology and Blockchain Projects Fostering Open Science. *Frontiers in Blockchain*, *2*. https://doi.org/10.3389/fbloc.2019.00016.
28. Luo, L. (2022). Application of Blockchain Technology in Intellectual Property Protection. *Mathematical Problems in Engineering*, *2022*, 1–12. https://doi.org/10.1155/2022/4641559.
29. Mettler, M. (2016). Blockchain technology in healthcare: The revolution starts here. In: *2016 IEEE 18th International Conference on E-Health Networking, Applications, and Services* (*Healthcom*), 1–3. https://doi.org/10.1109/HealthCom.2016.7749510.
30. Mohammadi, N. and Taylor, J.E. (2017). Smart city digital twins. In: *2017 IEEE Symposium Series on Computational Intelligence* (*SSCI*), 1–5. https://doi.org/10.1109/SSCI.2017.8285439.
31. Moroni, S., Alberti, V., Antoniucci, V. and Bisello, A. (2019). Energy communities in the transition to a low-carbon future: A taxonomical approach and some policy dilemmas. *Journal of Environmental Management*, *236*, 45–53. https://doi.org/10.1016/j.jenvman.2019.01.095.
32. Pandey, P. and Litoriya, R. (2020a). Ensuring elderly well being during COVID-19 by using IoT. *Disaster Medicine and Public Health Preparedness*, *16*(2), 763–766. https://doi.org/10.1017/dmp.2020.390.
33. Pandey, P. and Litoriya, R. (2020b). Implementing healthcare services on a large scale: Challenges and remedies based on blockchain technology. *Health Policy and Technology*, *9*(1), 69–78. https://doi.org/10.1016/j.hlpt.2020.01.004.
34. Pandey, P. and Litoriya, R. (2020c). Promoting Trustless Computation through Blockchain Technology. *National Academy Science Letters*, *44*, 225–231. https://doi.org/10.1007/s40009-020-00978-0.
35. Pandey, P. and Litoriya, R. (2020d). Securing and authenticating healthcare records through blockchain technology. *Cryptologia*, *44*(4), 341–356. https://doi.org/10.1080/01611194.2019.1706060.
36. Pandey, P. and Litoriya, R. (2020e). Securing E-health networks from counterfeit medicine penetration using Blockchain. *Wireless Personal Communications*, *117*, 7–25. https://doi.org/10.1007/s11277-020-07041-7.
37. Pandey, P. and Litoriya, R. (2021). Technology intervention for preventing COVID-19 outbreak. *Information Technology & People*, *34*(4), 1233–1251. https://doi.org/10.1108/ITP-05-2020-0298.
38. Parikshit Hooda. (2018). *Proof of Work* (*PoW*) *Consensus*. https://www.geeksforgeeks.org/proof-of-work-pow-consensus/.

39. Popović, T., Krčo, S., Maraš, V., Hakola, L., Radonjić, S., van Kranenburg, R. and Šandi, S. (2021). A novel solution for counterfeit prevention in the wine industry based on IoT, smart tags, and crowd-sourced information. *Internet of Things*, *14*, 100375. https://doi.org/10.1016/j.iot.2021.100375.
40. Reinecke, M. and Prinsloo, T. (2017). The influence of drone monitoring on crop health and harvest size. In: *2017 1st International Conference on Next Generation Computing Applications* (*NextComp*), 5–10. https://doi.org/10.1109/NEXTCOMP.2017.8016168.
41. Ren, Q., Man, K. L., Li, M. and Gao, B. (2019). Using Blockchain to Enhance and Optimize IoT-based Intelligent Traffic System. In: *2019 International Conference on Platform Technology and Service* (*PlatCon*), 1–4. https://doi.org/10.1109/PlatCon.2019.8669412.
42. Satoshi, N. (2008). *Bitcoin: A Peer-to-Peer Electronic Cash System*. SCIRP. https://bitcoin.org/bitcoin.pdf.
43. Sigwart, M., Borkowski, M., Peise, M., Schulte, S. and Tai, S. (2020). A secure and extensible blockchain-based data provenance framework for the Internet of Things. *Personal and Ubiquitous Computing*. https://doi.org/10.1007/s00779-020-01417-z.
44. Soner, S., Litoriya, R. and Pandey, P. (2021). Exploring Blockchain and Smart Contract Technology for Reliable and Secure Land Registration and Record Management. *Wireless Personal Communications*, *121*(4), 2495–2509. https://doi.org/10.1007/s11277-021-08833-1.
45. Srivastava, S., Gupta, R., Pandey, P. and Litoriya, R. (2021). Ethical and legal aspects of using blockchain technology for 5G-based health-care systems. In: *Blockchain for 5G Healthcare Applications: Security and Privacy Solutions*. Institution of Engineering and Technology, pp. 35–51. https://doi.org/10.1049/PBHE035E_ch2.
46. Sriyono, E. (2020). Digitizing water management: Toward the innovative use of blockchain technologies to address sustainability. *Cogent Engineering*, *7*(1), 1769366. https://doi.org/10.1080/23311916.2020.1769366.
47. Thakker, U., Patel, R., Tanwar, S., Kumar, N. and Song, H. (2021). Blockchain for Diamond Industry: Opportunities and Challenges. *IEEE Internet of Things Journal*, *8*(11), 8747–8773. https://doi.org/10.1109/JIOT.2020.3047550.
49. Upadhyay, N. (2020). Demystifying blockchain: A critical analysis of challenges, applications, and opportunities. *International Journal of Information Management*, *54*, 102120. https://doi.org/10.1016/j.ijinfomgt.2020.102120.
49. Wong, P.F., Chia, F.C., Kiu, M. S. and Lou, E.C.W. (2020). The potential of integrating blockchain technology into smart sustainable city development. *IOP Conference Series: Earth and Environmental Science*, *463*(1), 012020. https://doi.org/10.1088/1755-1315/463/1/012020
50. Ying, Z., Yi, L. and Ma, M. (2020). BEHT: Blockchain-Based Efficient Highway Toll Paradigm for Opportunistic Autonomous Vehicle Platoon. *Wireless Communications and Mobile Computing*, *2020*, 1–13. https://doi.org/10.1155/2020/8868656.
51. Yoon, H.-J. (2019). Blockchain Technology and Healthcare. *Healthcare Informatics Research*, *25*(2), 59. https://doi.org/10.4258/hir.2019.25.2.59.
52. Zhang, J., Xue, N. and Huang, X. (2016). A Secure System For Pervasive Social Network-Based Healthcare. *IEEE Access*, *4*, 9239–9250. https://doi.org/10.1109/ACCESS.2016.2645904.
53. Zheng, Z., Xie, S., Dai, H., Chen, X. and Wang, H. (2017). An Overview of Blockchain Technology: Architecture, Consensus, and Future Trends. In: *2017 IEEE International Congress on Big Data* (*BigData Congress*), 557–564. https://doi.org/10.1109/BigDataCongress.2017.85.

18 Blockchain and IoT: Advancing Smart Water Management for Sustainable Solutions

Aanandha Saravanan[1] and G. Sasikala[2*]

In recent times there has always been a problem with water scarcity throughout history. With the proliferation of IOT devices and technologies, the water supply chain has also been impacted. As a result of innovative research, blockchain has been developed as a reliable, auditable and transparent traceable system. A decentralized trustless system has been realized with blockchain technology, the distributed ledger technology. Due to its ability to keep records of all transactions, blockchain technology eliminates corruption. Block chains deliver integral chattels of culpability forbearance, immutability, limpidity, and full traceability of chronicles as well as cardinal depictions of possessions and communications. An IoT device to collect and add data along the setup of a Blockchain based water management system. We use Ethereum blockchain to represent a supplier-to-consumer use case in our water management system. As part of the water management system, we also discuss the roles of the participants. In the end, we demonstrated how blockchain-based water management could be a boon to modern society and should be adopted over current approaches. In order to be intelligent to accomplish water possessions more competently, actions must be recognized to encounter the growing demand for water. Contemporary machinery such as the Internet of Devices, Block chain, Resolution Sustenance Schemes, and Artificial Intellect have been used to progress numerous intelligent schemes with the objective of guaranteeing effective water management. Expertise developments nowadays permit the gathering, dispensation, and investigation of information from water dispersal systems so that resources can be enhanced and pronouncement creating can be amended. The artefact pronounces an assimilated podium for supervisory and observing inhabited grids in real-time expending blockchain machineries, with protected data managing. In order to assess the probability of storing real-time information in a dispersed ledger, a blockchain system was established exhausting hyper ledger Fabric.

1 Department of ECE, Vel Tech Rangarajan, Dr. Sagunthala R&D Institute of Science and Technology, Chennai, India.

2 Department of ECE, Vel Tech Rangarajan, Dr.Sagunthala R&D Institute of Science and Technology, Chennai, India.

* Corresponding author: gsasikala@veltech.edu.in

DOI: 10.1201/9781003686026-18

18.1 Introduction

There are presently no arrangements in place for tracing and storing water briefings and distributions. Auditability, transparency, and traceability are not provided by them. However, it can improve water management systems' safety by incorporating blockchain technology. IoT (Internet of Things) devices with cheaper connected devices are being used in water management systems with finer granularity by a number of research and development authorities. Nevertheless, most cloud solutions rely on centralized infrastructures with limited transparency, auditability, and confidentiality. Trust and reliability between all participants are crucial in some water management systems (Green & Eslamian, 2014). Due to the lack of centralized third parties, a distributed ledger system is an excellent solution since it is tamper-proof. No centralized servers are required with distributed ledger technology. There is consensus at the location of each record in the distributed ledger.

There are currently no systems in place for tracking and storing water orders and deliveries. Auditability, transparency, and traceability are not provided by them. However, we can improve water management systems' safety by incorporating blockchain technology. IoT devices with cheaper connected devices are being used in water management systems with finer granularity by a number of research and development authorities. Nevertheless, most cloud solutions rely on centralized infrastructures with limited transparency, auditability, and confidentiality. Trust and reliability between all participants are crucial in some water management systems. Due to the lack of centralized third parties, a distributed ledger system is an excellent solution since it is tamper-proof. No centralized servers are required with distributed ledger technology. The distributed ledger has a consensus at least by majority of peers at the location of each record. The blockchain-based solution would only require a stable connection to the nearest peer, unlike other solutions that require connectivity to a central cloud. Blockchain enables traceability in a wide range of views and domains. Here we present a blockchain-based system for decentralizing water supply chain management. Water management systems based on block chain use the Ethereum distributed ledger. IoT devices' information is stored directly on the blockchain, which enables participants to interact in a transparent manner. Participants in the network are assigned roles based on their roles in the network. Distributed ledger implementations are also shown to be necessary in water management systems. The role and process of each participant is also discussed. Using Blockchain technology to manage water can offer advantages over traditional methods.

An operative health fortification strategy must comprise admittance to safe imbibing water as an essential anthropological factual. In order to accomplish maintainable improvement, it is vital that water resources are managed in a way that ensures the quality of drinking water. Predictions are necessary aimed to progress strategies and formulate for convinced conceivable trials uniform once evidence nearby is not obtainable. IoT mentions an interconnected linkage of insolent things that are accomplished by pouring terminated and collaborating statistics around their interior circumstances and intermingling through the peripheral atmosphere. There is potential for the IoT explosion to transmute the manner of society's effort and sentience, as well as develop commotion, the situation, and the surroundings. Typical applications include an IoT network of sensors distributed geographically along a river connected to the Internet.

It may be possible to connect a grid of sensors through the Internet, which are geologically positioned alongside a river and monitor numerous constraints. As IoT has gained traction for isolated observing and computerization of water possessions, reassuring consequences have stayed and testified, but privacy and data security are major concerns (Angara & Saripalle, 2022). Using blockchain, the IoT network can create a distributed digital registry that shares transactions across nodes to overcome these security challenges. In addition to bitcoin, blockchain can be treated as a platform or a database, and developed into applications (Gurudas & Adarsh, 2020). As well as decision-making, blockchain technology can be applied to social governance (Hakak & Saqib, 2020). In addition to providing a distributed and decentralized database, it possesses two other properties that make it an essential platform (Hakak & Saqib, 2020).

***Transparency*:** Every node on the network shares the same data, and therefore contains the history of all decisions made until that time. New decisions will take previous decisions into account as a result of this property. Nodes in a network can be treated as crucial agents in decision-making, each having an integral role to play.

***Data Incorruptibility*:** A second important property is that data cannot be modified. A modification to the statistics of one chunk can cause a skirmish with the statistics of other blocks in the network because statistics is mutual to entire nodes. Owed to the circumstance that information cannot be corrupted at any time, decisions made in the future are based on accurate and reliable information. Based on this, we developed a lightweight blockchain architecture to store IoT sensor data in a water grid using IoT sensors.

18.2 Previous Works

Water consumption is constantly increasing (Imtiaz & Ahsan, 2023), which means that water distribution must be optimized to increase efficiency. Water demand is also increased by climate change. Moreover, IoT has created a vast array of intelligent systems that communicate through the Internet that have benefited a great deal. Water networks are monitored remotely using wireless sensors. Subsequently deliberating shrewd WDS (water distribution system) elucidations assimilated with Internet of grids, Block manacle, and managerial regulatory organisms, smart WDS solutions will be discussed more in detail. A simulation model is presented in (Joshi, 2023) that is used to model, design, and control WDS in real-time. Energy efficiency should be persistently enhanced in water dissemination schemes and propelling locations to diminish energy charges (Lin et al., 2017). The use of variable speed pumps can significantly improve the efficiency of the system, as shown in (Liu & Shang, 2022). Contemporary water delivery schemes progressively use capricious swiftness propels. An everyday demand monitoring system was constructed to track seasonal variations. The detection of leaks has been the subject of various research studies (Longo et al., 2022). This problem was solved by considering several factors, like accurateness, price, timing, and trustworthiness (Pincheira et al., 2023).

Numerous replicas must be integrated to achieve an efficient solution for complex WDNs (water distribution networks). WDN (water distribution network) pumping stations are controlled using a multimodel control structure. As part of the proposed solution, certain scenarios were tested with good results, though a standardized architecture can be developed that also integrates real-time data scientists and automated scenarios. The amalgamation of Blockchain skill with WDS schemes has stayed an additional methodology that has provided several benefits for augmenting water regulator structures. If water possessions are not accomplished proficiently, problems can arise in rustic regions, wherever cultivation produces a significant starring role in pecuniary constancy. Rupino Da Cunha et al. (2020) Proposes an architecture for controlling water using blockchain technology. Water quotas are utilized with the help of this network. Smart Contracts are used to implement all rules. Smart Contracts communicate with valves and pumps in the infrastructure (Satilmisoglu & Keskin, 2023). For instance, as a consequence, all rubrics are strictly observed and applied with decorum. In addition to observing and managing the water, policymaking processes can be realistic to the statistics stockpiled in the blockchain. Water springs and agronomic prosperities can distress the concert of these systems. A water governor structure was malformed into a disseminated conviction structure as a consequence of the learning. This broadside pronounces how a water propels arrangement mechanism in directive to increase an enhanced consideration of how irrigation organisms exertion.

Photovoltaic panels are used to irrigate agricultural land using green energy (Scott et al., 2021). A cryptocurrency called Solar Coin is also used to trade energy and water. Testing was initially conducted on a small farm to monitor soil moisture and pool temperature. In order to generate the necessary energy for the water pump, photovoltaic panels were proposed as the power source. A mobile application was developed that facilitates communication between farms through associations formed by several farms (Sosnowski & Sepczuk, 2023). Then, there are a number of trivial establishments and start-ups looking to transfigure water delivery by smearing contemporary machinery. Often, they face bureaucratic challenges, funding problems, and difficulties exchanging data with other regions/institutions. Blockchain is an innovative solution to reduce these bottlenecks because it uses shrewd indentures, shrewd disbursements, and the altercation of material among shareholders to eradicate the essential for correspondence. As part of the UN's 2035 Sustainable Development Goals (Waseem et al., 2023), a blockchain-based and IoT-based smart water management system is proposed. In addition, this paper (Xia, Wenjun, et al., 2022) provided a thorough examination of multiple applications (Zhang, 2023). Rendering to latest exploration in this extent, additional elucidations are tranquil desirable to progress water governor, contrivance Insolent Conventions, and stock information competently for decision-making processes in the imminent. Limitations include high implementation cost, security issues, standard architecture, integration, scalability, environmental factors, skilled workers.

18.2.1 Water Management: Blockchain Needs

There are certain requirements for a water management system:

- Water management systems are primarily concerned with water conservation. It is therefore necessary to maintain proper water supply tracing.
- Devices and other participants must be able to tamper with the data added to the records by IoT devices.
- Neither one entity nor a single authority should be responsible for managing the system.
- Depending on the role of the participants, the system should provide limited access.
- From producers to consumers, anyone within the system cannot commit any malpractice.
- There must be a secure system for keeping track of water quality and safety records.
- A transparent report should be provided to the customer based on usage.

18.3 Methodological Approach

Water management is becoming increasingly important as scarcity of water increases. In recent years, blockchain technology and its applications have been subject to an explosion of research. A transparent water management system has been lacking due to issues such as heterogeneity of the actors and confidentiality between the participants. Using blockchain technology, we use a resource-to-consumer approach to manage water resources. Since Ethereum is a single production blockchain framework, we use it. With Ethereum, the network nodes can only act when they reach consensus using the Proof of Work consensus algorithm. With REST APIs, we can abstract as well as integrate the systems well. Distributed ledger technology provides all the benefits of blockchain technology.

8.3.1 Defining Roles

Participants are assigned roles. One user can have multiple roles assigned to him or her. Participants have been assigned roles based on their tasks.

In our system, roles are categorized as follows:

a) Management of resources
b) Processing Center

c) Distribution Center
d) Consumer

Participants in water supply management are shown in Figure 18.1.

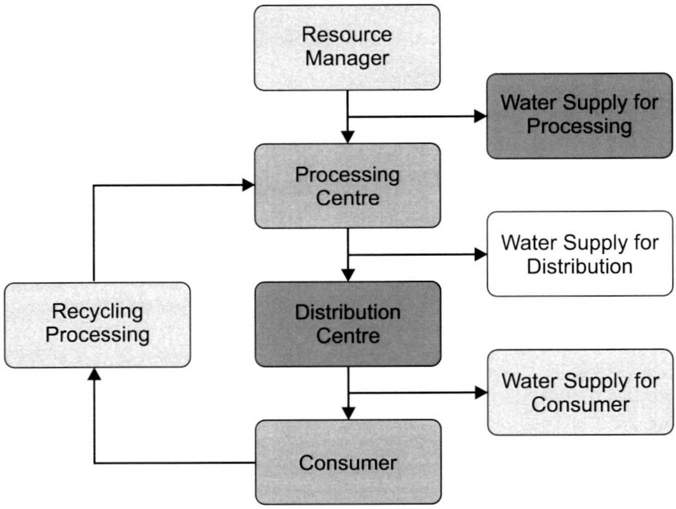

Fig. 18.1 Consignment of Roles.

Following the summary of the roles of the participants in a Blockchain-Based WDS, we are now able to display the respective tasks that the participants are expected to complete.

a) Management of Resources

Resource managers are at the forefront of water management. The role of these organizations is to act as an interface between natural resources and processing units. Various sources of water are supplied to the processor by the resource unit, including rivers, dams, lakes, reservoirs, etc. It is the resource unit's responsibility to manage data, such as water levels in reserves, analyze IoT devices, and develop smart contracts. Resource managers are responsible for managing water supply to processors.

b) Processing Center

Water management systems include a processing unit. Before water is distributed, it is processed in the processing unit. These processors purify water and make it usable for customers. Using pH meters or flow sensors, we can analyze the parameters of water quality and quantity. Consumers' waste water is also processed by them. Using IoT devices, water purification ingredients are accurately measured according to the amount added to the water. Through the use of Blockchain, every activity of the processing unit can be tracked. Each water processing activity carried out by the processing unit is transparent and safe for customers.

c) Distribution Centre

Customers and the processing unit are interconnected through the distributor unit. Upon receiving water from the processing unit, the distributor unit then distributes it to the customer. In the industrial sector, agricultural sector, commercial sector, or household sector, consumers can be large or small. The distribution unit is responsible for managing water distribution and checking the retail environment. Using IoT devices, they examine whether the charges are accurate according to the use by customers. During distribution, IoT sensors play a crucial role. It is also possible to determine how much water is wasted during the distribution phase through the output of the sensors. A GPS sensor on water container trucks, a flow sensor across pipelines, and pH meters in pipelines provide safety during water distribution. Smart contracts and IoT devices display errors if anomalies occur during distribution.

d) Consumer

In the water management system, consumers are the final phase. Blockchain-based water management systems are most beneficial to consumers. Every user report regarding their water usage can't be altered or deleted, so it provides a transparent experience. Water safety parameters are also ensured by the system by using IoT devices and smart contracts that display clear reports about water quality and quantity safety.

18.3.2 Issues of Security

Security must be ensured in every system. Security is ensured in water supply systems using blockchain technology. There are several security issues to consider, including:

I. Regularly checking IoT devices and their outputs is essential.
II. It is important that IoT devices add accurate data to the blockchain.
III. The blockchain must pair public and private keys properly.
IV. Digital signatures must be secured by the system.
V. A regular inspection of consumer feedback is necessary.

18.3.3 Benefits

Over a traditional centralized system, this decentralized system offers many advantages:

- Consumers receive transparent reports on their usages and charges.
- Due to the blockchain, malpractices cannot be committed.
- A traditional water management system relied on centralized servers, whereas a blockchain- based system creates a fully decentralized environment.
- By using blockchain technology, every transaction can be tracked transparently.
- In addition to providing auditability, traceability, and immutability, blockchains provide data security as well.

- The use of robust devices can ensure maximum security due to the increased capabilities of modern edge devices.
- High level participation is possible due to Ethereum's availability, reliability, and availability.

18.4 Conclusion

Earth's living creatures depend on water for their survival. A good water management system is a necessity in this environment of such scarcity. A blockchain-based water management system is an attempt to create a system that is transparent and tamper-proof. In a trust-less environment, blockchain-based water management systems act as the ideal solution to ensure customer satisfaction. As a result, each authority is relieved from a heavy burden by defining its roles. In comparison with traditional water management systems, distributed ledger technology and IoT devices make them smarter and advantageous. Using a private blockchain network can reduce the cost of using Ethereum from an economic perspective. The smart contract must also be written in the same language throughout the system as there is a limitation to their writing. However, the Ethereum implementation provides a CPU-intensive environment. Modern society will benefit from a blockchain-based water management system.

References

1. Angara, J.S.V. and Saripalle R.S. (2022). Towards a Virtual Water Currency for Industrial Products Using Blockchain Technology. *Water Policy*, *24*(6), 923–41. https://doi.org/10.2166/wp.2022.285.
2. Green, C. and Eslamian, S. (2014). Water Governance, In: Eslamian, S. (Ed.), *Handbook of Engineering Hydrology, Ch. 24* (Vol. 3): *Environmental Hydrology and Water Management*. Taylor and Francis, CRC Group, USA, pp. 461–483.
3. Gurudas, V. and Adarsh (2020). Blockchain Based Water Management System. *International Research Journal of Engineering and Technology*,7505–7507. www.irjet.net.
4. Hakak and Saqib. (2020). Industrial Wastewater Management Using Blockchain Technology: Architecture, Requirements, and Future Directions. *IEEE Internet of Things Magazine*, *3*, 38–43. https://doi.org/10.1109/iotm.0001.1900092.
5. Imtiaz and Ahsan (2023)."A Comprehensive Survey on Blockchain Technology Applications in Different Sectors of Pakistan to Develop Smart Cities. *Pakistan Journal of Emerging Science and Technologies*, *4*(3), 82–94, https://doi.org/10.58619/pjest.v4i3.122.
6. Joshi, P. (2023). Blockchain Technology for Sustainable Development: A Systematic Literature Review. *Journal of Global Operations and Strategic Sourcing*,. *16*(3), 683–717. https://doi.org/10.1108/JGOSS-06-2022-0054.
7. Lin, Y.P. et al. (2017). Blockchain: The Evolutionary Next Step for ICT e-Agriculture. *Environments - MDPI*, *4*(3), 1–13. https://doi.org/10.3390/environments4030050.
8. Liu, Y., Ch Shang. (2022). Application of Blockchain Technology in Agricultural Water Rights Trade Management. *Sustainability* (*Switzerland*), *14*(12), 1–10. https://doi.org/10.3390/su14127017.
9. Longo, F., et al. (2022). An Overview of Approaches and Methodologies for Supporting Smallholders: ICT Tools, Blockchain, Business Models, Sustainability Indicators,

Simulation Models. *Procedia Computer Science*, *217*, 1930–1939. https://doi.org/10.1016/j.procs.2022.12.393.

10. Pincheira, M., et al. (2023). An Infrastructure Cost and Benefits Evaluation Framework for Blockchain-Based Applications. *Systems*, *11*(4), 1–14. https://doi.org/10.3390/systems11040184.
11. Rupino Da Cunha, P., et al. (2020) *Association for Information Systems Association for Information Systems Blockchain for Development: Preliminary Insights from a Blockchain for Development: Preliminary Insights from a Literature Review*. https://aisel.aisnet.org/amcis2020.
12. Satilmisoglu, T.K. and Keskin, H. (2023). Smart Water Chain: Immutable, Distributed and Decentralized Water Transaction Ledgers. In: *IOP Conference Series: Earth and Environmental Science*, *1136*(1). https://doi.org/10.1088/1755-1315/1136/1/012014.
13. Scott, D.J., et al. (2021). Exploratory Literature Review of Blockchain in the Construction Industry. *Automation in Construction*, *132*, 103914. https://doi.org/10.1016/j.autcon.2021.103914.
14. Sosnowski, K. and Sepczuk, M. (2023) SURE: A Smart Failover Blockchain-Based Solution for the Recycling Reuse Process. *Electronics* (*Switzerland*), *12*(10). https://doi.org/10.3390/electronics12102201.
15. Waseem, M., et al. (2023). Incorporation of Blockchain Technology for Different Smart Grid Applications: Architecture, Prospects, and Challenges. *Energies*, *16*(2). https://doi.org/10.3390/en16020820.
16. Xia et al. (2022)."A Framework of Blockchain Technology in Intelligent Water Management." *Frontiers in Environmental Science*, *10*, 1–12. https://doi.org/10.3389/fenvs.2022.909606.
17. Zhang, J. (2023). *Application of Blockchain Technology in Water Management and New Energy Economic Development*. *10*, 4227–4241. IWA Publishing. https://doi.org/10.2166/ws.2023.258.

19 Blockchain for Hydrosystems: Transforming Water Management and Sustainability

Karkuzhali S.[1*] and Senthilkumar S.[2]

The rapid degradation of natural ecosystems and the urgent need for effective environmental conservation strategies have prompted innovative approaches to address these challenges. In recent years, blockchain technology has emerged as a potential tool for revolutionizing various sectors, and its application in environmental conservation holds promising possibilities. This paper explores the integration of blockchain technology into environmental conservation efforts, analyzing its potential benefits and limitations. The paper begins by providing an overview of blockchain technology, elucidating its fundamental principles and mechanisms that underpin its security, transparency, and immutability features. Subsequently, it delves into the current state of environmental conservation, highlighting the pressing issues faced by ecosystems and the shortcomings of existing conservation methods. The potential applications of blockchain in environmental conservation are then presented, ranging from enhancing transparency in supply chains to enabling more efficient carbon credit systems and fostering community engagement. To illustrate practical implementations, case studies of ongoing projects that leverage blockchain for environmental conservation are discussed. These cases encompass initiatives in tracking sustainable sourcing of goods, managing biodiversity data, and facilitating decentralized peer-to-peer energy trading. The benefits of using blockchain in these contexts, such as increased traceability, reduced fraud, and improved data accuracy, are examined alongside challenges like scalability, energy consumption, and regulatory concerns. Furthermore, the paper addresses the socioeconomic and policy implications of adopting blockchain technology for environmental conservation. It emphasizes the need for interdisciplinary collaboration between technology experts, environmental scientists, policymakers, and local communities to ensure effective implementation and equitable outcomes. The potential for blockchain to empower marginalized communities and indigenous groups in conservation efforts is explored. In conclusion, this paper underscores the transformative potential of blockchain

[1] Department of Computer Science and Engineering, Mepco Schlenk Engineering College, Sivakasi, Tamilnadu, India.

[2] Department of Chemistry, Ayya Nadar Janaki Ammal College, Sivakasi,Tamilnadu, India.

* Corresponding author: karkuzhali@mepcoeng.ac.in

DOI: 10.1201/9781003686026-19

technology in addressing critical environmental conservation challenges. While recognizing its nascent stage and associated challenges, it emphasizes the need for continued research and development to harness blockchain's capabilities fully. By leveraging blockchain's unique features, environmental conservation can be made more transparent, participatory, and accountable, thereby contributing to the global endeavor of safeguarding our planet's ecological balance.

19.1 Introduction

In the face of escalating environmental challenges, ranging from climate change and habitat loss to resource depletion and pollution, the need for innovative and effective conservation strategies has become more urgent than ever (Abdelkrim and Eslamian, 2024). Traditional methods of environmental protection often face limitations in terms of transparency, accountability, and inclusivity. Enter blockchain technology, a decentralized and tamper-resistant digital ledger that has captured global attention for its potential to revolutionize various sectors. In recent years, blockchain's unique attributes have led to explorations of its application in the realm of environmental conservation, offering novel solutions to enhance transparency, traceability, and community engagement. This paper delves into the promising intersection of blockchain technology and environmental conservation, examining its foundational principles, current challenges in conservation, and the transformative potential of blockchain-enabled solutions (Eslamian and Eslamian, 2025).

19.2 Blockchain Technology: Fundamentals and Features

This section provides an overview of the fundamental principles that underpin blockchain technology. It elucidates the decentralized nature of blockchain networks, highlighting their capacity to operate without a central authority, thus ensuring transparency and preventing single points of failure. The immutability of blockchain records, achieved through cryptographic mechanisms, is explored in terms of its relevance to maintaining a secure and tamper-resistant digital ledger. Additionally, the concept of smart contracts, self-executing code that automates agreements, is introduced as a tool that enhances the efficiency and reliability of transactions within blockchain systems.

19.3 Environmental Conservation: Challenges and Imperatives

The conservation of natural ecosystems faces an array of challenges, ranging from unsustainable resource extraction and habitat destruction to inadequate monitoring and ineffective regulatory mechanisms. This section outlines the critical issues and shortcomings within existing conservation practices. It highlights the complexity of global supply chains for commodities like timber, minerals, and seafood, often fraught with opacity and susceptibility to illegal and unsustainable practices.

Similarly, the intricacies of tracking carbon emissions and managing biodiversity data are explored, underscoring the need for more transparent and accountable systems.

19.4 The Promise of Blockchain in Conservation

This segment delves into the potential applications of blockchain technology in addressing the challenges outlined in the previous section. The transparency and traceability offered by blockchain have the potential to revolutionize supply chain management by enabling real-time monitoring of the origin, journey, and impact of commodities. Furthermore, the integration of blockchain in carbon credit systems promises enhanced accuracy and accountability in tracking emissions reductions. The engagement of local communities and indigenous groups through blockchain-enabled platforms is also discussed, emphasizing the empowerment of stakeholders in conservation efforts.

19.5 Practical Implementations and Case Studies

To illustrate the practicality of blockchain technology in environmental conservation, this section presents case studies of ongoing initiatives that utilize blockchain for sustainable practices. These examples range from tracking ethically sourced products to managing and sharing biodiversity data in a decentralized manner. Each case study highlights the specific benefits derived from integrating blockchain, such as increased trust, reduced fraud, and improved data accuracy. In conclusion, the introduction sets the stage for a comprehensive exploration of the potential of blockchain technology in environmental conservation. By examining the foundational principles of blockchain, the challenges faced by conservation efforts, and the promise offered by blockchain-enabled solutions, this paper aims to elucidate the transformative role that technology can play in safeguarding our planet's fragile ecosystems.

19.6 Principles of Blockchain Technology for Environmental Conservation

Blockchain technology, renowned for its decentralized and tamper-resistant nature, holds the potential to address critical challenges within environmental conservation. By leveraging its unique principles, blockchain can introduce transparency, traceability, and accountability to conservation efforts, ushering in a new era of sustainable practices. The following principles underpin the application of blockchain technology in environmental conservation:

19.6.1 Decentralization and Peer-to-Peer Networks

At the heart of blockchain lies its decentralized structure, where no single entity maintains control. Instead, a distributed network of participants, or nodes, collectively validates and records transactions. This principle eliminates the need for intermediaries and central authorities, facilitating direct peer-to-peer interactions. In the context of environmental conservation, decentralization enables transparent collaboration among stakeholders, empowering local communities, indigenous groups, researchers, policymakers, and businesses to engage directly in conservation efforts without relying on a centralized governing body.

19.6.2 Transparency and Immutability

Blockchain's transparency stems from its public and unalterable ledger. Once data is recorded on the blockchain, it cannot be retroactively modified, ensuring the integrity of information. In conservation, this principle is particularly relevant for tracking the origin and journey of commodities along complex supply chains. By providing an immutable record of transactions, blockchain allows stakeholders to verify the authenticity of sustainable sourcing claims, reducing the risk of illegal practices such as deforestation or overfishing.

19.6.3 Smart Contracts and Automation

Smart contracts are self-executing code segments that automate predefined actions when certain conditions are met. This feature streamlines processes and reduces the need for intermediaries. In conservation, smart contracts can be utilized to automate agreements between parties, ensuring adherence to sustainable practices. For instance, a contract could automatically trigger payments to local communities when specific conservation milestones are achieved, incentivizing proactive environmental stewardship.

19.6.4 Traceability and Supply Chain Management

Blockchain's traceability capabilities empower consumers and businesses to track the journey of products from origin to destination. Applied to conservation, this principle enables the creation of transparent supply chains for goods derived from natural resources. For instance, blockchain can be used to verify that timber originates from legal and sustainable sources, reducing the demand for illegal logging and promoting responsible forestry practices.

19.6.5 Data Integrity and Biodiversity Management

The secure and tamper-resistant nature of blockchain ensures the integrity of data stored within its ledgers. This attribute is vital for managing biodiversity information, where accurate and reliable data is essential for effective conservation planning. Blockchain can serve as a decentralized database for recording species data, habitat conditions, and research findings, enabling scientists and conservationists to collaborate and make informed decisions based on accurate information.

19.6.6 Community Engagement and Empowerment

Blockchain's transparency and direct engagement capabilities empower communities to actively participate in conservation efforts. By providing secure platforms for secure transactions and rewards, blockchain enables fair compensation for sustainable practices. This principle can elevate the role of local communities and indigenous groups, encouraging their active involvement in preserving ecosystems that they depend upon.

In summary, these principles demonstrate how blockchain technology aligns with the core goals of environmental conservation. By offering decentralization, transparency, traceability, automation, data integrity, and community empowerment, blockchain has the potential to revolutionize how we approach conservation, fostering a more sustainable and accountable relationship with our planet's natural resources.

19.7 Literature Review

Blockchain technology has garnered increasing attention for its potential to enhance transparency, traceability, and accountability in various sectors, including environmental conservation. This literature review examines key research papers that delve into the application of blockchain technology in the context of environmental conservation, highlighting their contributions, findings, and implications.

Peng and Huang (2022) explored the role of blockchain technology in promoting sustainable forest management within tribal communities. By leveraging the transparency and security features of blockchain, their study demonstrated how resource allocation and environmental monitoring could be enhanced. The integration of blockchain with traditional forest management practices facilitated accurate tracking and accountability, thereby contributing to more sustainable outcomes.

Kamble et al. (2023) investigated the antecedents for blockchain technology adoption in sustainable agricultural supply chains. Their research highlighted factors such as technological readiness, perceived benefits, and stakeholder collaboration as key enablers. Through empirical analysis, they demonstrated how blockchain enhances transparency, reduces fraud, and improves decision-making in agricultural supply chains.

Mubarik et al. (2021) examined the impact of blockchain technology on green supply chain practices in emerging economies. Their study provided evidence that blockchain applications can significantly enhance supply chain visibility, reduce carbon footprints, and ensure responsible sourcing. They emphasized the importance of supportive regulatory frameworks and technological infrastructure for successful implementation.

Kouhizadeh et al. (2022) critically analyzed the relationship between blockchain and the circular economy. They identified potential tensions, including technological complexity and interoperability issues. However, their research also underscored blockchain's potential to improve resource efficiency, waste management, and material traceability.

Kechagias et al. (2023) conducted a systematic literature review on blockchain's role in transitioning towards a circular economy. They identified blockchain applications in waste management, recycling, and product lifecycle tracking. Their findings suggested that blockchain fosters stakeholder collaboration and supports the creation of closed-loop supply chains.

Benítez-Martínez and Toledo (2023) explored blockchain's contribution to environmental economics through holistic and integrated life cycle sustainability assessments. They proposed that blockchain-enabled transparency could enhance decision-making processes in environmental management, particularly in tracking carbon emissions and monitoring resource utilization.

Thanasi-Boçe and Hoxha (2024) examined blockchain as a catalyst for sustainable development, focusing on economic, social, and environmental synergies. Their research demonstrated how blockchain applications in carbon credit trading, waste management, and sustainable finance contribute to achieving the United Nations Sustainable Development Goals (SDGs).

Zhang (2019) presented a case study on the Yitong System, utilizing blockchain technology to incentivize efficient use of rural waste. Through smart contracts and token-based incentives, the system successfully promoted waste recycling and resource optimization, offering a scalable model for rural waste management.

Song et al. (2022) proposed a blockchain-based framework for hazardous waste transfer management. By ensuring traceability and accountability across the waste management lifecycle, their system minimized environmental risks and improved regulatory compliance. They highlighted the importance of integrating blockchain with IoT sensors for real-time data capture.

Saad et al. (2023) introduced a blockchain-enabled Vehicular Ad Hoc Network (VANET) for smart solid waste management. Their research demonstrated how blockchain's decentralized architecture facilitated real-time data sharing, route optimization, and secure transactions among waste collection vehicles.

França et al. (2020) suggested the application of blockchain in improving solid waste management in small municipalities. They argued that blockchain could streamline waste collection processes, reduce illegal dumping, and encourage community participation through transparent waste tracking systems.

Damadi and Namjoo (2021) presented a smart waste management solution using blockchain technology. Their model incorporated IoT sensors to monitor waste levels, while blockchain ensured data integrity and transparency. The study highlighted the potential for reducing operational costs and optimizing waste collection routes.

Dua et al. (2020) explored blockchain-based e-waste management in 5G smart communities. Their proposed system used blockchain to enhance accountability in e-waste disposal processes, reduce illegal dumping, and ensure proper recycling of electronic waste.

Gupta and Bedi (2018) examined the use of blockchain-based smart contracts for e-waste management. They demonstrated how automated contracts could enforce regulatory compliance, track e-waste throughout its lifecycle, and facilitate responsible recycling practices.

Kshetri (2018) analyzed blockchain's role in achieving key supply chain management objectives. By ensuring data transparency and traceability, blockchain enhanced operational efficiency, reduced fraud, and supported sustainable supply chain practices. The study emphasized the importance of stakeholder collaboration for successful adoption.

Saberi et al. (2019) explored blockchain technology's relationships with sustainable supply chain management. They identified blockchain's potential to reduce information asymmetry, enhance traceability, and improve decision-making. Their findings underscored the need for robust governance frameworks to ensure blockchain's effective implementation.

Wang et al. (2019) provided insights into the transformative impact of blockchain on supply chains. Their research demonstrated how blockchain applications improve visibility, optimize inventory management, and foster trust among stakeholders. The study concluded that blockchain adoption requires addressing interoperability challenges and regulatory uncertainties.

Table 19.1 Comparison of Existing methods in Blockchain Technology for Environmental Conservation.

Author(s) and Year	*Focus Area*	*Key Findings*
Peng & Huang (2022)	Sustainable Forest Management	Blockchain promotes transparency, accountability, and resource management in tribal communities.
Kamble et al. (2023)	Agricultural Supply Chains	Enhances transparency, reduces fraud, and improves decision-making.
Mubarik et al. (2021)	Green Supply Chain	Blockchain ensures visibility, reduces carbon footprints, and supports responsible sourcing.
Kouhizadeh et al. (2022)	Circular Economy	Identified tensions in implementation, but highlighted benefits in resource efficiency and waste management.
Kechagias et al. (2023)	Circular Economy	Blockchain supports waste management, recycling, and lifecycle tracking.
Benítez-Martínez & Toledo (2023)	Environmental Economics	Blockchain enables lifecycle sustainability assessments and enhances decision-making.
Thanasi-Boçe & Hoxha (2024)	Sustainable Development	Supports SDGs through applications in carbon credit trading, waste management, and sustainable finance.
Zhang (2019)	Rural Waste Management	Blockchain incentivizes waste recycling and resource optimization using token-based systems.

Contd...

Table 19.1 *Contd...*

Author(s) and Year	*Focus Area*	*Key Findings*
Song et al. (2022)	Hazardous Waste Management	Ensures traceability, accountability, and regulatory compliance using blockchain.
Saad et al. (2023)	Solid Waste Management	Blockchain-enabled VANET supports real-time data sharing and route optimization.
França et al. (2020)	Municipal Waste Management	Streamlines waste collection, reduces illegal dumping, and enhances community participation.
Damadi & Namjoo (2021)	Smart Waste Management	Integrates blockchain and IoT sensors for data integrity and optimized waste collection.
Dua et al. (2020)	E-waste Management	Blockchain ensures accountability, reduces illegal dumping, and enhances recycling.
Gupta & Bedi (2018)	E-waste Management	Smart contracts automate compliance, track e-waste, and ensure responsible recycling.
Kshetri (2018)	Supply Chain Management	Blockchain enhances transparency, operational efficiency, and fraud reduction.
Saberi et al. (2019)	Sustainable Supply Chain	Blockchain reduces information asymmetry, improves traceability, and supports decision-making.
Wang et al. (2019)	Supply Chain Transformation	Blockchain optimizes inventory management, increases visibility, and fosters stakeholder trust.

In conclusion, these research papers contribute to the ongoing exploration of how blockchain technology can drive environmental conservation efforts. By examining applications in the circular economy, renewable energy, carbon trading, agriculture, and nature conservation, these studies showcase the diverse ways in which blockchain can foster sustainability. However, they also acknowledge the need for addressing challenges related to adoption, scalability, and regulation to fully harness the potential of blockchain for advancing environmental conservation objectives.

19.8 Methodology

19.8.1 Research Design

The research employs a qualitative approach to explore the potential applications of blockchain technology in the context of energy conservation. It involves a comprehensive review of existing literature, case studies, and expert opinions to identify key insights and trends related to the integration of blockchain in energy conservation strategies.

19.8.2 Case Studies and Use Cases

A collection of real-world case studies and use cases is analyzed to showcase how blockchain technology has been implemented to promote energy conservation. These cases span different sectors, including renewable energy generation, energy trading, grid management, and demand response. The analysis includes the identification of success factors, challenges faced, and lessons learned from each case.

19.8.3 Expert Interviews and Surveys

Expert interviews and surveys are conducted with professionals, researchers, and practitioners in the fields of blockchain technology and energy conservation. These engagements provide valuable insights into the current state of blockchain integration in energy conservation, potential barriers, and future directions. The collected data is qualitatively analyzed to identify common themes, opinions, and recommendations.

19.8.4 Framework Development

Based on the insights gained from the literature review, case studies, and expert opinions, a conceptual framework is developed to outline the potential applications of blockchain technology for energy conservation. This framework categorizes the various use cases, highlights the benefits and challenges, and proposes strategies for successful implementation.

19.8.5 Challenges and Opportunities Analysis

A comprehensive analysis is conducted to identify the challenges and opportunities associated with adopting blockchain technology for energy conservation. These challenges may include technological limitations, regulatory hurdles, data privacy concerns, and scalability issues. Conversely, opportunities may involve increased transparency, decentralized energy trading, enhanced grid management, and consumer engagement.

19.8.6 Ethical and Environmental Considerations

The research addresses ethical considerations related to energy conservation and blockchain technology, such as the potential for unequal access to blockchain-based energy systems and the environmental impact of blockchain's energy consumption. This analysis provides a balanced perspective on the implications of blockchain integration for sustainable energy practices.

19.8.7 *Synthesis and Recommendations*

The findings from the literature review, case studies, expert interviews, and challenges/opportunities analysis are synthesized to provide recommendations for stakeholders interested in leveraging blockchain technology for energy conservation. These recommendations encompass strategies for addressing challenges, maximizing benefits, fostering collaboration, and ensuring a sustainable and equitable implementation.

In conclusion, the methodology combines literature analysis, case studies, expert insights, and ethical considerations to provide a comprehensive exploration of the potential and challenges of integrating blockchain technology for energy conservation. Through this approach, the research aims to contribute valuable insights to the ongoing dialogue on sustainable energy practices and innovative technological solutions.

19.9 Architecture

19.9.1 *System Architecture and Components*

a. **Blockchain Infrastructure:** Choose a suitable blockchain platform (public or private), consensus mechanism (e.g., Proof of Work, Proof of Stake), and protocol (e.g., Ethereum, Hyperledger).
b. **Smart Contracts:** Design and develop smart contracts to automate and enforce energy conservation transactions and agreements.
c. **Nodes and Participants:** Define the roles and responsibilities of participants, such as energy producers, consumers, regulators, and grid operators.
d. **Decentralized Applications (DApps):** Develop user-friendly DApps for participants to interact with the blockchain system.

19.9.2 *Data Collection and Integration*

a. **IoT Integration:** Integrate IoT devices to collect real-time energy consumption data from smart meters, sensors, and devices.
b. **Data Oracles:** Use data oracles to securely feed external data into the blockchain, ensuring accurate and tamper-proof data.

19.9.3 *Energy Trading and Conservation Mechanisms*

a. **Peer-to-Peer Energy Trading:** Design a mechanism for participants to directly trade excess energy with each other using smart contracts.
b. **Demand Response Programs:** Implement automated demand response programs to encourage energy conservation during peak periods.
c. **Energy Tokens:** Create a tokenized system where energy is represented by digital tokens, facilitating easy trading and incentivizing energy-saving behaviors.

19.9.4 Verification and Validation

a. **Energy Provenance:** Establish a mechanism to trace the origin of energy sources, ensuring transparency and reducing reliance on non-renewable sources.
b. **Smart Meter Integration:** Integrate smart meters to verify energy consumption and trading data, minimizing fraud and inaccuracies.

19.9.5 Incentive Mechanisms

a. **Token-based Rewards:** Design a reward system where participants receive tokens for energy conservation actions.
b. **Carbon Credits:** Implement a blockchain-based carbon credit system to incentivize participants to reduce their carbon footprint.

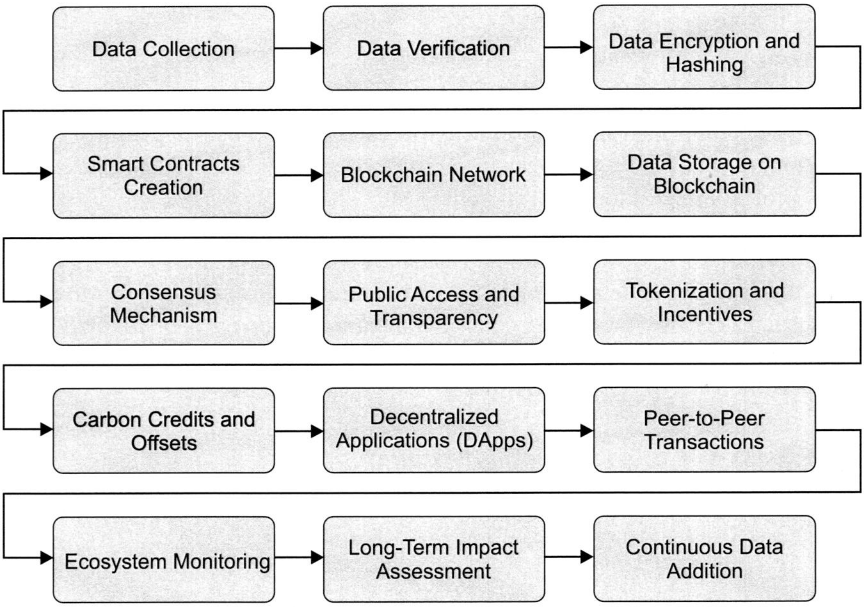

Fig. 19.1 Block diagram for Blockchain Technology for Environmental Conservation.

19.9.6 Regulatory Compliance and Governance

a. **Regulatory Framework:** Ensure the system adheres to energy regulations and standards in the respective jurisdiction.
b. **Identity Management:** Develop a secure identity management system to ensure compliance with KYC (Know Your Customer) and AML (Anti-Money Laundering) regulations.

19.9.7 Security and Privacy

a. **Data Encryption:** Implement strong encryption techniques to protect sensitive energy consumption and trading data.
b. **Private Transactions:** Incorporate mechanisms for private transactions, allowing participants to engage in confidential energy trading.

19.9.8 User Experience and Accessibility

a. **User-Friendly Interface:** Design intuitive user interfaces for participants to monitor energy consumption, trading activities, and rewards.
b. **Mobile Apps:** Develop mobile applications for participants to access and interact with the blockchain system conveniently.

19.9.9 Scalability and Interoperability

a. **Off-Chain Solutions:** Consider implementing off-chain solutions (such as state channels) to improve scalability and reduce transaction costs.
b. **Integration with Existing Systems:** Ensure compatibility and integration with existing energy management and grid systems.

19.9.10 Monitoring and Analytics

a. **Real-time Analytics:** Implement real-time analytics to monitor energy consumption trends, trading patterns, and conservation outcomes.
b. **Predictive Analysis:** Utilize predictive analytics to anticipate energy demands and optimize conservation strategies.

By carefully designing and implementing such a system, blockchain technology can play a pivotal role in revolutionizing energy conservation efforts by enabling secure, transparent, and efficient energy management and trading practices.

19.10 Challenges

Implementing blockchain technology for energy conservation presents several challenges that need to be addressed to ensure the success and effectiveness of the system. Here are some key challenges:

1. **Scalability:** Blockchain networks can face scalability issues as the number of participants and transactions increase. Energy conservation systems involve a large number of users, devices, and real-time transactions, requiring a blockchain solution that can handle the scale without compromising efficiency.
2. **Energy Consumption:** The energy consumption of blockchain networks, especially those using Proof of Work consensus, can be counterproductive to energy conservation goals. Striking a balance between the energy saved

through conservation efforts and the energy used by the blockchain itself is crucial.

3. **Interoperability:** Integrating blockchain with existing energy infrastructure and management systems can be complex. Achieving interoperability between various systems, protocols, and standards requires careful planning and coordination.
4. **Regulatory Compliance:** Energy markets are heavily regulated, and blockchain systems need to comply with energy and financial regulations. Overcoming legal and regulatory barriers while maintaining transparency can be challenging.
5. **Data Privacy:** While blockchain provides transparency, ensuring data privacy is equally important, especially when dealing with sensitive energy consumption and trading data. Balancing transparency with privacy concerns is a challenge.
6. **Adoption and Education:** Introducing blockchain technology to energy consumers, producers, and regulators requires education and awareness efforts. Many stakeholders may be unfamiliar with blockchain, which can slow down adoption.
7. **Smart Contract Complexity:** Developing and auditing complex smart contracts for energy conservation mechanisms can be challenging. Errors or vulnerabilities in smart contracts can lead to unintended consequences.
8. **Token Valuation and Stability:** If a token-based reward system is employed, the valuation and stability of tokens can impact participant motivation. Fluctuating token values may affect the effectiveness of incentive mechanisms.
9. **Energy Source Verification:** Blockchain can enhance the tracking of renewable energy sources, but verifying the authenticity of these sources on the blockchain can be challenging, particularly in the case of energy trading.
10. **Initial Setup Costs:** Implementing a blockchain system requires initial investment in infrastructure, development, and integration. For some stakeholders, the upfront costs might outweigh the immediate benefits.
11. **User Experience:** Designing user-friendly interfaces and experiences for participants is crucial. Ensuring that users can easily interact with the blockchain system is essential for its success.
12. **Network Security:** Blockchain networks are vulnerable to attacks, such as 51% attacks and hacking attempts. Ensuring robust network security to protect the integrity of energy conservation data is vital.
13. **Technological Maturity:** Blockchain technology is still evolving, and the tools and platforms available may lack some necessary features for specific energy conservation applications.
14. **Resistance to Change:** Traditional energy systems and practices may face resistance to change from established stakeholders. Convincing all parties to adopt blockchain-based solutions can be challenging.

Addressing these challenges requires collaboration among stakeholders, innovation, research, and the development of tailored solutions that consider the unique aspects of energy conservation within the context of blockchain technology.

19.11 Benefits

Blockchain technology offers several benefits that can contribute to more effective energy conservation strategies. Here are some key advantages:

1. **Transparency and Traceability:** Blockchain provides a transparent and immutable ledger where energy transactions and conservation efforts are recorded. This transparency allows participants to track the source and usage of energy, fostering accountability and trust among stakeholders.
2. **Decentralization:** Decentralized blockchain networks eliminate the need for intermediaries, such as energy providers or brokers. This enables direct peer-to-peer energy trading and reduces reliance on centralized energy grids.
3. **Smart Contracts for Automation:** Smart contracts on the blockchain can automate energy conservation processes. These self-executing contracts can automatically trigger actions based on predefined conditions, such as adjusting energy usage during peak hours or rewarding users for reducing consumption.
4. **Energy Tokenization:** Tokenizing energy on the blockchain allows for efficient trading and exchange of energy units. This tokenization simplifies energy transactions, making them more accessible and enabling participants to directly buy and sell energy.
5. **Real-time Monitoring:** IoT devices and sensors can collect real-time energy consumption data and feed it into the blockchain. This data can be used to monitor energy usage patterns, identify inefficiencies, and optimize conservation strategies.
6. **Incentive Mechanisms:** Blockchain-based systems can offer incentives to encourage energy conservation. Participants can earn rewards, tokens, or discounts for reducing energy consumption during peak periods or using renewable sources.
7. **Enhanced Demand Response:** With blockchain, demand response programs become more effective. Smart contracts can automatically adjust energy usage or pricing based on real-time demand and supply conditions, helping balance the energy grid.
8. **Secure Data Sharing:** Blockchain ensures data security and integrity through encryption and decentralized storage. This allows secure sharing of energy consumption data among stakeholders without compromising privacy.
9. **Efficient Grid Management:** Blockchain can streamline grid management by enabling real-time monitoring and optimization of energy distribution. It helps prevent overloads and blackouts by maintaining a balanced energy supply.

10. **Carbon Credit Management:** Blockchain can simplify the tracking and verification of carbon emissions reductions, making it easier to issue and trade carbon credits. This encourages businesses and individuals to participate in emission reduction efforts.
11. **Reduction of Energy Loss:** Blockchain's decentralized nature reduces the need for multiple intermediaries, reducing energy loss during transmission and distribution.
12. **Renewable Energy Integration:** Blockchain can enhance the integration of renewable energy sources into the grid. It enables tracking and verification of renewable energy generation, promoting its adoption.
13. **Immutable Records:** Once data is added to the blockchain, it cannot be altered or tampered with, ensuring the accuracy and reliability of energy consumption records and conservation efforts.
14. **Greater Consumer Engagement:** Blockchain-based systems can empower consumers by giving them more control over their energy usage. This engagement can lead to increased awareness and participation in conservation activities.
15. **Resilience and Reliability:** Blockchain's distributed nature makes it resistant to single points of failure and attacks, increasing the reliability and resilience of energy conservation systems.

Overall, blockchain technology offers a robust foundation for enhancing energy conservation efforts through increased transparency, automation, security, and direct participation of consumers and producers in shaping the energy landscape.

19.12 Conclusion

In conclusion, blockchain technology holds immense promise for revolutionizing energy conservation practices and contributing to a more sustainable and efficient energy ecosystem. The synergy between blockchain's inherent features and the goals of energy conservation paves the way for innovative solutions that address challenges in energy management, distribution, and consumption. By leveraging the transparency, traceability, and automation offered by blockchain, energy conservation efforts can be amplified and streamlined, leading to a range of positive outcomes.Blockchain's transparency empowers participants with real-time access to accurate energy consumption data, enabling informed decision-making and encouraging responsible energy usage. The traceability of energy sources ensures that renewable and sustainable energy generation can be reliably tracked, verified, and integrated into the energy grid. This not only promotes the adoption of cleaner energy sources but also supports the reduction of carbon emissions.The automation capabilities of blockchain, facilitated through smart contracts, automate energy trading, demand response mechanisms, and reward systems. These automated processes enhance the efficiency of energy conservation initiatives, allowing for timely adjustments based on real-time data and conditions. Moreover, the integration of Internet of Things (IoT) devices with blockchain enables the collection of granular energy consumption data, facilitating the identification of inefficiencies and the optimization of energy usage patterns.However, the implementation of blockchain for energy conservation is not without its challenges. Scalability, energy consumption of blockchain networks, regulatory compliance, and interoperability with existing systems remain areas of concern. Overcoming these challenges requires collaborative efforts among governments, industries, researchers, and technology developers. As the world increasingly focuses on sustainable development and reducing its carbon footprint, blockchain technology emerges as a catalyst for driving energy conservation initiatives forward. Its potential to create a decentralized, transparent, and efficient energy ecosystem aligns well with the global goals of mitigating climate change and ensuring the availability of energy resources for future generations. To fully harness the benefits of blockchain for energy conservation, a multidisciplinary approach involving energy experts, blockchain developers, policymakers, and stakeholders from various sectors is essential.In conclusion, blockchain's transformative capabilities, when integrated into energy conservation strategies, have the potential to redefine how energy is managed, traded, and consumed. As we continue to explore innovative solutions for a more sustainable future, blockchain technology stands poised to play a pivotal role in shaping the energy landscape and driving us towards a more energy-efficient and environmentally conscious world.

References

1. Abdelkrim Z. and Eslamian S. (2024). Potential Impacts of Climate Change. In: Eslamian, S., Huda M.B., Rather N.A., and Eslamian F.A. (Eds.), *Handbook of Climate Change Impacts on River Basin Management, Ch. 7* (Vol. 3): *Technology, Innovations, and Management Practices*. Taylor and Francis, CRC Group, USA.
2. Benítez-Martínez, J. and Toledo, M. (2023). Blockchain technology in environmental economics: A service for a holistic and integrated life cycle sustainability assessment. *Journal of Risk and Financial Management, 16*(3), 209. https://doi.org/10.3390/jrfm16030209.
3. Damadi, H. and Namjoo, M. (2021). Smart waste management using blockchain. *IT Professional, 23*(3), 81–87. https://doi.org/10.1109/MITP.2021.3062980.
4. Dua, A., Dutta, A., Zaman, N. and Kumar, N. (2020). Blockchain-based e-waste management in 5G smart communities. In: *Proceedings of the IEEE INFOCOM 2020—IEEE Conference on Computer Communications Workshops* (*INFOCOM WKSHPS*), pp. 195–200. https://doi.org/10.1109/INFOCOMWKSHPS50562.2020.9162783
5. Eslamian, S. and F., Eslamian (2025). *Blockchain Technology in Water and Environmental Systems*. CRC Press, Taylor and Francis, USA, 282 p.
6. França, A., Neto, J.A., Gonçalves, R. and Almeida, C. (2020). Proposing the use of blockchain to improve solid waste management in small municipalities. *Journal of Cleaner Production, 244*, 118529. https://doi.org/10.1016/j.jclepro.2019.118529.
7. Gupta, N. and Bedi, P. (2018). E-waste management using blockchain-based smart contracts. In: *Proceedings of the 2018 International Conference on Advances in Computing, Communications and Informatics* (*ICACCI*), (pp. 915–921). https://doi.org/10.1109/ICACCI.2018.8554552.
8. Kamble, S.S., Gunasekaran, A. and Sharma, R. (2023). Antecedents for blockchain technology-enabled sustainable agriculture supply chain. *Annals of Operations Research, 327*, 293–337. https://doi.org/10.1007/s10479-021-04423-3.

9. Kechagias, E., Kotsialos, A. and Vlachos, D. (2023). The role of blockchain technology in the transition toward the circular economy: Findings from a systematic literature review. *Resources, Conservation, and Recycling Advances*, *17*, 200126. https://doi.org/10.1016/j.rcradv.2022.200126.
10. Kouhizadeh, M., Zhu, Q. and Sarkis, J. (2022). Blockchain and the circular economy: Potential tensions and critical reflections from practice. *Production Planning and Control*, *33*(11-12), 991–1007. https://doi.org/10.1080/09537287.2021.1934584.
11. Kshetri, N. (2018). Blockchain's roles in meeting key supply chain management objectives. *International Journal of Information Management*, *39*, 80–89. https://doi.org/10.1016/j.ijinfomgt.2017.12.005.
12. Mubarik, M., Raja Mohd Rasi, R.Z., Mubarak, M.F. and Ashraf, R. (2021). Impact of blockchain technology on green supply chain practices: Evidence from emerging economy. *Management of Environmental Quality*, *32*(5), 1023–1039. https://doi.org/10.1108/MEQ-11-2020-0277.
13. Peng, Y. and Huang, W. (2022). Using blockchain technology and sharing culture to promote sustainable forest management in tribal communities. *Journal of Environmental and Public Health*, *2022*, 1529407. https://doi.org/10.1155/2022/1529407.
14. Saberi, S., Kouhizadeh, M., Sarkis, J. and Shen, L. (2019). Blockchain technology and its relationships to sustainable supply chain management. *International Journal of Production Research*, *57*(7), 2117–2135. https://doi.org/10.1080/00207543.2018.1533261.
15. Saad, M., Ahmad, M.B., Asif, M., Khan, M.K., Mahmood, T. and Mahmood, M.T. (2023). Blockchain-enabled VANET for smart solid waste management. *IEEE Access*, *11*, 5679–5700. https://doi.org/10.1109/ACCESS.2023.3242078.
16. Song, G., Lu, Y., Feng, H., Lin, H. and Zheng, Y. (2022). An implementation framework of blockchain-based hazardous waste transfer management system. *Environmental Science and Pollution Research*, *29*, 36147–36160. https://doi.org/10.1007/s11356-022-19265-0.
17. Thanasi-Boçe, M. and Hoxha, J. (2024). Blockchain technology as a catalyst for sustainable development: Exploring economic, social, and environmental synergies. *Academic Journal of Interdisciplinary Studies*, *13*(2), 151. https://doi.org/10.36941/ajis-2024-0041.
18. Wang, Y., Singgih, M., Wang, J. and Rit, M. (2019). Making sense of blockchain technology: How will it transform supply chains? *International Journal of Production Economics*, *211*, 221–236. https://doi.org/10.1016/j.ijpe.2019.02.002.
19. Zhang, D. (2019). Application of blockchain technology in incentivizing efficient use of rural wastes: A case study on Yitong System. *Energy Procedia*, *158*, 6707–6714. https://doi.org/10.1016/j.egypro.2019.01.1154.

20 Blockchain Technology for Environmental Conservation

Pulkit Kumar[1*] and Harpreet Kaur Channi[2]

Initially designed to support cryptocurrencies like Bitcoin, blockchain technology can solve many problems, including environmental protection. This abstract explores how blockchain technology might promote sustainable practices, reduce ecological impact, and enable transparent stakeholder engagement in environmental conservation. Blockchain's immutability, decentralization, and transparency provide novel environmental conservation options. Blockchain verifies environmental data like carbon emissions and animal monitoring by creating a tamper-proof, traceable record. This helps regulatory agencies, organizations, and people check ecological compliance. Blockchain's environmental protection benefits from intelligent contracts and programmable self-executing agreements. These contracts may automate sustainable practices and incentivize eco-friendly behavior. This lets stakeholders contribute to conservation and be paid securely and auditable. Blockchain's ability to change supply chain management is an environmental benefit. Blockchain may help identify environmentally responsible suppliers and promote responsible procurement by generating transparent and irreversible product origin and route records. Transparency inhibits illicit logging, overfishing, and unsustainable mining and promotes supply chain accountability. Blockchain enables stakeholders to share and collaborate across borders due to its decentralization. This allows international cooperation on climate change and deforestation, which need cross-border collaboration. Finally, blockchain technology in environmental conservation offers a possible paradigm change. Its capacity to safeguard and openly handle data, automate sustainable practices, and enable collaboration might revolutionize ecological preservation. Technology has many benefits, but its application must overcome technological, regulatory, and ethical issues to help the globe.

20.1 Introduction

"Blockchain Technology for Environmental Conservation" represents a cutting-edge approach to addressing the pressing environmental challenges of our time.

[1] Electrical Engineering Department, University Institute of Engineering, Chandigarh University, Mohali, Punjab, India.

[2] Electrical Engineering Department, University Institute of Engineering, Chandigarh University, Mohali, Punjab, India.

* Corresponding author: 22mev10001@cuchd.in

DOI: 10.1201/9781003686026-20

As we stand at a critical juncture in our planet's history, with issues like climate change, biodiversity loss, and resource depletion becoming increasingly urgent, the need for innovative solutions is paramount. Blockchain technology, originally developed as the foundation for cryptocurrencies like Bitcoin, has emerged as a potent tool with the potential to revolutionize how we approach environmental conservation. At its core, Blockchain is a decentralized and immutable digital ledger that records transactions across a network of computers. This inherent structure brings a range of attributes that can be harnessed to benefit our environment. Firstly, blockchain ensures data integrity and transparency. Every entry in the ledger is cryptographically sealed, making tampering with or falsifying records practically impossible. In an era where trust and accurate data are crucial for conservation efforts, this feature can significantly enhance the reliability of environmental data, such as emissions tracking, wildlife monitoring, and supply chain sustainability.

Wildlife conservation, another pivotal aspect of environmental preservation, can benefit significantly from blockchain technology. By tracking the movements and interactions of endangered species using blockchain, we can create a secure, unalterable ledger of their behavior. This can aid in combating illegal wildlife trade and protecting vulnerable animal populations. The global supply chain is another arena where blockchain's transformative potential shines. Sustainable and responsible sourcing of products has become an ethical imperative, and blockchain can provide an unbroken chain of custody for goods, verifying their origins and ensuring they meet environmental and ethical standards. This transparency empowers consumers to make informed choices and encourages businesses to adopt more eco-friendly practices. As we embark on this journey into the world of blockchain technology for environmental conservation, it's essential to explore its promises and challenges. Issues such as scalability, regulatory compliance, and ethical concerns demand careful consideration. Additionally, we must look to the future, exploring how blockchain can synergize with emerging technologies like artificial intelligence and the Internet of Things to create a holistic and data-driven approach to environmental stewardship. Figure. 20.1 shows the six ways in which blockchain technology helps in the construction of a sustainable environment.

20.2 Literature Review

To this end (Oberhauser, 2019) presents a proof-of-concept of a blockchain-based wildlife conservation performance payments scheme in Namibia: the habitat integrity of an elephant corridor is assessed by remote sensing algorithms, which in turn trigger fictitious blockchain innovative contract payments to surrounding communities. This commentary (Howson et al., 2019) explores how blockchain is being leveraged to address the fundamental problems with market-based forest protection globally. Despite the increasing importance of blockchain in improving supply chain efficiencies and bringing societal changes, research investigating its potentialities from the lens of sustainability is scarce. The primary aim of Rejeb & Rejeb (2020) is to fill this knowledge gap and synthesize the literature from leading

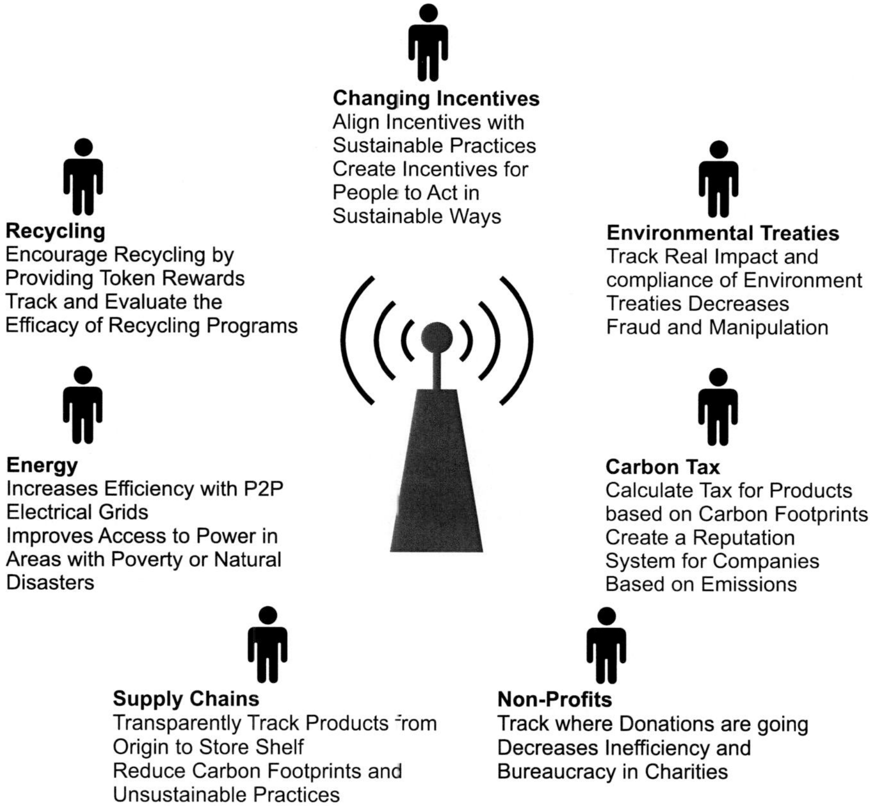

Fig. 20.1 6-Ways–How-Blockchain-Technology-Helps-Environmental-Sustainability.

journals on blockchain and its relation to supply chain sustainability. Tsao et al. (2021) address the sustainable microgrid design problem by leveraging blockchain technology to provide real-time-based demand response programs. This study investigates the impact of blockchain technology on green supply chain practices to promote pro-environmental settings in the supply chains of manufacturing firms (Mubarik et al., 2021). Abramyan et al. (2021) provide an interconnection diagram of organizational, technological, and environmental blockchains. Future research should explore whether blockchain initiatives may translate into concrete environmental outcomes and contribute to the well-being of natural resource managers (Stuit et al., 2022). In response to rising environmental concerns, blockchain technology (BCT) is transforming green innovation, culminating in green economic practices and well-established business models. Recognizing this, Polas et al. (2022) investigate the role of blockchain technology in green innovation practices and its impact on green economic sustainability, which impacts green environmental sustainability. The primary purpose of Munir et al.

(2022) is to explore the literature relevant to blockchain for sustainable supply chain management. Zhang & Zhu (2022) propose an environmental accounting system model based on artificial intelligence blockchain and embedded sensors.

Kouhizadeh and Sarkis (2018) seek to help advance the discussion and motivate additional practice and research related to green supply chains and blockchain technology. Kouhizadeh et al. (2019) study at the nexus of blockchain technology, the circular economy, and product deletion. Many nuances of relationships are introduced. Kouhizadeh et al. (2021) study blockchain technology and the sustainable supply chain: theoretically exploring adoption barriers. The technology-organization-environment framework and force field theories are utilized to investigate blockchain adoption barriers. This commentary (Howson et al., 2019) explores how blockchain is being leveraged to address the fundamental problems with market-based forest protection globally. In doing so, Howson (2020) considers the technical and political challenges of building trust and equity for various stakeholders. Howson (2021) argues that if blockchain is ever to prove beneficial for the degrowth movement, it would need to overcome challenges in three important areas: 1) building democratic and (re)distributive economies, 2) regenerating the environment without commodifying it, and 3) facilitating international alliances without imposing a particular set of values. The aim of Clohessy and Acton (2019) is to elucidate the impact of organizational factors on the adoption of blockchain and the adoption of blockchain in companies based in Ireland. Other influential work includes Saberi et al. (2018). Xie et al. (2019) provide a comprehensive survey of the literature involving blockchain technology applied to smart cities. Wong et al. (*2020) aim to investigate the effects of relative advantage, complexity, upper management support, cost, market dynamics, competitive pressure, and regulatory support on blockchain adoption for operations and supply chain management among Small-Medium Enterprises (SMEs) in Malaysia. Truby (2018) examine government intervention choices to dissocialize negative environmental externalities caused by high-energy consuming Blockchain technology designs. The subject of Esmaeilian et al. (2020) is to provide an overview of Blockchain technology and Industry 4.0 for advancing supply chains towards sustainability. The aim of Clohessy et al. (2019) is to elucidate the impact of organizational factors on the adoption of blockchain and the adoption of blockchain in companies based in Ireland.

20.3 Understanding Blockchain Technology

Understanding blockchain technology is fundamental to appreciating its potential impact on various sectors, including finance, supply chain management, healthcare, and environmental conservation. At its core, blockchain is a decentralized and immutable digital ledger that records transactions across a network of computers in a way that ensures transparency, security, and trust without the need for a central authority (Eslamian and Eslamian, 2025).

20.3.1 *Fundamentals of Blockchain*

The fundamentals of blockchain technology are the foundational concepts and principles that underpin its operation. Understanding these fundamentals is essential for grasping blockchain and its potential applications. Below are the key fundamentals of blockchain:

- **Decentralization:** Blockchain is a decentralized technology that operates on a network of computers (nodes) spread across the globe rather than relying on a central authority, such as a bank or government. This decentralization ensures that no single entity controls the entire network, making it resistant to censorship and single points of failure.
- **Distributed Ledger:** A blockchain is essentially a distributed ledger, which is a record-keeping system that is maintained and synchronized across all nodes in the network. Transactions are grouped into blocks, and each block is linked to the previous one, creating a chain of blocks. This structure ensures the transaction history is transparent, secure, and unchangeable (Khang et al., 2022).
- **Cryptographic Security:** Blockchain relies on cryptographic techniques to secure transactions and control access to data. Participants in the network have a pair of cryptographic keys: a private key (known only to the owner) and a public key (shared with others). Transactions are signed with the private key to prove authenticity and are verified using the public key. This cryptography ensures the integrity and confidentiality of data.
- **Consensus Mechanisms:** Blockchain networks use consensus mechanisms to agree on the validity of transactions and the order in which they are added to the ledger. The most well-known consensus mechanisms are Proof of Work (PoW) and Proof of Stake (PoS), but several others exist. These mechanisms ensure that all participants reach a standard agreement without needing a central authority (Singhal et al., 2018).
- **Immutability:** Once data is added to the blockchain, it becomes challenging, if not impossible, to alter or delete. Each block contains a cryptographic hash of the previous block, creating a chain. Changing any data in a block would require changing all subsequent blocks, which is computationally infeasible. This immutability ensures the integrity of the blockchain's historical data.
- **Transparency:** Transactions recorded on the blockchain are visible to all participants in the network. This transparency promotes accountability and trust among users. Anyone can view the transaction history and verify the authenticity of transactions.
- **Smart Contracts:** Smart contracts are self-executing contracts with predefined rules and conditions written in code. These contracts automatically execute actions when specific conditions are met without intermediaries. Smart contracts enable trustless automation of processes and agreements.
- **Public vs. Private Blockchains:** Public and private are two leading blockchains. Public blockchains are open to anyone and are maintained by a

distributed network of volunteers. Private blockchains are permissioned and operated by a single entity or a consortium of entities. They are typically used for specific business applications (Julie et al., 2020).

Understanding these fundamentals is a crucial first step in comprehending how blockchain technology operates and its potential for transforming various industries, from finance to supply chain management.

20.3.2 Types of Blockchain

- **Public Blockchains (Open Blockchains) Characteristics:** Public blockchains are open to anyone and are maintained by a distributed network of participants who act as nodes. They are characterized by decentralization, transparency, and immutability. Examples include Bitcoin and Ethereum.
  Use Cases: Public blockchains are ideal for applications that require censorship resistance, transparency, and trust without relying on a central authority. They are commonly used for cryptocurrencies, decentralized applications (DApps), and digital asset management.
- **Private Blockchains (Permissioned Blockchains) Characteristics:** Private blockchains are restricted to a specific group of participants who have permission to join the network and validate transactions. They are more centralized than public blockchains, with a select group of trusted validators or nodes. These blockchains often have faster transaction processing times.
  Use Cases: Private blockchains are suitable for organizations that want to leverage blockchain technology for internal purposes while maintaining control over access and data. Everyday use cases include supply chain management, interbank transactions, and enterprise resource planning (ERP) systems (Laurence, 2023).
- **Consortium Blockchains (Semi-Private Blockchains) Characteristics:** Consortium blockchains are a hybrid between public and private blockchains. They are governed by a consortium or organizations that collaboratively maintain and validate transactions. While they are not open to the public, they involve multiple trusted entities rather than a single organization.
  Use Cases: Consortium blockchains are often used when multiple organizations want to collaborate and share a blockchain network while maintaining some degree of control and privacy. Use cases include industry-specific consortia, supply chain partnerships, and financial consortia for cross-border transactions (Saranya & Murugan, 2023).

20.3.3 Advantages of Types of Blockchain

Certain types of Blockchains are readily being used in the latest technology. Each Type of Blockchain has its particular advantages and tradeoffs that are explained below:

- **Public Blockchains:** They offer maximum decentralization, security, and transparency, but may have slower transaction processing times and higher energy consumption. They are well-suited for applications where trust among participants is essential and censorship resistance is desired.
- **Private Blockchains:** They provide greater control, faster transaction speeds, and privacy, making them suitable for enterprise use. They are beneficial when an organization wants to experiment with blockchain technology internally or when regulatory compliance concerns them (Paul et al., 2021).
- **Consortium Blockchains:** These strike a balance between decentralization and control. They are helpful when multiple entities want to collaborate while maintaining a certain level of trust. Consortia can benefit from shared infrastructure and data without exposing everything to the public (Roman-Belmonte et al., 2018). The choice of blockchain type depends on the specific requirements of the application or organization as shown in Figure 20.2. Some projects may even integrate multiple blockchain types to meet various needs within a single ecosystem.

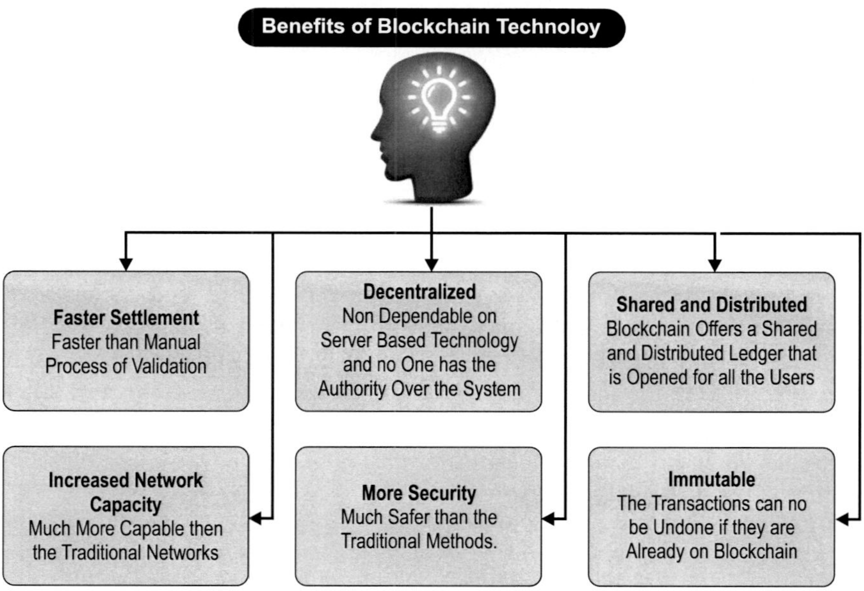

Fig. 20.2 Benefits of Blochchain Technology.

20.4 Environmental Conservation Challenges

Environmental conservation faces many challenges in the modern era, each presenting unique obstacles to preserving our planet's delicate ecosystems. Climate change is a paramount concern, with rising global temperatures, extreme weather events, and sea-level rise threatening the stability of ecosystems and human societies. Habitat destruction, driven by urbanization, deforestation, and agriculture expansion, erodes biodiversity, endangering countless species and destabilizing entire ecosystems. Pollution, whether in the form of air pollutants, plastics in the oceans, or toxic chemicals, poses severe threats to terrestrial and aquatic environments, impairing the health of wildlife and contaminating essential resources (Truong & Clayton, 2020).

Overexploitation of natural resources, such as overfishing and deforestation, disrupts delicate balances in ecosystems and threatens long-term sustainability. Additionally, the invasive species introduced through global trade can wreak havoc on native flora and fauna, causing cascading ecological impacts. Moreover, the lack of global cooperation and effective governance mechanisms hinders international efforts to address environmental challenges that often transcend borders. Finally, the rapid pace of technological advancement and its unintended consequences, such as electronic waste and energy consumption, pose further complications. Addressing these challenges requires concerted local, national, and international efforts involving governments, businesses, communities, and individuals. Solutions will demand innovative approaches, sustainable resource management, and a commitment to protecting the planet for current and future generations (Beng & Corlett, 2020).

20.4.1 Overview of Environmental Issues

The modern world is confronted by many pressing environmental issues that pose significant threats to the planet's ecological balance and human well-being. Climate change, driven by the accumulation of greenhouse gases in the atmosphere, is a paramount concern. It is causing rising global temperatures, resulting in more frequent and severe heatwaves, melting polar ice caps, and disruptions to weather patterns, ultimately contributing to devastating events like wildfires, hurricanes, and floods (Balogh et al., 2020). Habitat destruction and deforestation are exacerbating the loss of biodiversity, endangering countless species, and diminishing the resilience of ecosystems. Air and water pollution poses severe risks to human and environmental health. Air pollution from industrial emissions and vehicle exhaust harms respiratory systems and contributes to climate change. Meanwhile, water pollution from chemicals, plastics, and contaminants imperils aquatic ecosystems and the availability of clean freshwater (Ide et al., 2023).

Overexploitation of natural resources, including overfishing and excessive logging, threatens the sustainability of vital ecosystems and disrupts fragile ecological balances. Invasive species introduced through global trade further disrupt native habitats and species. The lack of practical international cooperation and governance exacerbates these challenges, as environmental problems often transcend political borders. Addressing these complex issues necessitates a concerted global effort emphasizing sustainability, conservation, and protecting our planet's natural resources for future generations (Glicksman et al., 2023).

20.4.2 Need for Innovative Solutions

The need for innovative solutions in environmental conservation has never been more pressing as we grapple with complex and interconnected challenges threatening our planet's ecosystems and biodiversity. These challenges demand fresh approaches and technologies that can effectively address the following key areas:

- **Climate Change Mitigation:** Innovative solutions are vital to combat climate change, including developing renewable energy sources, carbon capture technologies, and sustainable agriculture practices. These innovations can help reduce greenhouse gas emissions and promote climate resilience.
- **Biodiversity Preservation:** As species face extinction rates unseen in millions of years, innovative conservation strategies such as genetic rescue, habitat restoration, and assisted migration are essential to protect and restore biodiversity.
- **Habitat Restoration:** Innovative techniques for restoring degraded habitats, such as reforestation and coral reef restoration using 3D printing, are essential for maintaining ecosystem health and supporting wildlife.
- **Pollution Control:** To combat pollution of air, water, and soil, innovative approaches like advanced filtration systems, nanotechnology-based water treatment, and sustainable packaging materials can mitigate environmental harm.
- **Resource Efficiency:** Innovations in resource management, including circular economy models, can reduce waste and resource depletion. Advanced recycling technologies and sustainable urban planning can help create more resource-efficient societies.
- **Conservation Technology:** The application of technology, such as satellite monitoring, drones, and AI-driven data analysis, allows for more effective monitoring and protection of wildlife and natural habitats.
- **Community Engagement:** Innovative social and educational programs can raise awareness and empower local communities to participate in conservation efforts, ensuring sustainable and inclusive conservation initiatives.
- **Policy and Governance:** Innovative policy frameworks, international agreements, and financial mechanisms can incentivize environmentally friendly practices and promote cross-border cooperation.

Innovation in environmental conservation is not just desirable; it is imperative. These innovations can help address our planet's multifaceted challenges while fostering a sustainable and harmonious relationship between humans and the natural world. Encouraging research, investment, and collaboration in these areas is crucial for the well-being of both current and future generations.

20.5 Applications of Blockchain in Environmental Conservation

Initially conceived for digital currencies, blockchain technology has emerged as a powerful tool with transformative potential across various sectors. In recent years, it has garnered attention for its applications in environmental conservation, offering innovative solutions to combat climate change, biodiversity loss, and unsustainable resource management. These applications are poised to revolutionize how we approach conservation efforts on a global scale. One of the most impactful applications of blockchain in environmental conservation is carbon credits and offset tracking. Carbon credits are pivotal in reducing greenhouse gas emissions (Kaur, 2023). Blockchain provides a transparent and immutable ledger for documenting carbon emissions and removals. This ensures the credibility and authenticity of carbon credits, allowing companies and individuals to track their carbon footprint and invest in offsetting initiatives with confidence. Through blockchain-based systems, participants can verify the environmental impact of their contributions, creating a more accountable and effective carbon market (Dondjio & Themistocleous, 2021).

Supply chain transparency is another critical area where blockchain drives positive change in conservation efforts. Industries like agriculture and forestry often grapple with deforestation, illegal logging, and unethical practices. By recording every step of the supply chain on a blockchain, stakeholders and consumers can trace the origin of products, ensuring they come from sustainable and responsible sources. This transparency empowers consumers to make informed choices while incentivizing businesses to adopt sustainable practices, ultimately reducing the environmental footprint of goods and commodities. Wildlife conservation benefits immensely from blockchain technology. The illegal wildlife trade significantly contributes to biodiversity loss and threatens numerous endangered species. Conservationists are leveraging blockchain to create digital identities for these species. This facilitates monitoring wildlife populations, tracks poaching incidents, and enforces protective regulations. By storing this information on a tamper-proof blockchain, stakeholders can work collectively to safeguard vulnerable ecosystems and wildlife, curbing illegal activities that jeopardize biodiversity (Dondjio & Themistocleous, 2021).

Blockchain is also playing a pivotal role in promoting sustainable energy solutions. Decentralized energy grids powered by blockchain allow for peer-to-peer energy trading, making it more accessible for individuals and businesses to generate and sell excess renewable energy. This reduces reliance on fossil fuels and encourages the adoption of clean energy sources. Additionally, blockchain can certify the origin of renewable energy, ensuring consumers can trust the environmental claims of their energy providers. This transparency is fundamental in accelerating the global transition towards a more sustainable and renewable energy landscape. Conservation finance is transforming blockchain-based crowdfunding and tokenization of projects. By tokenizing conservation initiatives, blockchain allows fractional ownership, making investing in reforestation, marine conservation, or wildlife protection easier for individuals and organizations. This democratization of conservation finance generates additional funding for critical initiatives. It establishes a transparent mechanism for tracking the impact of these investments, ensuring that funds are allocated efficiently and effectively (Mattila et al., 2022).

Furthermore, blockchain technology facilitates secure data sharing and collaboration among diverse stakeholders involved in environmental conservation. Governments, researchers, non-profit organizations, and private companies can securely share and access environmental data. By ensuring data integrity and privacy, blockchain encourages the open sharing of information, leading to better-informed conservation strategies and policies. This collaborative approach enables more effective conservation efforts and fosters international cooperation in addressing global environmental challenges. In conservation agreements, blockchain's smart contracts hold immense promise. These self-executing contracts with terms written into code can automate and enforce agreements related to environmental protection. For example, contracts can trigger payments or penalties based on predefined conditions, such as achieving specific conservation goals. This ensures that funds allocated for conservation efforts are used efficiently and that outcomes are measurable and accountable, enhancing the effectiveness and transparency of conservation finance (Mattila et al., 2022).

20.5.1 Challenges of Blockchain in Environmental Conservation

While blockchain technology holds immense promise for environmental conservation, it also faces several challenges and barriers that must be addressed to unlock its full potential. Following are some of the key challenges as shown in Figure 20.3.

- **Energy Consumption:** One of the most significant criticisms of blockchain, particularly public blockchains like Bitcoin and Ethereum, is their high energy consumption due to the PoW consensus mechanism. This energy-intensive process can have negative environmental impacts and is at odds with conservation goals.
- **Scalability:** Public blockchains often struggle with scalability issues, which can become slower and less efficient as more users join the network. This can hinder the real-time monitoring and management of environmental data, which requires fast and responsive systems (Dogo et al., 2019).
- **Data Privacy:** Blockchain's transparency, while valuable, can also present privacy concerns, especially in environmental conservation efforts that involve sensitive or personal data. Striking a balance between transparency and privacy is a challenge.

- **Regulatory Uncertainty:** The regulatory landscape for blockchain and cryptocurrencies is still evolving. Unclear or inconsistent regulations can create uncertainty for organizations and individuals looking to use blockchain in conservation projects.
- **Technological Adoption:** Widespread adoption of blockchain technology in the conservation sector requires not only technological infrastructure but also the willingness and capacity of organizations and governments to adopt and integrate blockchain solutions.
- **Interoperability:** Different blockchain networks may use different protocols and standards, making it challenging to ensure interoperability and data sharing among various conservation projects and stakeholders (Senthil et al., 2022).
- **Data Accuracy:** While blockchain can provide data integrity, it cannot guarantee the accuracy of the data input into the system. Garbage in, garbage out (GIGO) remains a concern, as inaccurate or fraudulent data can compromise the effectiveness of conservation efforts.
- **Costs and Funding:** Implementing blockchain solutions can be expensive, particularly for smaller organizations or conservation projects with limited budgets. Funding challenges may hinder the adoption of blockchain technology.
- **User Education:** Users and stakeholders in conservation projects may require education and training to effectively understand how to interact with blockchain systems. The learning curve can be steep, slowing down adoption.
- **Legacy Systems Integration:** Integrating blockchain with legacy systems can be complex and requires significant resources and expertise.

Despite these challenges, ongoing research and innovation in blockchain technology, coupled with increased awareness of its potential benefits in environmental conservation, are gradually paving the way for solutions to these obstacles. Collaboration among stakeholders, including governments, NGOs, and the private sector, will be crucial in addressing these challenges and harnessing blockchain's capabilities to advance conservation efforts (Jin & Chang, 2023).

20.5.2 Considerations of Blockchain in Environmental Conservation

When implementing blockchain technology in environmental conservation, several important considerations must be considered to ensure conservation efforts' success and sustainability. Following are key considerations:

- **Data Accuracy and Validation**
  Consideration: Ensuring the accuracy and validation of data input into the blockchain is crucial. GIGO can compromise the effectiveness of conservation initiatives.
  Mitigation: Implement data validation mechanisms and leverage IoT devices or sensors for real-time data collection. Encourage data verification through multiple trusted parties.

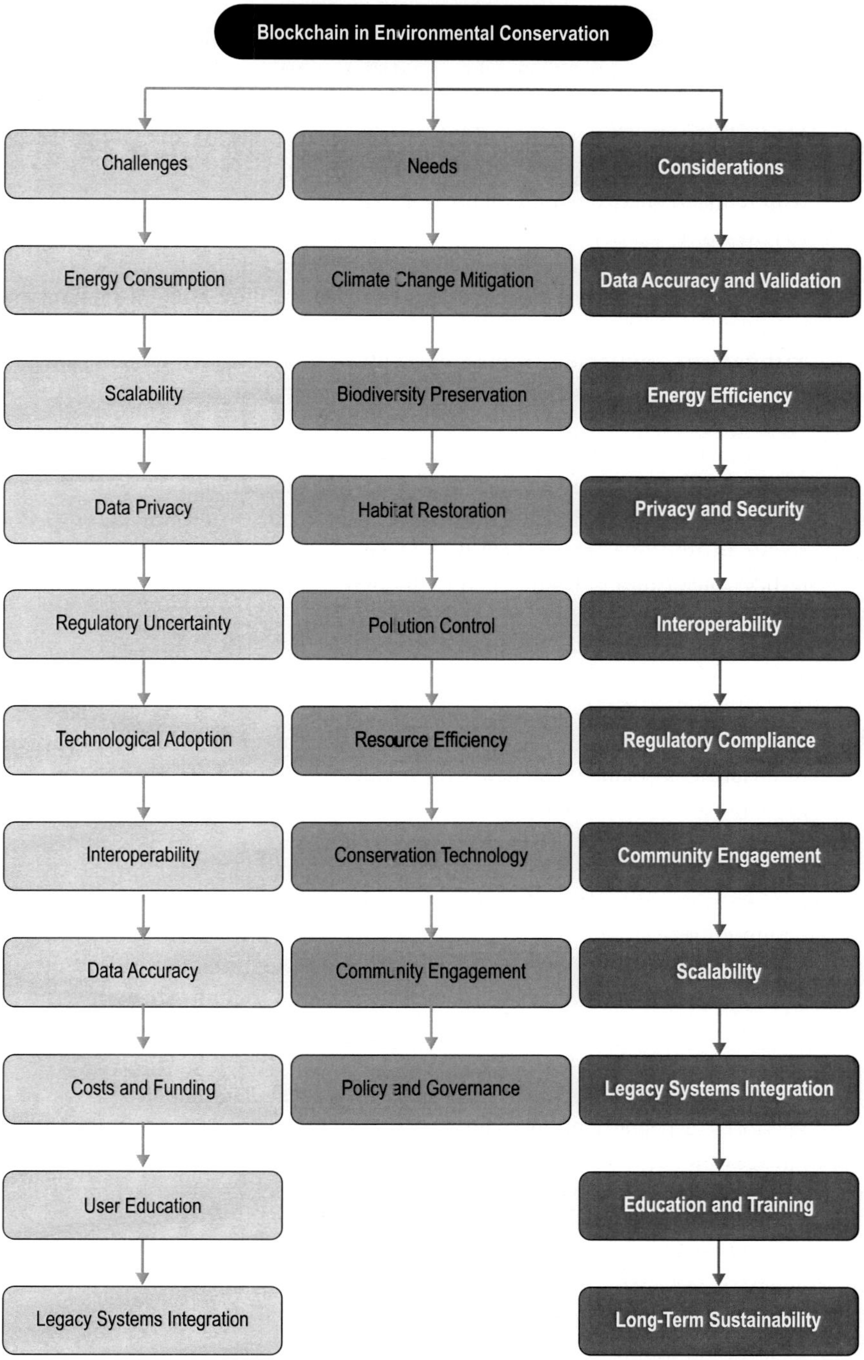

Fig. 20.3 Challenges, Needs, and Considerations for Blockchain in Environment Conservation.

- **Energy Efficiency**
  Consideration: Blockchain's energy consumption, especially in proof-of-work systems, can have environmental impacts that contradict conservation goals.
  Mitigation: Explore energy-efficient consensus mechanisms like proof-of-stake or hybrid solutions. Consider using private or consortium blockchains with reduced energy requirements (Oberhauser, 2019).
- **Privacy and Security**
  Consideration: Balancing transparency with data privacy and security is essential, particularly when handling sensitive environmental or personal information.
  Mitigation: Implement robust encryption and access controls. Explore privacy-focused blockchain solutions or off-chain solutions for sensitive data.
- **Interoperability**
  Consideration: Different blockchain networks may have varying protocols and standards, challenging data sharing and collaboration.
  Mitigation: Standardize data formats and protocols within the conservation sector to improve interoperability. Develop cross-blockchain solutions or utilize consortium blockchains for collaborative projects.
- **Regulatory Compliance**
  Consideration: The regulatory environment for blockchain and cryptocurrencies is evolving, and compliance with relevant laws and regulations is critical.
  Mitigation: Stay informed about regulatory changes and seek legal advice to ensure compliance. Engage with regulatory authorities to establish clear guidelines for blockchain use in conservation.
- **Community Engagement**
  Consideration: Effective community engagement is vital for successful conservation initiatives. Blockchain technology should not alienate local communities.
  Mitigation: Involve local communities in decision-making processes and ensure that benefits from blockchain-enabled conservation efforts are shared with them. Promote transparency in community interactions (Li et al., 2022).
- **Scalability**
  Consideration: Scalability issues can impede the performance of blockchain networks, especially public ones.
  Mitigation: Choose a blockchain platform that suits the project's needs and consider layer-2 scaling solutions for public blockchains.
- **Legacy Systems Integration**
  Consideration: Integrating blockchain with existing legacy systems can be complex and costly.
  Mitigation: Develop a clear integration plan, allocate resources, and seek expertise to minimize disruptions during integration.

- **Education and Training**
  Consideration: Users and stakeholders may need education and training to use blockchain systems effectively.
  Mitigation: Provide comprehensive training and support to ensure that users are proficient in using blockchain technology for conservation.
- **Long-Term Sustainability**
  Consideration: The sustainability of blockchain-enabled conservation initiatives beyond the initial implementation phase is vital.
  Mitigation: Develop a long-term sustainability plan that includes ongoing maintenance, funding strategies, and community involvement (Zhang et al., 2020).

Addressing these considerations strategically and thoughtfully can help maximize the positive impact of blockchain technology on environmental conservation while minimizing potential pitfalls and challenges. Collaboration among stakeholders and a commitment to the overarching goals of conservation are essential for success.

20.5.3 *Real-World Examples of Blockchain in Environmental Conservation*

Several real-world examples showcase the application of blockchain technology in environmental conservation, highlighting its potential to address critical environmental challenges. Below are some notable examples:

- **Wildlife Conservation: Wildchain**
  Use Case: Wildchain, a blockchain-based platform, is designed to combat wildlife poaching by tracking and verifying wildlife data.
  How it Works: Users can upload photos and videos of animals, which are time-stamped and recorded on the blockchain. The data is used for research, monitoring, and anti-poaching efforts.
  Impact: Wildchain provides a transparent and immutable record of wildlife sightings, aiding conservationists and researchers in tracking and protecting endangered species.
- **Carbon Credit Market: Verra**
  Use Case: Verra, formerly the Verified Carbon Standard (VCS), is a blockchain-based platform certifying and tracking carbon offset projects.
  How it Works: Verra verifies and issues carbon credits for projects that reduce greenhouse gas emissions. These credits are recorded on a blockchain to ensure transparency and prevent double-counting.
  Impact: Verra's blockchain-based system helps mitigate climate change by providing a trustworthy and transparent way to trade carbon credits and support emission reduction projects.
- **Supply Chain Traceability: IBM Food Trust**
  Use Case: While primarily focused on food supply chains, IBM Food Trust's blockchain platform can also be applied to sustainable and ethical sourcing in the seafood industry.

How it Works: The platform allows consumers to trace the origin and journey of seafood products, including details about the catch location and method.
Impact: By promoting transparency and traceability in seafood supply chains, IBM Food Trust helps combat illegal fishing and supports sustainable fisheries management.

- **Forest Conservation: Rainforest Foundation US**
  Use Case: Rainforest Foundation US uses blockchain technology to protect rainforests and indigenous land rights.
  How it Works: The organization uses blockchain to create a transparent and immutable record of land titles and land-use agreements, empowering indigenous communities to secure their land rights.
  Impact: Rainforest Foundation US's blockchain-based system supports forest conservation and indigenous rights by preventing illegal land grabs and deforestation.
- **Waste Management: Plastic Bank**
  Use Case: Plastic Bank uses blockchain to incentivize plastic recycling in developing countries.
  How it Works: Individuals can collect plastic waste, which is weighed and recorded on a blockchain. They are rewarded with digital tokens that can be used for various purposes.
  Impact: Plastic Bank's blockchain-based platform encourages plastic recycling, reduces plastic pollution, and provides economic opportunities in marginalized communities.

These real-world examples demonstrate the diverse applications of blockchain technology in environmental conservation, from wildlife protection and carbon credit tracking to sustainable supply chains and waste management. These initiatives leverage blockchain's transparency, security, and accountability to address critical environmental issues and promote conservation efforts worldwide.

20.6 Future Directions and Innovation

The future of blockchain in environmental conservation holds excellent promise as technological advancements and innovative approaches continue to emerge. There are some future directions and innovations expected in the field: Blockchain technology, designed initially to underpin cryptocurrencies like Bitcoin, has transcended its initial purpose and found applications across various domains. Environmental conservation is one of the most promising and impactful areas where blockchain is making significant strides. Innovative solutions are urgently needed in the face of mounting global challenges such as climate change, deforestation, biodiversity loss, and pollution. Blockchain, with its inherent features of transparency, security, and decentralization, offers a set of emerging trends that can revolutionize how we address these environmental crises and conserve our planet's natural resources (Karaszewski et al., 2021).

One of the most prominent applications of blockchain in environmental conservation is the tracking and trading of carbon credits. Carbon credits are critical in mitigating climate change by incentivizing carbon reduction activities. Blockchain provides an immutable ledger for documenting carbon emissions and removals, offering greater transparency and trust in the carbon credit market. This technology allows companies and individuals to track their carbon footprint, purchase offsets, and verify the authenticity of credits through blockchain-based systems. This aligns with global efforts to reduce greenhouse gas emissions and ensures that the traded carbon credits genuinely contribute to the fight against climate change (Rana et al., 2019). Another exciting application of blockchain in environmental conservation is enhancing supply chain transparency, particularly in industries like agriculture and forestry. In these sectors, unethical practices and unsustainable resource management can have devastating environmental consequences. By recording each step of the supply chain on a blockchain, consumers and stakeholders can trace the origin of products, ensuring that they come from sustainable and environmentally responsible sources. This transparency helps combat deforestation, illegal logging, and unethical labor practices. It empowers consumers to make informed choices about the products they purchase and encourages companies to adopt more sustainable practices throughout their supply chains (Esmaeilian et al., 2020).

Wildlife conservation is yet another area where blockchain is making a significant impact. The illegal wildlife trade severely threatens biodiversity, and blockchain technology is being harnessed to combat this crisis. Conservationists are utilizing blockchain to create digital identities for endangered species. These digital profiles enable monitoring of wildlife populations, tracking poaching incidents, and enforcing regulations. By making this information publicly accessible and tamper-proof, blockchain contributes to preserving vulnerable ecosystems and wildlife. Sustainable energy is also benefiting from blockchain's transformative potential. Decentralized energy grids powered by blockchain technology enable peer-to-peer energy trading, making it easier for individuals and businesses to generate and sell excess renewable energy. This reduces dependence on fossil fuels and incentivizes the adoption of clean energy solutions. Blockchain can also certify the origin of renewable energy, ensuring that consumers can trust the environmental claims of their energy providers. This transparency is crucial in promoting the growth of renewable energy sources and transitioning to a more sustainable energy landscape (Esmaeilian et al., 2020).

In the realm of conservation finance, blockchain is creating innovative solutions. Crowdfunding and tokenization of conservation projects are becoming more accessible through blockchain-based platforms. By tokenizing conservation initiatives, blockchain allows fractional ownership, making investing in reforestation, marine conservation, or wildlife protection easier for individuals and organizations. This provides additional funding for critical conservation efforts and creates a transparent mechanism for tracking the impact of these investments, ensuring that funds are used efficiently and effectively. Furthermore, blockchain technology facilitates secure data sharing and collaboration among stakeholders involved in environmental conservation. Governments, researchers, non-profit organizations, and private companies can use blockchain to share and access environmental data securely. By ensuring data integrity and privacy, blockchain encourages more open information sharing, leading to better-informed conservation strategies and policies. This collaborative approach enables more effective conservation efforts and fosters international cooperation in addressing global environmental challenges. Smart contracts, self-executing agreements with contract terms directly written into code, are another innovative aspect of blockchain technology. In conservation, intelligent contracts can automate and enforce agreements related to environmental protection. For instance, contracts can trigger payments or penalties based on predefined conditions, such as achieving specific conservation goals. This ensures that funds are used efficiently and that conservation efforts are held accountable.

20.6.1 Emerging Trends in Blockchain for Environmental Conservation

Blockchain technology, originally designed to support cryptocurrencies, is now emerging as a powerful tool for environmental conservation. As the world grapples with the urgent need to address climate change and protect the environment, blockchain offers innovative solutions that are gaining momentum. One of the most notable applications of blockchain in environmental conservation is tracking and trading carbon credits. Carbon credits are critical to global efforts to reduce greenhouse gas emissions and mitigate climate change. Blockchain technology provides an immutable ledger for recording carbon emissions and removals, creating a transparent and trustworthy system for documenting these crucial environmental metrics (Giungato et al., 2017). This transparency is precious in a market where the authenticity of carbon credits can be a source of concern. Through blockchain-based systems, companies and individuals can easily track their carbon footprint, purchase offsets, and verify the authenticity of credits, ensuring that their contributions to carbon reduction efforts are meaningful and verifiable.

Moreover, blockchain technology is revolutionizing supply chain transparency, a crucial aspect of environmental conservation. Blockchain is significantly impacting industries like agriculture and forestry, where unethical practices and unsustainable resource management can have devastating environmental consequences. By recording every step of the supply chain on a blockchain, consumers and stakeholders can trace the origins of products, ensuring that they come from sustainable and environmentally responsible sources. This transparency helps combat deforestation, illegal logging, and other harmful practices and empowers consumers to make more informed choices, promoting sustainable resource management and reducing the environmental footprint of everyday products (Jin & Chang, 2023). In wildlife conservation, blockchain is creating digital identities for endangered species. The illegal wildlife trade is a significant threat to biodiversity, and blockchain technology is crucial in combatting this crisis. Conservationists use blockchain to track and monitor wildlife populations, record poaching incidents, and enforce regulations to protect endangered species. By making this information publicly accessible and tamper-proof, blockchain contributes to preserving vulnerable ecosystems and wildlife, offering hope for the future of our planet's biodiversity.

Sustainable energy is another area where blockchain is making a substantial impact on environmental conservation. Decentralized energy grids powered by blockchain technology enable peer-to-peer energy trading, making it easier for individuals and businesses to produce and sell excess renewable energy. This reduces reliance on fossil fuels and incentivizes the adoption of clean energy solutions, furthering the global transition to sustainable and environmentally friendly energy sources. Additionally, blockchain can certify the origin of renewable energy, ensuring that consumers can trust their energy providers' environmental claims and promote greater confidence in green energy options (Mishra & Kaushik, 2023). Conservation finance is also being transformed by blockchain technology. Crowdfunding and tokenization of conservation projects are emerging trends that make investing in environmental conservation efforts easier for individuals and organizations. By tokenizing conservation initiatives, blockchain allows for fractional ownership and a more accessible way for people to support projects like reforestation, marine conservation, or wildlife protection. This provides additional funding for critical conservation efforts and creates a transparent mechanism for tracking the impact of these investments, ensuring that funds are used efficiently and effectively (Bai et al., 2020).

Furthermore, blockchain technology facilitates secure data sharing and collaboration among stakeholders involved in environmental conservation. Governments, researchers, non-profit organizations, and private companies can use blockchain to share and access environmental data securely. By ensuring data integrity and privacy, blockchain encourages more open information sharing, leading to better-informed conservation strategies and policies. This collaborative approach enables more effective conservation efforts and facilitates international cooperation in addressing global environmental challenges. In conservation agreements, blockchain's smart contracts are gaining traction. Smart contracts are self-executing agreements with the contract terms written into code. They can automate and enforce conservation agreements by triggering payments or penalties based on predefined conditions, such as achieving specific conservation goals. This ensures that funds allocated for conservation efforts are used efficiently and that the outcomes are measurable and accountable, making conservation finance more effective and transparent (Rejeb & Rejeb, 2020).

20.7 Conclusion

The burgeoning intersection of blockchain technology and environmental conservation promises a transformative era in our approach to safeguarding the planet. These innovative applications address various challenges, from tracking wildlife movements to transparent carbon markets, empowering local communities in conservation efforts, and revolutionizing sustainable finance. The emergence of eco-friendly blockchain protocols and decentralized autonomous organizations in conservation underscore the sector's adaptability and embrace of cutting-edge technologies. Moreover, as global initiatives to combat climate change gain momentum, blockchain's role in verifying carbon offsets and supporting green bonds exemplifies its pivotal role in transitioning to a more sustainable future. The emphasis on transparency, community engagement, and cross-border collaboration through blockchain-powered platforms marks a paradigm shift in how we share and manage critical environmental data. The trends shaping this landscape are diverse and dynamic. From enabling individuals to directly participate in renewable energy trading to combatting illegal fishing through transparent supply chains, blockchain's versatility knows few bounds. Its adoption in environmental education and awareness initiatives highlights its potential to mobilize communities and individuals towards more eco-conscious behaviors. As conservation projects increasingly adopt blockchain, the technology's potential to democratize decision-making and funding allocation is a powerful force for change. In the coming years, as blockchain technology matures, we anticipate even greater integration and innovation in environmental conservation. As regulatory frameworks solidify and sustainability practices become more ingrained in our societies, blockchain's role is poised to expand. By harnessing the power of decentralization, transparency, and secure data management, we embark on a more resilient and harmonious coexistence with the natural world. The future of environmental conservation, enriched by the possibilities of blockchain, holds immense promise for the preservation of our planet and the well-being of future generations. Through collaborative efforts and a shared commitment to a sustainable future, we stand at the precipice of a new era in environmental stewardship.

References

1. Abramyan, S., Klenina, E., Peskov, A. and Burlachenko, A. (2021). Monitoring of Organizational, Technological and Environmental Solutions in Construction with Blockchain Technologies. *IOP Conference Series: Materials Science and Engineering.*
2. Bai, C.A., Cordeiro, J. and Sarkis, J. (2020). Blockchain technology: Business, strategy, the environment, and sustainability. *Business Strategy and the Environment*, *29*(1), 321–322.
3. Balogh, J.C., Gibeault, V.A., Walker, W.J., Kenna, M.P., & Snow, J.T. (2020). Background and overview of environmental issues. In: *Golf Course Management and Construction*). CRC Press, pp. 1–37.
4. Beng, K.C. and Corlett, R.T. (2020). Applications of environmental DNA (eDNA) in ecology and conservation: Opportunities, challenges, and prospects. *Biodiversity and Conservation*, *29*, 2089–2121.
5. Clohessy, T., & Acton, T. (2019). Investigating the influence of organizational factors on blockchain adoption: An innovation theory perspective. *Industrial Management and Data Systems*, *119*(7), 1457–1491.

6. Clohessy, T., Acton, T. and Rogers, N. (2019). Blockchain adoption: Technological, organisational, and environmental considerations. *Business Transformation through Blockchain: Volume I*, 47–76.
7. Dogo, E.M., Salami, A.F., Nwulu, N.I. and Aigbavboa, C.O. (2019). Blockchain and internet of things-based technologies for intelligent water management system. *Artificial Intelligence in IoT*. MDPI, pp. 129–150.
8. Dondjio, I. and Themistocleous, M. (2021). Blockchain technology and waste management: A systematic literature review. *European, Mediterranean, and Middle Eastern Conference on Information Systems*.
9. Eslamian, S. and Eslamian, F. (2025). *Blockchain Technology in Water and Environmental Systems*. CRC Press, Taylor and Francis, USA. p. 282.
10. Esmaeilian, B., Sarkis, J., Lewis, K. and Behdad, S. (2020). Blockchain for the future of sustainable supply chain management in Industry 4.0. *Resources, Conservation, and Recycling*, *163*, 105064.
11. Giungato, P., Rana, R., Tarabella, A. and Tricase, C. (2017). Current trends in sustainability of bitcoins and related blockchain technology. *Sustainability*, *9*(12), 2214.
12. Glicksman, R.L., Buzbee, W.W., Mandelker, D.R., Hammond, E. and Camacho, A. (2023). *Environmental Protection: Law and Policy*. Aspen Publishing.
13. Howson, P. (2020). Building trust and equity in marine conservation and fisheries supply chain management with blockchain. *Marine Policy*, *115*, 103873.
14. Howson, P. (2021). Distributed degrowth technology: Challenges for blockchain beyond the green economy. *Ecological Economics*, *184*, 107020.
15. Howson, P., Oakes, S., Baynham-Herd, Z. and Swords, J. (2019). Cryptocarbon: The promises and pitfalls of forest protection on a blockchain. *Geoforum*, *100*, 1–9.
16. Ide, T., Johnson, M.F., Barnett, J., Krampe, F., Le Billon, P., Maertens, L., Von Uexkull, N. and Vélez-Torres, I. (2023). The future of environmental peace and conflict research. *Environmental Politics*, 1–27.
17. Jin, S. and Chang, H. (2023). The trends of blockchain in environmental management research: A bibliometric analysis. *Environmental Science and Pollution Research*, *30*(34), 81707–81724.
18. Julie, E.G., Nayahi, J.J.V. and Jhanjhi, N.Z. (2020). *Blockchain Technology: Fundamentals, Applications, and Case Studies*. CRC Press.
19. Karaszewski, R., Modrzyński, P., Müldür, G.T. and Wójcik, J. (2021). Blockchain Technology in Life Cycle Assessment—New Research Trends. *Energies*, *14*(24), 8292.
20. Kaur, G. (2023). Application of Blockchain for Sustaining Green Finance. In: *Perspectives on Blockchain Technology and Responsible Investing*. IGI Global, pp. 226–235.
21. Khang, A., Chowdhury, S. and Sharma, S. (2022). *The Data-Driven Blockchain Ecosystem: Fundamentals, Applications, and Emerging Technologies*. CRC Press.
22. Kouhizadeh, M., Saberi, S. and Sarkis, J. (2021). Blockchain technology and the sustainable supply chain: Theoretically exploring adoption barriers. *International Journal of Production Economics*, *231*, 107831.
23. Kouhizadeh, M. and Sarkis, J. (2018). Blockchain practices, potentials, and perspectives in greening supply chains. *Sustainability*, *10*(10), 3652.
24. Kouhizadeh, M., Sarkis, J. and Zhu, Q. (2019). At the nexus of blockchain technology, the circular economy, and product deletion. *Applied Sciences*, *9*(8), 1712.
25. Laurence, T. (2023). *Blockchain for Dummies*. John Wiley & Sons.
26. Li, X., Zhou, Y. and Yuen, K.F. (2022). Blockchain implementation in the maritime industry: critical success factors and strategy formulation. *Maritime Policy and Management*, 1-19.
27. Mattila, V., Dwivedi, P., Gauri, P. and Ahbab, M. (2022). Blockchain for environmentally sustainable economies: case study on 5irechain. *International Journal of Social Sciences and Management Review*, *5*, 50–62.

28. Mishra, L. and Kaushik, V. (2023). Application of blockchain in dealing with sustainability issues and challenges of financial sector. *Journal of Sustainable Finance and Investment, 13*(3), 1318–1333.
29. Mubarik, M., Raja Mohd Rasi, R.Z., Mubarak, M.F. and Ashraf, R. (2021). Impact of blockchain technology on green supply chain practices: Evidence from emerging economy. *Management of Environmental Quality: An International Journal, 32*(5), 1023–1039.
30. Munir, M.A., Habib, M.S., Hussain, A., Shahbaz, M.A., Qamar, A., Masood, T., Sultan, M., Mujtaba, M., Imran, S. and Hasan, M. (2022). Blockchain adoption for sustainable supply chain management: Economic, environmental, and social perspectives. *Frontiers in Energy Research, 10*, 899632.
31. Oberhauser, D. (2019). Blockchain for environmental governance: can smart contracts reinforce payments for ecosystem services in Namibia? *Frontiers in Blockchain, 2*, 21.
32. Paul, P., Aithal, P., Saavedra, R. and Ghosh, S. (2021). Blockchain Technology and its Types: A Short Review. *International Journal of Applied Science and Engineering* (*IJASE*), *9*(2), 189–200.
33. Polas, M.R.H., Kabir, A.I., Sohel-Uz-Zaman, A.S.M., Karim, R. and Tabash, M.I. (2022). Blockchain technology as a game changer for green innovation: Green entrepreneurship as a roadmap to green economic sustainability in Peru. *Journal of Open Innovation: Technology, Market, and Complexity, 8*(2), 62.
34. Rana, R.L., Giungato, P., Tarabella, A. and Tricase, C. (2019). Blockchain applications and sustainability issues. *Amfiteatru Economic, 21*(13), 861–870.
35. Rejeb, A. and Rejeb, K. (2020). Blockchain and supply chain sustainability. *Logforum, 16*(3).
36. Roman-Belmonte, J. M., De la Corte-Rodriguez, H. and Rodriguez-Merchan, E.C. (2018). How blockchain technology can change medicine. *Postgraduate Medicine, 130*(4), 420–427.
37. Saberi, S., Kouhizadeh, M. and Sarkis, J. (2018). Blockchain technology: A panacea or pariah for resources conservation and recycling? *Resources, Conservation, and Recycling, 130*(March), 80–81.
38. Saranya, R. and Murugan, A. (2023). A systematic review of enabling blockchain in healthcare system: Analysis, current status, challenges, and future direction. *Materials Today: Proceedings, 80*, 3010–3015.
39. Senthil, G., Prabha, R., Divya, B. and Sathya, A. (2022). An Empirical Analysis and Challenging Era of Blockchain in Green Society. In: *International Conference on Communication and Intelligent Systems.*
40. Singhal, B., Dhameja, G. and Panda, P.S. (2018). *Beginning Blockchain: A Beginner's Guide to Building Blockchain Solutions* (Vol. 1). Springer.
41. Stuit, A., Brockington, D. and Corbera, E. (2022). Smart, commodified, and encoded. *Conservation and Society, 20*(1), 12–23.
42. Truby, J. (2018). Decarbonizing Bitcoin: Law and policy choices for reducing the energy consumption of Blockchain technologies and digital currencies. *Energy Research and Social Science, 44*, 399–410.
43. Truong, M.-X. A. and Clayton, S. (2020). Technologically transformed experiences of nature: A challenge for environmental conservation? *Biological Conservation, 244*, 108532.
44. Tsao, Y.-C., Thanh, V.-V. and Wu, Q. (2021). Sustainable microgrid design considering blockchain technology for real-time price-based demand response programs. *International Journal of Electrical Power and Energy Systems, 125*, 106418.
45. Wong, L.-W., Leong, L.-Y., Hew, J.-J., Tan, G. W.-H. and Ooi, K.-B. (2020). Time to seize the digital evolution: Adoption of blockchain in operations and supply chain management among Malaysian SMEs. *International Journal of Information Management, 52*, 101997.
46. Xie, J., Tang, H., Huang, T., Yu, F. R., Xie, R., Liu, J. and Liu, Y. (2019). A survey of blockchain technology applied to smart cities: Research issues and challenges. *IEEE Communications Surveys & Tutorials, 21*(3), 2794–2830.

47. Zhang, A., Zhong, R.Y., Farooque, M., Kang, K. and Venkatesh, V.G. (2020). Blockchain-based life cycle assessment: An implementation framework and system architecture. *Resources, Conservation and Recycling*, *152*, 104512.
48. Zhang, W. and Zhu, M. (2022). Environmental accounting system model based on artificial intelligence blockchain and embedded sensors. *Computational Intelligence and Neuroscience*.

21 Blockchain and Environmental Sustainability: Addressing Global Challenges and Advancing SDGs

Harpreet Kaur Channi[1], Parminder Singh[2] and Saeid Eslamian[3*]

Blockchain technology across several sectors has inspired creative solutions to complicated problems, including hydro systems. This detailed review covers blockchain research in hydrosystems management. Water shortages, contamination, and poor management plague hydrosystems, which include water resources, distribution networks, and environmental concerns. The decentralization, transparency, and data immutability of Blockchain have led to studies on its potential to transform hydrosystem management. The chapter begins with blockchain basics and hydrosystem contexts. Blockchain provides hydrosystems with crucial benefits, including data security, verifiable recordkeeping, and transparent, traceable data-sharing ecosystems. A thorough examination of current research highlights hydrosystem management aspects under blockchain integration scrutiny. These include water quality monitoring, resource optimization, intelligent metering, water distribution supply chain transparency, and collaborative water governance. The chapter provides real-world implementation ideas and integration issues via case studies and prototype projects. The chapter also discusses hydrosystem blockchain adoption challenges. Scalability, energy efficiency, legacy system compatibility, regulatory compliance, and data protection are these problems. The iterative nature of blockchain integration in this environment is highlighted by exploring forward-looking tactics and research pathways to mitigate these problems. This study summary summarizes Blockchain's impact on hydrosystem applications. It shows Blockchain's potential to change water management and promote sustainability. This exposition helps scholars, practitioners, and policymakers use Blockchain to improve global hydrosystem management by synthesizing potential obstacles and current improvements.

[1] Department of Electrical Engineering, Guru Nanak Dev Engineering College, Ludhiana, Punjab, India.

[2] Department of Electrical Engineering, Chandigarh University, Gharuan, Mohali, Punjab.

[3] Department of Water Science and Engineering, College of Agriculture, Isfahan University of Technology, Isfahan, Iran.

* Corresponding author: saeid@iut.ac.ir

DOI: 10.1201/9781003686026-21

21.1 Introduction

Blockchain technology, characterized by its nascent adoption stage, holds immense potential to revolutionize the management and operation of hydrosystems. Within this critical sector encompassing water resources, distribution networks, and environmental concerns, effective management is paramount to mitigate challenges such as water shortages, contamination, and suboptimal administration (Pandey et al., 2022). Blockchain, albeit in its early application stages, offers a promising avenue for redefining conventional methodologies and fostering a paradigm shift in hydrosystem management. A particularly promising application of blockchain technology in this domain lies in water data management. By leveraging Blockchain, a secure and tamper-proof repository of all water-related data, ranging from sensor readings to water quality measurements, can be established. This reservoir of data can be seamlessly shared in real-time among stakeholders, facilitating a more efficient and collaborative decision-making process (Mardiani & Iswahyudi, 2023).

Transparency and accountability are integral facets of effective hydrosystem management, and Blockchain offers a compelling solution. Tracking the movement of water throughout the hydrosystem, from source to end user, using Blockchain holds the potential to curb water theft and contamination, ensuring reasonable and equitable utilization of this precious resource (Hu and Chuang, 2023). Furthermore, Blockchain presents an opportunity to cultivate novel business models within the hydro sector. For instance, the creation of peer-to-peer water trading platforms or the development of micro-financing schemes for small-scale hydro projects can be facilitated through blockchain technology (Wu et al., 2022).

As blockchain technology advances and matures, its applications within the hydrosystem management sector are expected to evolve and diversify, fostering a future characterized by innovative solutions and heightened operational efficiency. This chapter aims to shed light on the trajectory of this evolution, offering insights into the burgeoning potential of Blockchain to shape the future of hydrosystem management (Wu et al., 2022). This chapter presents notable ongoing projects within the sector to provide a tangible perspective on implementing Blockchain in hydrosystem management. While still in its early stages of adoption within the hydrosystems sector, blockchain technology is witnessing a surge of innovative projects worldwide. One notable initiative is HydroChain, a blockchain-based platform facilitating the management and trade of water data, gaining traction among various water utilities globally (Nithish et al., 2023).

Similarly, Water Ledger is employed by the World Bank to monitor water usage in irrigation projects within developing nations. The Water Neutrality platform addresses the water footprint offset for businesses, collaborating with industries like food and beverage, agriculture, and technology. Initiatives like WaterChain and H20Chain also make strides in managing and trading water resources, embracing a blockchain framework. These projects showcase the potential of Blockchain in revolutionizing water data management, enhancing transparency, supporting dynamic pricing models, and promoting sustainable water resource management (Parhamfar et al., 2023).

Moreover, broader trends are emerging, emphasizing Blockchain's potential to enhance transparency in water governance, develop sustainable pricing models, and augment accountability within the hydrosystem. By providing a secure and tamper-proof record of water data, Blockchain facilitates real-time data sharing and collaboration among stakeholders for informed decision-making (Koedel et al., 2022). Additionally, it aids in tracking water movement throughout the hydrosystem, preventing theft contamination, and ensuring equitable usage. The technology also supports dynamic pricing models, reducing water consumption, and enhancing efficiency. Blockchain promises to propel hydrosystem management towards greater efficiency, sustainability, transparency, and equitability, setting the stage for transformative applications in future years (Vera, 2022).

21.1.1 Background and Motivation

Hydro systems are critical in providing essential resources such as drinking water, irrigation, and energy generation. However, managing hydrosystems is complex, demanding a delicate balance between competing demands, including water quality maintenance, flood control, and environmental preservation. Recently, interest has emerged in leveraging blockchain technology to enhance hydrosystem management. Blockchain, recognized as a distributed ledger technology, can create secure and unalterable transaction records (Ezeh et al., 2023). This characteristic renders Blockchain highly suitable for multiple applications within hydrosystem management.

Blockchain's potential applications in hydrosystem management include robust water data management, where it establishes an immutable record of all water-related data, spanning sensor readings to water quality assessments. This data can be shared among stakeholders in real-time, fostering efficiency and collaboration in decision-making processes. Moreover, Blockchain can enhance transparency and accountability within hydrosystems. Blockchain can mitigate water theft and contamination by tracking water movement throughout the hydrosystem, from the source to end user while ensuring equitable and efficient water usage (Kumar,et al., 2023). Another compelling aspect is facilitating new business models and opportunities within the hydro sector. Blockchain can pave the way for peer-to-peer water trading platforms and novel micro-financing schemes for small-scale hydro projects, thereby democratizing access to hydrosystems.

The motivations for integrating blockchain technology in hydrosystem management are multifaceted. Firstly, it aims to enhance operational efficiency and transparency. By providing a secure and unchangeable record of water data, Blockchain minimizes the risks associated with errors and fraudulent activities, thereby improving communication and collaboration among stakeholders. Secondly, Blockchain offers a means to bolster the sustainability of hydrosystems. Through improved water data management and transparency, it aids in reducing water wastage and pollution, aligning with sustainability goals (Mejías, 2022).

Additionally, Blockchain can support the development of renewable energy sources like solar and wind power, reducing greenhouse gas emissions associated with the hydro sector. Lastly, Blockchain can fuel the creation of innovative business models and opportunities, democratizing access to hydrosystems. By enabling peer-to-peer water trading platforms and innovative financing mechanisms, Blockchain enhances the accessibility and affordability of hydro resources (Coban et al., 2022).

This overview delineates its objectives into four primary dimensions. Firstly, it aims to furnish a comprehensive and current status overview regarding integrating blockchain technologies within hydrosystem management. Secondly, it endeavors to engage in a detailed discourse concerning the potential advantages and hurdles pertinent to the application of blockchain technology in this sector. Furthermore, the overview seeks to exemplify specific blockchain projects presently underway within the hydro industry. Lastly, it highlights the transformative potential of blockchain technology, underscoring its capacity to fundamentally revolutionize extant methodologies and operational frameworks integral to the management of hydrosystems (Palominos et al., 2022). The discussion is strategically structured to address pivotal inquiries, including the mechanisms through which Blockchain can be employed to enhance hydrosystem management, the prospective benefits attributable to blockchain integration, the challenges that necessitate resolution for the widespread adoption of Blockchain within the hydro sector, notable ongoing projects serving as exemplars, and the envisaged trajectory for blockchain-driven transformation within the purview of hydrosystem management. In amalgamating these facets, this overview aspires to furnish a comprehensive academic resource for stakeholders seeking profound insights into Blockchain's prospective role in shaping the trajectory of hydrosystem management (Paniyil, 2023).

21.1.2 Scope and Structure

The overarching objective of this comprehensive overview is to thoroughly examine the current landscape of blockchain technologies within hydrosystem management. Through an in-depth exploration, we aim to elucidate the potential advantages and obstacles associated with integrating Blockchain in hydrosystem management, shedding light on how this technology could revolutionize operations in this sector. To achieve this, the overview structure is carefully organized, starting with an introduction that provides a foundational understanding of hydrosystems, their management challenges, and an introduction to blockchain technology (Wu et al., 2022). Subsequent sections will delve into the potential benefits and challenges of ongoing blockchain projects, culminating in a discussion on Blockchain's immense potential to reshape hydrosystems' management and operation. A formal tone will be maintained to ensure the content is academically rigorous and accessible to a broad audience, including hydrosystem professionals, policymakers, researchers, and the general public. Additionally, proper citations will be included to substantiate all presented information (Hu and Chuang, 2023).

21.2 Blockchain Technology Fundamentals

Blockchain, categorized as a distributed ledger technology (DLT), is a foundational innovation enabling secure, transparent, and tamper-proof transactions. At its core, it functions as a shared database distributed across a network of computers, with each node possessing a replicated ledger. This unique structure not only fortifies the system's security but also makes unauthorized alterations or hacking exceedingly challenging without the consensus of most network participants (Khang et al., 2022).

20.2.1 Comprehending Blockchain Principles

Blockchain, categorized as a DLT, represents an innovative mechanism ensuring secure, transparent, and unalterable transactions. Essentially, it functions as a shared database distributed across a network of computers, each node possessing a copy of the ledger. This shared structure grants blockchain high security, making it remarkably resistant to hacking or unauthorized modifications, requiring consensus from most of the network for any changes (Lacity & Lupien, 2022) (Figure 21.1).

- **Decentralization:** Empowering Security and Resilience:-Decentralization is a fundamental principle of blockchain technology, distinguishing it from traditional centralized systems. In a typical centralized system, a single entity or governing body controls the entire network and its operations. Conversely, Blockchain operates decentralized, eliminating the need for a central authority. It functions on a network of computers, termed nodes, with each node containing an identical copy of the ledger. This replication across numerous nodes ensures that no single entity holds complete control, enhancing the security and resilience of the system. In hydrosystems, decentralization can have a profound impact (Namasudra &Akkaya, 2023). By decentralizing the management and monitoring of water-related data, various stakeholders, including government bodies, environmental agencies, and local communities, can collectively participate in overseeing and safeguarding the integrity of water-related information. This distributed oversight can enhance trust and credibility within the hydrosystem management, ultimately promoting more secure and efficient water resource management.
- **Transparency:** Fostering Trust and Collaboration: Transparency is a core attribute of Blockchain, ensuring that every transaction executed on the network is visible to all participants. This transparency brings a high degree of openness and accessibility to the system, fostering stakeholder trust and collaboration. In a blockchain, all transactions are recorded in a public ledger that anyone can access and verify, creating a sense of accountability. In the context of hydrosystems, transparency is crucial for building trust among stakeholders involved in water management (Namasudra & Akkaya, 2023). By utilizing Blockchain to record and share water-related data, such as quality assessments, consumption patterns, and distribution, stakeholders, including

water authorities and local communities, can gain real-time visibility into the state of water resources. This transparent data sharing can lead to informed decision-making and collaborative efforts to address water scarcity, contamination, and other pertinent challenges within the hydrosystem.

- **Immutability:** Ensuring Data Integrity and Reliability: Immutability is a defining feature of Blockchain, ensuring that once a transaction is recorded, it cannot be altered or deleted. This characteristic guarantees the permanence and integrity of the data stored on the Blockchain. Any changes to the data would require consensus from the majority of the network, providing a high level of security against unauthorized modifications (Alam, 2022). In hydrosystems, immutability is paramount for preserving the accuracy and credibility of water-related data. Water quality assessments, supply records, and environmental impact data stored on the Blockchain are tamper-proof, assuring stakeholders that the information is accurate and reliable. This immutable data record supports evidence-based decision-making and regulatory compliance within the hydrosystem domain (Zhan et al., 2023).

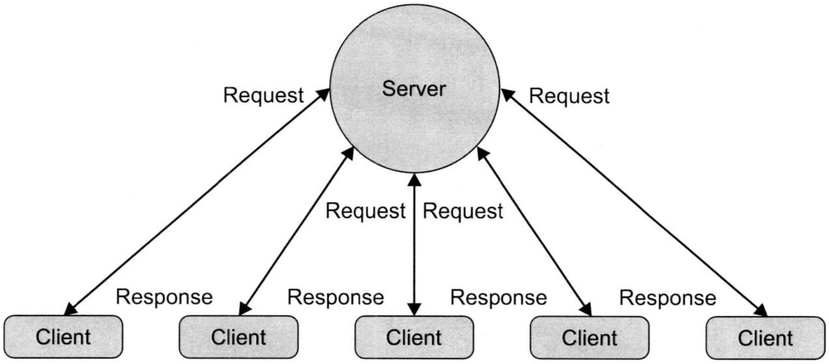

Fig. 21.1 Blockchain Framework.

Incorporating these blockchain principles into hydro systems management can significantly enhance the security, transparency, and reliability of data crucial for efficient water resource management. Applying Blockchain in hydrosystems can contribute to better decision-making, regulatory compliance, and stakeholder collaboration, ultimately promoting sustainable and resilient water management practices (Sunny et al., 2022).

21.2.2 Components and Terminology of Blockchain

Blockchain, a foundational technology underpinning cryptocurrencies and an array of applications beyond, consists of intricate components and terminologies integral to its structure and functionality. Understanding these fundamental elements is paramount for grasping Blockchain's architecture and potential applications.

Fig. 21.2 Overview of Blockchain Processing.

At its core, a blockchain is a decentralized, distributed ledger characterized by its security and transparency. It operates through a series of blocks, each containing a set of transactions, forming a connected chain (Yang et al., 2020). These blocks, the building units of the Blockchain, serve as repositories for various transactions.

- **Block:** A block is a fundamental unit of a blockchain, grouping a set of transactions. It encapsulates a collection of data about transactions and possesses a unique identifier. Moreover, each block references the preceding block in the chain, creating a linked structure.
- **Node:** Nodes are integral components of the blockchain network, representing individual computers connected to the system (Yang et al., 2020). They perform crucial functions, including transaction verification, consensus building, and storing and maintaining a complete copy of the blockchain ledger. In a decentralized network, nodes play a pivotal role in upholding the security and integrity of the entire system.
- **Hash:** Hashes are unique cryptographic identifiers generated for each block or transaction within the Blockchain. These identifiers, produced through specific algorithms, ensure the integrity and security of the data. Hashes are vital for validating the content within a block, and even a minor alteration to the data would result in a completely different hash.
- **Smart Contract:** A smart contract, a standout feature of Blockchain, is a self-executing contract with the terms of the agreement between the buyer and the seller directly written into code. These contracts are stored on the Blockchain and automatically execute predefined actions when the specified conditions are met. Smart contracts can revolutionize various sectors, including finance, real estate, and supply chain management, by automating and streamlining complex processes (Livieris et al., 2020).

In essence, comprehending these fundamental components and terminologies is critical to unraveling the intricate world of Blockchain, offering a glimpse into its potential to revolutionize various industries and processes.

21.2.3 Significance of Blockchain in Hydrosystems

Blockchain technology presents immense potential to revolutionize the management of hydrosystems, offering transformative capabilities in crucial domains (Eslamian and Eslamian, 2024)

In water data management, Blockchain can establish an unchangeable record encompassing a spectrum of vital water-related data, spanning sensor readings

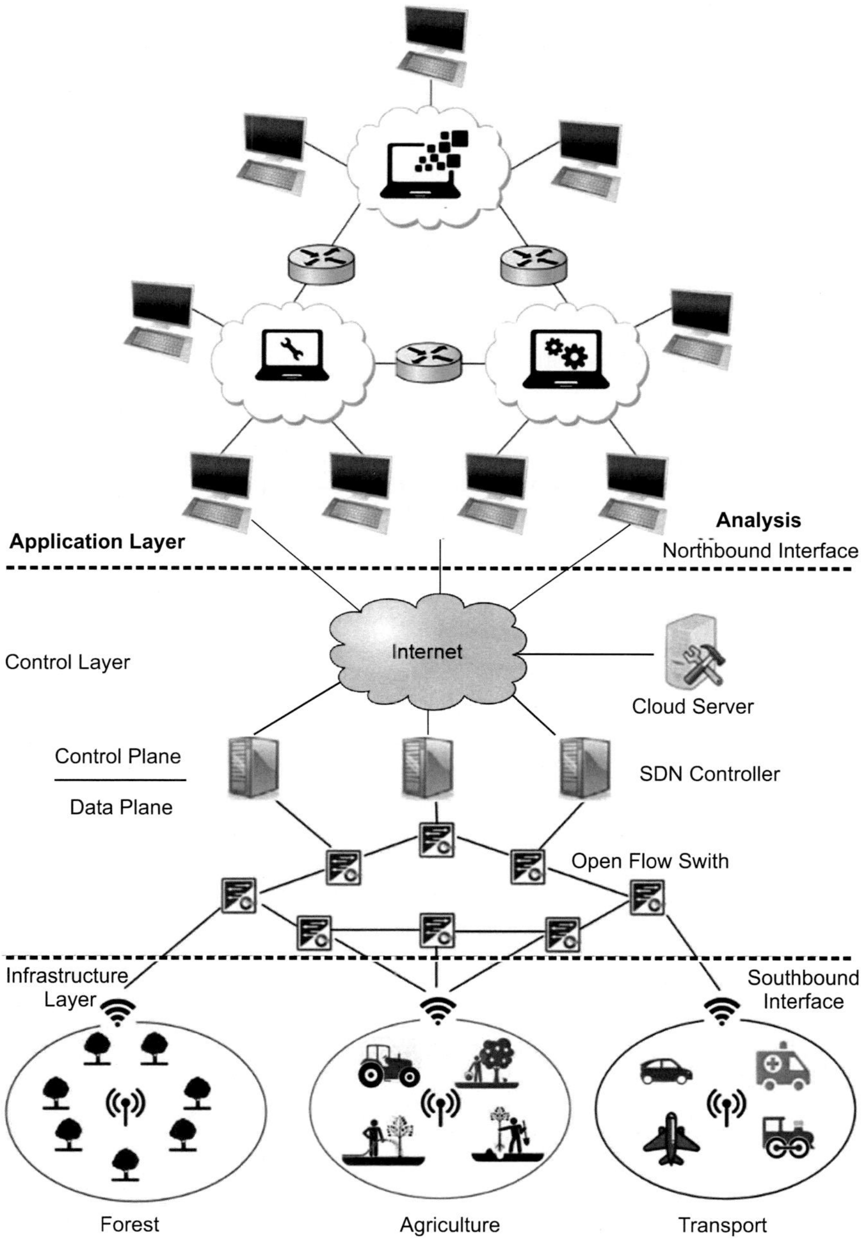

Fig. 21.3 Significance of Blockchain.

and precise water quality measurements (Figure 21.3). By creating a secure and easily shareable database, this technology dramatically facilitates influential and collaborative decision-making processes among stakeholders involved in hydrosystem management. Furthermore, blockchain integration contributes significantly to enhancing transparency and accountability within hydrosystems (Yin et al., 2021). By diligently tracking water movement at every stage of the system, from source to end-user, Blockchain mitigates risks of water theft and contamination, ensuring an equitable and efficient utilization of this vital resource. Additionally, Blockchain opens doors to innovative business models and opportunities within the hydro sector. Notable examples include establishing peer-to-peer water trading platforms and the introduction of pioneering micro-financing schemes tailored for small-scale hydro projects. Overall, Blockchain is a promising technology with the potential to optimize hydrosystem management, offering advancements in data reliability and system transparency and fostering novel business prospects critical for a sustainable water future (Shari & Malip, 2022). In conclusion, blockchain technology's potential to enhance the efficiency, transparency, sustainability, and accessibility of hydrosystems underscores its promising prospects for the water management sector. While still in its nascent stages of development, understanding Blockchain's fundamental principles and potential applications is pivotal in preparing for the transformative future of water management.

21.3 Advantages of Blockchain in Hydrosystems

Blockchain technology presents a transformative potential for revolutionizing the management and operation of hydrosystems. The immutable and transparent nature of Blockchain, a decentralized distributed ledger technology, offers a range of advantages in water resource management. Its applications span from securing critical water data to enhancing the efficiency of resource allocation. This section delves into these advantages, detailing how the Blockchain stands poised to improve hydrosystems' effectiveness, transparency, and sustainability, which is critical for ensuring a reliable and equitable supply of water resources (Pandey et al., 2022).

21.3.1 Enhanced Data Security and Immutability

A pivotal advantage of Blockchain is its capability to ensure secure and immutable data storage. Data within a blockchain is stored in a distributed ledger, spread across a network of computers, making it exceedingly resistant to tampering or unauthorized modifications. Alterations to the data would necessitate replication across the entire network, providing a robust level of security. In hydrosystems, Blockchain is a secure repository for diverse critical data, including water sensor readings, water quality measurements, and water usage data. This stored data can significantly enhance hydrosystem management and facilitate informed decision-making. Although blockchain data security and immutability are still in the early stages of development, ongoing projects, such as the HydroChain project, are actively working towards implementing blockchain-based solutions for hydrosystem data management (Wu et al., 2022).

21.3.2 Transparency and Traceability

Another fundamental advantage of blockchain technology is its transparency and traceability. All transactions conducted on a blockchain are public and accessible to all participants, enabling high transparency and traceability. This feature simplifies tracking the movement of data and assets on the Blockchain.In hydrosystems, Blockchain can enhance transparency and traceability in water management. For instance, it can effectively track the water journey through a hydrosystem from its source to the end user. This functionality helps prevent water theft contamination and promotes efficient and equitable water utilization (Thaeer Hammid, Awad et al., 2020). While blockchain transparency and traceability are still evolving, projects like the Water Ledger project are actively developing blockchain-based platforms to track water usage in irrigation projects in developing nations.

21.3.3 Decentralization and Trust Enhancement

Blockchain operates on a decentralized model devoid of central control, as it relies on a network of computers, each housing a copy of the ledger. This decentralization feature significantly fortifies the technology against censorship and fraudulent activities. In the context of hydrosystems, Blockchain can play a vital role in fostering stakeholder trust. For instance, it can create a transparent and tamper-proof record of water transactions between farmers and irrigation districts, reducing disputes and enhancing water management efficiency. Although blockchain decentralization and trust enhancement are in the early stages of development, projects like the Water Neutrality project are actively working on blockchain-based platforms to assist businesses and organizations in offsetting their water footprint (Thaeer Hammid et al., 2020).

21.3.4 Efficiency Gains and Data Integrity

Blockchain can markedly enhance the efficiency of hydrosystem management by automating tasks and minimizing the need for manual intervention. For instance, it can automate the billing and payment process for water usage, leading to cost reduction and improved customer satisfaction. Furthermore, Blockchain aids in ensuring the integrity of hydrosystem data by preventing data tampering and fraud. For instance, it can create a tamper-proof record of water quality measurements, ensuring compliance with water quality standards and providing consumers access to safe drinking water (Giudicianni et al., 2020).

Although efficiency gains and data integrity through Blockchain are in the early stages of development, projects like the Water Chain project are actively developing blockchain-based platforms to manage and trade water resources, showcasing the technology's potential to revolutionize hydro system management. In conclusion, blockchain technology holds immense potential to offer many advantages for hydro system management. However, it is essential to acknowledge that blockchain technology is in its nascent stages, and overcoming challenges is imperative for its widespread adoption in the hydro sector. Continued research, development, and collaboration are crucial to fully realizing Blockchain's transformative potential in hydrosystems (Parhamfar et al., 2023).

21.4 Research Domains in Hydrosystems

The management and optimization of hydrosystems are vital components in ensuring sustainable water resources for a growing population. Within this dynamic domain, several critical research areas demand attention to address pressing challenges and drive advancements. This section examines the significant research domains within hydrosystems to shed light on the ongoing endeavors to enhance water management, quality, and efficiency. From water quality monitoring to innovative allocation strategies, exploring these research domains provides valuable insights into hydro system management's present and future landscape (Kilic et al., 2023).

21.4.1 Water Quality Monitoring

Water quality monitoring plays a pivotal role in ensuring the safety and dependability of water supplies, encompassing a spectrum of parameters such as chemical composition, microbial content, and physical characteristics. Traditional water quality monitoring methods often involve a laborious and time-intensive process, where samples are collected manually, transported to laboratories, and analyzed through complex procedures. This conventional approach is resource-intensive and may delay obtaining critical information regarding water safety. However, with the advent of blockchain technology, the landscape of water quality monitoring is evolving. Blockchain, a decentralized and distributed ledger, offers a promising avenue to enhance the efficiency and precision of water quality monitoring through automation. Blockchain can be leveraged to automate collecting and analyzing water quality data obtained from sensors deployed in various water system locations (Flipo et al., 2023). These sensors continuously monitor water quality parameters and generate data, which is then automatically recorded on the Blockchain. This automation reduces the reliance on manual intervention, accelerates the data analysis process, and provides real-time insights into water quality.

Moreover, Blockchain's decentralized nature is crucial for data storage and processing in water quality monitoring. Traditional centralized storage systems are vulnerable to single points of failure and potential security breaches. In contrast, Blockchain decentralizes the storage and processing of water quality data across a network of computers (nodes), ensuring each node has a copy of the complete data. This redundancy enhances the security and resilience of the monitoring system, making it highly resistant to unauthorized alterations or tampering. Furthermore, Blockchain's transparency and immutability features contribute to water quality data's reliability and trustworthiness. Each entry (block) in the Blockchain contains a unique cryptographic hash of the previous block, creating an unbroken chain of records. This characteristic provides transparency and ensures that once the water quality data is recorded on the Blockchain, it cannot be altered or deleted. Stakeholders, including regulatory bodies, utility providers, and consumers, can access and verify the data, instilling trust in the accuracy and authenticity of the information (Schuite et al., 2019).

Finally, blockchain technology promises to revolutionize water quality monitoring by automating data collection and analysis, decentralizing data storage and processing, and providing transparent and tamper-proof data records. These advancements enhance the efficiency, accuracy, and reliability of water quality monitoring practices, ensuring the safety and trustworthiness of water supplies for communities and ecosystems.

21.4.2 Efficient Water Resource Allocation

Adequate and equitable allocation of water resources is a critical concern, especially in regions grappling with limited water availability due to various factors, including population growth, climate change, and inefficient water management practices. In such contexts, blockchain technology emerges as a potential solution to optimize the allocation of water resources efficiently and transparently (Boyer, 2022).

Firstly, Blockchain's capability to establish transparent and tamper-proof water usage records is instrumental in addressing water allocation challenges. Every water usage transaction through Blockchain is recorded in an immutable and transparent ledger. This ledger can be accessed by relevant stakeholders, including water authorities, consumers, and governing bodies. Consequently, the evident nature of Blockchain ensures that water usage records are accurate and indisputable, allowing for fair and efficient water allocation. Additionally, Blockchain facilitates peer-to-peer water trading, which holds promise in enhancing water resource allocation. Traditional water allocation systems are often centralized and bureaucratic, leading to delays and inefficiencies (de Assis Souza Filho et al., 2023) . However, Blockchain allows direct user transactions, eliminating intermediaries and streamlining the water transfer process. Through smart contracts on the Blockchain, water can be securely and instantly transferred from surplus areas to regions facing shortages. This peer-to-peer trading feature enhances the overall efficiency of water resource allocation and provides a mechanism to address sudden fluctuations in demand or supply.

Furthermore, Blockchain's potential to support innovative water pricing models is critical to efficient water allocation. Dynamic pricing and demand-response pricing are examples of such models. Dynamic pricing involves adjusting water prices based on real-time demand and supply conditions, incentivizing consumers to use water during off-peak hours, thus optimizing water usage. Demand-response pricing applies varying prices to encourage consumers to reduce usage during peak periods. Implementing these models through Blockchain makes the entire process automated and transparent, ensuring fair pricing and efficient water allocation (Ollivier, 2019).

21.4.3 Smart Metering and Consumption Tracking

Efficient water management, underpinned by precise, smart metering and consumption tracking, is crucial in modern water resource management. While the benefits of such systems are undeniable, the traditional implementation of intelligent metering systems often poses challenges, primarily in terms of cost and complexity. Blockchain technology offers innovative solutions to mitigate these challenges and enhance the efficiency and affordability of intelligent metering and consumption tracking. Firstly, Blockchain's role in reducing the cost of smart meters is a significant advantage. Traditional intelligent metering systems involve substantial costs associated with hardware, data management, and maintenance. Blockchain can disrupt this cost structure by enabling peer-to-peer trading of smart meter data. In a blockchain-based system, smart meters can securely and autonomously record consumption data, which can be traded directly between users and utilities. This eliminates the need for costly intermediaries and reduces the cost of implementing and operating intelligent metering systems (Bondarenkom et al., 2021).

Moreover, blockchain technology contributes to intelligent meter data's enhanced security and privacy. Security and privacy concerns have been notable barriers to adopting innovative metering systems. Blockchain addresses these concerns through robust data encryption mechanisms and decentralized data storage. In a blockchain-based system, smart meter data is encrypted and stored in a tamper-proof and transparent ledger. This ensures that data remains secure and immune to unauthorized access or manipulation, thus bolstering consumer trust and regulatory compliance. Furthermore, Blockchain's ability to facilitate real-time data sharing and analysis between water utilities and consumers is a transformative aspect of intelligent metering and consumption tracking. Traditional systems often delay data transmission and analysis, hindering timely decision-making and water conservation efforts. With Blockchain, real-time data sharing becomes seamless and instantaneous. Water utilities can access consumption data in real time, allowing for rapid responses to changes in demand or supply. Consumers can also monitor their water usage in real-time, enabling them to make informed decisions about their consumption habits. This real-time interaction fosters improved water efficiency, reduced water wastage, and enhanced communication between utilities and consumers (Gillefalk et al., 2022).

In summary, blockchain technology offers innovative solutions to the challenges associated with efficient smart metering and consumption tracking. By reducing the cost of smart meters, enhancing security and privacy, and enabling real-time data sharing and analysis, Blockchain contributes to more accessible and efficient water management practices, benefiting utilities and consumers (Gawusu et al., 2022).

21.4.4 Transparency in Water Distribution Supply Chains

Transparency and accountability in water distribution supply chains are critical elements in safeguarding the quality and safety of water supplies. Traditional supply chain systems often encounter challenges in maintaining this transparency due to their complex and multitiered structures. With its inherent features, blockchain technology offers promising mechanisms to address these challenges and significantly enhance transparency within water distribution supply chains. Firstly, Blockchain's ability to track water movement throughout the supply chain is fundamental to ensuring transparency. Traditional supply chains may face difficulties in precisely tracing the journey of water from its source to the end user. Blockchain, however, enables a comprehensive and transparent record of this movement. Each transaction relating to water movement is securely recorded in a tamper-proof and publicly accessible ledger. This feature empowers stakeholders to trace the entire path of water, identify potential contamination points, and promptly take necessary actions to ensure the safety and quality of water supplies (Anderson, 2020).

Additionally, Blockchain provides a platform for maintaining transparent records of water quality testing. In traditional supply chains, water quality testing records may not always be easily accessible or face authenticity and data integrity challenges. Blockchain addresses these concerns by creating an immutable record of each water quality test conducted at various supply chain stages. This transparent and tamper-proof record instills public confidence by assuring consumers and regulatory authorities that water quality standards are diligently maintained throughout the supply chain, ultimately guaranteeing safe and reliable water supplies. Furthermore, blockchain technology enables consumers to track their water footprint effectively. By incorporating Blockchain into water distribution supply chains, consumers can access their water usage data, providing insights into their consumption patterns and allowing them to monitor and manage their water consumption more responsibly. This feature promotes informed decision-making, encouraging consumers to reduce their water usage, adopt sustainable practices, and contribute to the overall conservation of water resources.

Blockchain technology offers a transformative potential to enhance transparency within water distribution supply chains (Azaroual, 2019). By tracking water movement, maintaining transparent water quality testing records, and enabling consumer water footprint tracking, Blockchain fosters a new era of accountability, confidence, and sustainability in water management, benefiting stakeholders and the environment.

21.4.5 Collaborative Water Governance

Collaborative water governance is critical for achieving sustainable and effective water management. However, traditional governance structures often grapple with issues related to fragmentation and inefficiency. Blockchain technology emerges as a potential solution to enhance collaborative water governance by revolutionizing how stakeholders interact, make decisions, and contribute to the management of water resources. Firstly, Blockchain is a powerful tool providing a secure and transparent platform for communication and collaboration among water stakeholders. Traditional communication mechanisms can be vulnerable to data breaches, lack transparency, or face challenges ensuring all stakeholders access accurate and up-to-date information. Conversely, Blockchain leverages its decentralized and immutable nature to establish a secure and transparent platform. All interactions and communications within this platform are securely recorded and can be accessed by relevant stakeholders. This enhanced transparency leads to improved coordination, informed decision-making, and heightened trust among stakeholders, ultimately improving the efficiency and effectiveness of collaborative water governance (Compère et al., 2021).

Furthermore, Blockchain supports establishing new water governance models that promote inclusivity and transparency. Traditional governance models often concentrate decision-making power within a limited group of stakeholders, potentially excluding others and hindering broad participation. Blockchain introduces decentralized and participatory water management models, distributing decision-making across participants. This inclusivity ensures that a broader spectrum of stakeholders can actively engage, voice their concerns, and contribute to decision-making processes. Additionally, the transparency of Blockchain ensures that decisions are made in an accountable and transparent manner, fostering a sense of trust and collaboration among all stakeholders (Heck et al., 2020).

Moreover, Blockchain's capability to facilitate the development of new water management solutions is a fundamental aspect of advancing collaborative water governance. Blockchain allows new solutions to be established, such as peer-to-peer water trading and water markets. These solutions promote efficient resource allocation, encourage responsible consumption, and enable stakeholders to manage water resources actively. Peer-to-peer water trading, for example, empowers stakeholders to exchange water resources directly, optimizing resource allocation and reducing inefficiencies in the system.

Overall, blockchain technology offers a transformative potential to enhance collaborative water governance. By providing a secure and transparent communication platform, supporting innovative governance models, and enabling the development of new water management solutions, Blockchain can revolutionize the governance of water resources, fostering sustainability, inclusivity, and efficiency in water management (Thaeer Hammid et al., 2020).

21.5 Case Studies and Real-World Implementations

This section is a repository of tangible examples showcasing how Blockchain addresses practical challenges within hydro system management. By presenting diverse applications, ranging from water quality monitoring to efficient resource allocation, this section offers a holistic view of Blockchain's versatility in enhancing hydrosystems. Furthermore, it substantiates the advantages of Blockchain by providing empirical evidence of its benefits, encouraging stakeholders to consider its adoption.

Moreover, this section is a valuable source of best practices and lessons learned from early adopters' experiences. This knowledge aids other organizations in navigating potential obstacles and implementing Blockchain effectively. Ultimately, through real-world case studies, confidence in Blockchain's potential to revolutionize hydrosystem management is bolstered, promoting its broader integration across the industry (Tagnon et al., 2020).

Case Study 1: Blockchain-Enhanced Water Quality Monitoring

Blockchain has revolutionized the process by enabling the seamless automatic collection, analysis, and storage of water quality data obtained from various sensors. This automation has significantly reduced the dependence on manual intervention, amplifying the overall efficiency of the monitoring process. One of the core strengths lies in decentralizing data storage and processing within the blockchain framework. Blockchain inherently bolsters security measures and resilience in handling water quality data by dispersing this critical function across the network. This decentralization addresses paramount concerns surrounding the safety and integrity of the data collected in water quality monitoring, establishing a robust foundation for reliable decision-making processes.

Furthermore, creating transparent and tamper-proof records of water quality data is a pinnacle achievement in blockchain-enhanced water quality monitoring. These incorruptible records foster heightened trust among stakeholders, including regulatory bodies, policymakers, and the public. Such faith is vital for shaping informed and credible decisions, ultimately ensuring the provision of safe and secure water supplies for communities (Pahonțu et al., 2020).

Case Study 2: Optimized Water Allocation Via Blockchain

This insightful case study illuminates the paradigm shift enabled by blockchain technology in the domain of efficient water resource allocation. The revolutionary impact lies in Blockchain's ability to establish transparent and tamper-proof water usage records, providing a foundation for equitable and effective distribution of precious water resources. Integrating peer-to-peer water trading through Blockchain presents a pivotal advancement, significantly enhancing the efficiency of water resource allocation. This innovation directly tackles the pressing issue of water scarcity by enabling a streamlined, peer-driven approach to water transactions. It breaks down barriers and inefficiencies associated with traditional allocation systems, promoting optimal utilization of water resources (Pahonțu et al., 2020).

Moreover, Blockchain's support for innovative water pricing models, such as dynamic pricing and demand-response pricing, adds another layer of efficiency to the allocation process. These models actively influence consumer behavior, encouraging judicious water usage and ultimately contributing to a more sustainable and balanced distribution of water resources. In this context, Blockchain emerges as a catalyst for transformative change, empowering better decision-making and fostering a future where water allocation is efficient, environmentally conscious, and economically viable (Liu & Shang, 2022).

Case Study 3: Transparent Water Distribution Tracking
This case study offers a comprehensive examination of how blockchain technology catalyzes augmenting transparency within water distribution supply chains, a pivotal aspect for ensuring the safety and quality of water supplies. Blockchain's inherent characteristics enable precise and unalterable tracking of water movement, effectively tracing its journey from the source to the ultimate end-user. This meticulous tracking serves as a robust mechanism to identify and eliminate potential sources of contamination, fortifying the safety standards associated with water distribution. Furthermore, Blockchain's capability to furnish a transparent record of water quality testing significantly contributes to building public trust and confidence in the safety of water distribution networks. The immutable and accessible nature of this record assures stakeholders and the public of the authenticity and accuracy of water quality assessments, aligning with paramount health and safety considerations (Elashmawy, 2020) .

In addition to its pivotal role in industrial applications, Blockchain empowers consumers by allowing them to monitor their water footprint actively. This not only cultivates a sense of responsibility among consumers but also accentuates the transparency and accountability fostered by blockchain technology. Blockchain amplifies the collective effort towards sustainable and conscientious water consumption practices by actively involving end-users. Thus, this case study illuminates the instrumental role of Blockchain in revolutionizing water distribution supply chains, championing enhanced transparency, and bolstering confidence in the safety of our vital water resources (Gore et al., 2023).

Case Study 4: Decentralized Collaborative Water Governance
This case study sheds light on how Blockchain supports collaborative water governance. Blockchain provides a secure and transparent platform for communication and collaboration among water stakeholders. This innovation promotes enhanced coordination and decision-making, which is essential for sustainable water management. By supporting decentralized and participatory water management models, Blockchain increases the inclusivity and transparency of water governance. Additionally, by enabling the development of novel water management solutions like peer-to-peer water trading and water markets, Blockchain significantly contributes to the efficiency and sustainability of water governance and management (Ibrahim et al., 2022).

21.6 Challenges and Mitigation Strategies

While promising, the integration of Blockchain technology into hydrosystems is not without its challenges. These challenges encompass various aspects, from technical hurdles to regulatory and environmental concerns. This section will delve into these challenges and explore potential mitigation strategies. By addressing these obstacles head-on, we can pave the way for the effective and responsible utilization of Blockchain in the management of hydrosystems.

21.6.1 Scalability and Performance Concerns

Blockchain technology exhibits immense potential, yet it grapples with infancy issues, prominently among scalability and performance. As Blockchain strives to find extensive application in hydrosystems, it encounters challenges regarding transaction speed and the ability to handle a burgeoning amount of data efficiently. The consensus mechanisms, designed to ensure security and decentralization, often pose a bottleneck regarding speed and scalability. To mitigate these concerns, innovative solutions like sharding, layer-2 protocols such as Lightning Network, and advancements in consensus algorithms like Proof-of-Stake (PoS) are being explored. Sharding involves breaking the Blockchain into smaller, manageable segments, allowing for parallel transaction processing (Boroomandnia et al., 2022). Layer-2 solutions like Lightning Network enable faster transactions by conducting most transactions off-chain, reducing the load on the main Blockchain. Advancements in consensus algorithms, particularly PoS, significantly enhance scalability by minimizing the energy-intensive nature of traditional proof-of-work (PoW) consensus.

21.6.2 Energy Efficiency and Sustainability

A pressing concern with blockchain technology is its notable energy consumption, which raises valid questions about its sustainability, especially in energy-sensitive sectors like hydrosystems. The energy-intensive nature primarily stems from the consensus mechanisms employed, such as PoW. The extensive computational power required to validate transactions necessitates substantial energy usage. However, ongoing research and development within the blockchain sphere have given rise to promising alternatives. Transitioning towards more energy-efficient consensus mechanisms like PoS, which does not entail the same level of computational competition as PoW, holds considerable potential (Zakari et al. 2022). Additionally, integrating Blockchain with renewable energy sources or exploring consensus mechanisms that promote energy conservation and sustainable practices could further mitigate these concerns.

21.6.3 Interoperability With Legacy Systems

Blockchain's seamless integration and interoperability with legacy systems are pivotal for its widespread adoption in hydrosystems. Achieving this harmony is challenging due to the intrinsic differences in the architecture and functioning of traditional methods versus Blockchain. Legacy systems often operate in silos and may not be compatible with Blockchain's decentralized and distributed nature. Employing Application Programming Interfaces (APIs) and gateways is a potential mitigation strategy. These tools bridge Blockchain and traditional systems, allowing smooth and secure data exchange. Standardizing data formats and structures can facilitate seamless interoperability, ensuring Blockchain can cohesively operate within the existing ecosystem (de Oliveira et al., 2021).

21.6.4 Regulatory Compliance and Legal Considerations

The nascent stage of blockchain technology introduces uncertainty regarding its regulatory compliance and legal implications, particularly in sectors like hydrosystems. Regulators actively navigate this landscape to establish frameworks that balance innovation with compliance and protection. Collaboration between industry stakeholders and regulators is crucial to developing comprehensive and consistent regulations that foster innovation while ensuring compliance with legal and ethical standards (Nižetić et al., 2019). Proactive engagement with regulatory bodies, participation in policy discussions, and compliance with existing laws are integral components of a strategic approach to address these concerns.

21.6.5 Data Privacy and Security in Hydrosystems

Preserving data privacy and security is paramount, especially within sensitive domains like hydrosystems. While inherently secure due to its cryptographic nature, Blockchain faces challenges in protecting sensitive information regarding public and permissionless blockchains. Privacy-preserving techniques, such as zero-knowledge proofs and homomorphic encryption, are actively explored to enable selective data sharing and confidential transactions. Additionally, private or permissioned blockchains, where access to data and validation of transactions is restricted to authorized entities, offer a promising solution to enhance data security and privacy in the context of hydrosystems. Collaborative efforts involving industry experts, regulators, and technologists are essential to strengthen the security features of blockchain technology (Abdelkader et al., 2021).

21.7 Future Directions and Research Opportunities

In this ever-evolving landscape of hydrosystem management, the potential for integrating blockchain technology is immense and promising. As many stand on the precipice of a new era in technology and environmental stewardship, exploring the uncharted territories of Blockchain's applications within hydrosystems becomes increasingly vital. This includes but is not limited to delving into improved consensus mechanisms, enhancing scalability, devising energy-efficient protocols, and addressing interoperability challenges. Moreover, there is a pressing need to investigate the socioeconomic implications and legal frameworks that will govern the integration of Blockchain into existing hydro system structures. This opening sets the stage for a comprehensive exploration of future directions and research opportunities at the intersection of Blockchain and hydrosystems (Shih et al., 2022).

21.7.1 Integration of IoT and Sensor Technologies

The transformative potential is achieved through the seamless integration of the Internet of Things (IoT) and sensor technologies with Blockchain in hydrosystem management. This convergence holds promise to revolutionize the traditional approaches to overseeing and governing hydrosystems. The subsequent discourse delineates the compeling arguments substantiating that integrating IoT and sensor technologies with Blockchain constitutes a paradigm shift with far-reaching implications in enhancing efficiency, transparency, and sustainability within hydrosystem management.

Integrating IoT devices and sensors into hydrosystem infrastructure enables real-time data collection and monitoring, significantly enhancing data accuracy and transparency. When this data is securely and immutably recorded on a blockchain, it provides stakeholders with unparalleled credibility and up-to-date information. This authenticated data becomes the foundation for critical decisions, ultimately fostering a more informed and efficient hydrosystem management process. Moreover, the marriage of Blockchain's decentralized ledger with IoT-generated data streamlines data management and analysis by minimizing fragmentation and data silos. This seamless integration offers decision-makers a comprehensive understanding of the hydrosystem's dynamics, empowering them to optimize resource allocation and predictive maintenance strategies (Srinidhi et al., 2020).

Furthermore, the immutable nature of Blockchain ensures secure recordkeeping, which is crucial for maintaining trustworthy records of hydrosystem operations and environmental data, demonstrating compliance with regulations, and building trust among stakeholders and regulatory bodies. Additionally, the integration with Blockchain facilitates the implementation of smart contracts, automating various aspects of hydro system management like billing, maintenance scheduling, and compliance adherence. Smart contracts execute agreements and operations precisely as intended, reducing disputes and human errors, thus promoting operational efficiency and cost-effectiveness in hydrosystem management.

Subsequently, blockchain technology is employed for storage and management to enhance the credibility and security of this wealth of data. Blockchain's decentralized and immutable ledger provides an ideal platform for securely recording and storing the collected data. Each piece of data is encrypted and linked to the previous transaction, forming a chain of blocks that ensures the information remains unaltered and traceable. Immutability guarantees that once recorded, the data cannot be tampered with or deleted, reinforcing its integrity and trustworthiness. Moreover, the transparency inherent to Blockchain ensures that stakeholders, including regulatory bodies, can access the data and verify its authenticity without any intermediaries. This transparency fosters trust and confidence among stakeholders, encouraging collaborative efforts for effective hydro system management. Additionally, the decentralized nature of Blockchain minimizes the risk of a single point of failure and unauthorized access, further enhancing the security and reliability of the stored data (Taha et al., 2022).

In summary, the synergy between IoT sensors for real-time data collection and Blockchain for secure and transparent data storage represents a significant advancement in hydro system management. This integration facilitates timely and accurate monitoring of water quality and quantity. It ensures the integrity and credibility of the data, ultimately contributing to informed decision-making and the sustainable management of water resources.

21.7.2 Hybrid Solutions: Blockchain and Emerging Technologies

The emergence of hybrid solutions, which amalgamate blockchain technology with other cutting-edge innovations like artificial intelligence (AI) and machine learning (ML), introduces a compelling dimension to hydrosystem management. These synergistic combinations hold immense potential to bolster the capabilities of blockchain technology in this domain. Hydrosystem management can benefit from advanced data analytics, predictive modeling, and automation by integrating AI and ML algorithms with Blockchain. AI-powered algorithms can analyze the vast datasets stored on the Blockchain, uncovering insights and patterns that might otherwise remain hidden. These insights can inform decision-makers about potential issues, resource allocation optimization, and even predictive maintenance, enhancing overall operational efficiency and sustainability (Bagrecha et al., 2020).

Integrating blockchain technology with other burgeoning advancements, particularly AI and ML, represents a transformative approach poised to elevate the landscape of hydrosystem management significantly. The potential for enhancing predictive analytics tools within the hydrosystem sector is substantial by fusing Blockchain's decentralized ledger capabilities with AI and ML algorithms. AI and ML technologies can systematically analyze extensive datasets harnessed by the Blockchain, discerning intricate patterns and trends that might elude conventional analytical methods (Laledia & Channi, 2023). These predictive analytics tools, empowered by AI and ML, enable hydro system managers to anticipate potential issues, deviations, or anomalies in the system's behavior well in advance. Such foresight allows for proactive decision-making, facilitating timely interventions to mitigate impending problems before they escalate. This amalgamation embodies a data-driven and forward-thinking approach, underscoring the symbiotic relationship between Blockchain and AI/ML in augmenting hydro system management practices for improved operational efficiency, resource allocation, and sustainability (Channi & Chowdhary, 2023).

Furthermore, by enabling smart contracts to interact with AI-driven systems, Blockchain fosters automation in billing, compliance monitoring, and anomaly detection tasks. Combining Blockchain with AI and ML technologies in this holistic approach creates a robust framework for more proactive and data-driven hydrosystem management, ultimately leading to more efficient resource utilization and improved environmental sustainability (Sharma & Park, 2018).

21.7.3 Addressing Societal and Environmental Water Challenges

Blockchain technology is a powerful tool capable of addressing critical societal and environmental water challenges. In the face of escalating issues like water scarcity, pollution, and climate change, Blockchain offers a promising solution. Its decentralized and transparent nature facilitates secure and immutable recordkeeping of vital water-related data. This characteristic ensures trust and accountability among stakeholders involved in managing water resources. Blockchain's ability to enable efficient and traceable transactions can revolutionize water management systems by facilitating optimized allocation and distribution of water resources (Cosgrove and Loucks, 2015).

One of its critical applications is the establishment of decentralized water markets, fundamentally altering how water resources are allocated and utilized. By leveraging Blockchain's decentralized ledger, stakeholders can transparently and securely participate in water trading, promoting efficient resource allocation based on demand and availability. This fosters a dynamic, market-driven approach to water distribution, ensuring equitable access while incentivizing water conservation. Furthermore, Blockchain's potential in tracking and reducing water pollution is remarkable. The immutable nature of blockchain records creates an indelible trail of water quality data. Starting from the source and encompassing various touchpoints, this trail ensures transparency and accountability in monitoring water quality. Authorities and stakeholders can trace the origin and impact of pollutants, enabling more targeted and effective pollution mitigation strategies (Allan et al., 2013).

Moreover, by integrating IoT sensors with Blockchain, real-time water quality monitoring can be facilitated, offering timely alerts and enabling proactive measures to prevent pollution events. In summary, blockchain technology significantly contributes to resolving societal and environmental water challenges by fostering decentralized water markets for optimized resource allocation and providing robust tools for tracking and mitigating water pollution. This integration underscores the potential to build a more sustainable and responsive water management system, crucial for addressing the growing concerns of water scarcity, corruption, and climate change (Pahl-Wostl et al., 2011).

21.7.4 Advancements in Blockchain Consensus Mechanisms

The progression of blockchain consensus mechanisms holds paramount importance in elevating the efficacy of blockchain technology within hydro system management, focusing specifically on scalability, performance, and energy efficiency. While robust, the conventional PoW consensus mechanism has faced criticism due to its energy-intensive nature and limitations in scaling. These limitations are particularly relevant in applications like hydrosystem management, which demand swift and high-volume transactions. Specific challenges within hydrosystems, such as the necessity for real-time data processing and precise resource allocation, underscore the urgency for tailored consensus mechanisms. Addressing these challenges calls for concerted research into alternative consensus mechanisms prioritizing energy efficiency and high scalability. Promising research directions encompass exploring and implementing PoS and delegated proof-of-stake (DPoS) mechanisms, significantly reducing blockchain networks' energy footprint while preserving robust security and transaction validation (Oyinloye et al., 2021).

Additionally, advancements in Byzantine fault tolerance (BFT) and practical Byzantine fault tolerance (PBFT) mechanisms offer potential avenues for achieving consensus in a more efficient and scalable manner. Furthermore, customizing consensus mechanisms to the unique needs of hydrosystem management is a burgeoning area of research involving the development of consensus protocols aligned with the real-time monitoring requirements of water quality, quantity, and distribution—a critical aspect for effective hydrosystem operation. These customized consensus algorithms can ensure rapid decision-making, resource optimization, and timely responses to emerging issues within the hydrosystem domain. Overall, ongoing research and development efforts aimed at refining and innovating consensus mechanisms are pivotal in unlocking the full potential of blockchain technology to address the specific demands and challenges of hydrosystem management (Sharma, Pant et al., 2020).

21.8 Conclusion

In summation, integrating blockchain technology into the domain of hydrosystem management presents a compelling potential to address the formidable challenges prevailing in water resource management. At its core, Blockchain embodies a robust framework characterized by security, transparency, and decentralization, pivotal in redefining the management and optimization of water resources. The immutability of records and the introduction of innovative consensus mechanisms within Blockchain pave the way for a transformative approach to combating water scarcity, pollution, and the impacts of climate change. By ensuring an unalterable ledger of water-related transactions and activities, Blockchain establishes a foundation of trust and credibility. It instills transparency into operations, enabling stakeholders to make informed decisions concerning resource allocation and utilization.

Furthermore, the efficiency brought about by Blockchain in managing and distributing water resources signifies a notable shift towards a sustainable future. The technology facilitates streamlined processes, reduces administrative complexities, and enhances the traceability of resources. Blockchain contributes significantly to fostering a balanced and equitable water distribution by establishing decentralized water markets and accurate resource usage tracking. Consequently, it stands as a promising catalyst for achieving a sustainable and considered approach to water management, essential for ensuring the well-being of the environment and society.

21.8.1 Key Insights and Takeaways

Throughout this comprehensive analysis of blockchain technology and its implications in hydro system management, crucial insights have come to the forefront. A cornerstone takeaway is the remarkable potential to establish decentralized water markets, allowing for a more equitable and efficient distribution of this vital resource. Concurrently, the opportunity to optimize resource allocation, a fundamental requirement in sustainable water management, is illuminated. Integrating IoT sensor data is a powerful tool for data-driven decision-making, enhancing operational efficiency, and informed strategic planning. Equally paramount is recognizing the necessity to refine consensus mechanisms, a critical step toward achieving scalability, improving energy efficiency, and enabling real-time data processing. These insights collectively underscore the transformative capacity of Blockchain in revolutionizing hydrosystem management, offering innovative solutions to age-old challenges in water resource allocation and sustainability.

21.8.2 Implications for Hydrosystems Management

The integration of Blockchain into hydrosystem management ushers in a transformative era, and its implications reverberate profoundly across several dimensions. First and foremost, there is a notable enhancement in data accuracy. The inherent features of Blockchain, such as immutability and transparency, engender a secure and unchangeable ledger. Consequently, the critical water-related data stored within the Blockchain is reliable and trustworthy. This is pivotal in informing decision-makers with accurate and real-time information, leading to more precise judgments in water resource management.

In addition to improved data accuracy, Blockchain ensures streamlined data management. The decentralized and distributed nature of Blockchain allows for efficient data storage and retrieval, reducing the bureaucratic bottlenecks associated with centralized systems. This streamlined approach translates to expedited processes and enhanced operational efficiency, crucial for dynamic hydrosystem management.

Equally critical is the aspect of enhanced security in recording water-related data. Blockchain's cryptographic algorithms and decentralized structure safeguard the integrity and confidentiality of the recorded data. This fortified security is vital when dealing with sensitive water quality, quantity, and distribution data. It prevents unauthorized access, tampering, or malicious alterations, instiling confidence in stakeholders and regulators regarding data integrity. Moreover, integrating Blockchain significantly augments the ability to promptly predict, mitigate, and respond to water-related challenges. The real-time data recorded and processed on the Blockchain facilitates proactive decision-making. Predictive modeling and data analytics become more accurate, enabling timely responses to emerging challenges. This predictive capability is fundamental in maintaining sustainable hydrosystems, ensuring resilience in evolving environmental and societal dynamics.

The implications of integrating blockchain technology into hydrosystem management encompass enhanced data accuracy, streamlined data management, fortified security, and proactive responsiveness to challenges. These implications collectively chart a course towards sustainable and efficient management of hydrosystems, promising a future where water resources are judiciously utilized and managed, mitigating potential crises and fostering environmental harmony.

21.8.3 Collaborative Research and Adoption

Given the potential demonstrated by Blockchain technology, a call for collaborative research and widespread adoption within the hydrosystem management domain is warranted. Interdisciplinary collaboration is essential to further explore and innovate the application of Blockchain, refine consensus mechanisms, and tailor solutions to the specific needs of hydrosystem management. Additionally, stakeholders across the water management sector must actively engage in adopting and implementing blockchain solutions to realize a more sustainable and efficient future for hydrosystems. We can only unlock the transformative potential of blockchain technology in managing our water resources effectively and responsibly through concerted efforts and collaboration.

References

1. Abdelkader, M.M., Gaplaev, M.S., Terekbaev, A.A. and Puchkov, M.Y. (2021). The influence of biostimulants on tomato plants cultivated under hydroponic systems. *Journal of Horticultural Research, 29*(2), 107–116.
2. Alam, A. (2022). Platform utilising blockchain technology for eLearning and online education for open sharing of academic proficiency and progress records. In: *Smart Data Intelligence: Proceedings of ICSMDI 2022*. Springer, pp. 307–320.
3. Allan, C., Xia, J. and Pahl-Wostl, C. (2013). Climate change and water security: challenges for adaptive water management. *Current Opinion in Environmental Sustainability*, 5(6), 625–632.
4. Anderson, J.J. (2020). 30 Decadal Climate Cycles and Declining Columbia River Salmon. In: *Sustainable Fisheries Management: Pacific Salmon*. Routledge, p. 467.
5. Azaroual, M. (2019). Predictive Geosciences for Georesources Exploration and Management, Underground Exploitation, and Environmental Issues. In: *2019-Sustainable Industrial Processing Summit, Flogen Star Outreach*.
6. Bagrecha, N.R., Polishwala, I.M., Mehrotra, P.A., Sharma, R. and Thakare, B. (2020). Decentralised blockchain technology: Application in banking sector. In: *2020 International Conference for Emerging Technology*. IEEE.
7. Bondarenko, V.L., Khetsuriani, E.D., Yliasov, A.I., Semenva, E.A. and Azarov, V.N. (2021). Basics of systems approach to methodology of assessing waterworks facilities on natural environment. *Nexo Revista Científica, 34*(02), 872–885.
8. Boroomandnia, A., Rismanchi, B. and Wu, W. (2022). A review of micro hydrosystems in urban areas: Opportunities and challenges. *Renewable and Sustainable Energy Reviews, 169*, 112866.
9. Boyer, A.-L. (2022). HYDECO Connections, disconnections, and reconnections in socio-hydrosystems: Does (in) visibility make the difference? In: *International Symposium of LabEx DRIIHM 2022*, Nantes, France.

10. Channi, H.K. and Chowdhary, C.L. (2023). Blockchain-Based IoT E-Healthcare. In: *Handbook of Research on Solving Societal Challenges through Sustainability-Oriented Innovation*. IGI Global, pp. 56–73.
11. Coban, H.H., Rehman, A. and Mousa, M. (2022). Load frequency control of microgrid system by battery and pumped-hydro energy storage. *Water*, *14*(11), 1818.
12. Compère, F., Kern, G. and Bellenfant, G. (2021). *A 3D Feflow Hydrogeological Uranium Underground Mine Model, France*. International Mine Water Association.
13. Cosgrove, W.J. and Loucks, D.P. (2015). Water management: Current and future challenges and research directions. *Water Resources Research*, *51*(6), 4823–4839.
14. de Assis Souza Filho, F., de Carvalho Studart, T.M., Filho, J.D.P., Martins, E.S.P.R., Ayrimoraes, S.R., Pessoa, C.A.P., Rolim, L.Z.R., de Araujo Junior, L.M., da Silva, S.M.O. and Carvalho, T.M.N. (2023). Integrated proactive drought management in hydrosystems and cities: Building a nine-step participatory planning methodology. *Natural Hazards*, *115*(3), 2179–2204.
15. de Oliveira, K.B., dos Santos, E.F., Neto, A.F., de Mello Santos, V.H. and de Oliveira, O.J. (2021). Guidelines for efficient and sustainable energy management in hospital buildings. *Journal of Cleaner Production*, *329*, 129644.
16. Elashmawy, M. (2020). Improving the performance of a parabolic concentrator solar tracking-tubular solar still (PCST-TSS) using gravel as a sensible heat storage material. *Desalination*, *473*, 114182.
17. Eslamian, S. and Eslamian, F. (2024). *Hydrosystem Restoration Handbook: Streamflow Recharge and Lake Rehabilitation*. Elsevier, p. 322.
18. Ezeh, M., Fidel-Anekwe, T. and Ikpabi, P. (2023). Evaluating the Potential of Renewable Energy to Address Energy Accessibility, Affordability, and Sustainability Challenges in Africa. In: *SPE Nigeria Annual International Conference and Exhibition*, OnePetro.
19. Flipo, N., Gallois, N. and Schuite, J. (2023). Regional coupled surface–subsurface hydrological model fitting based on a spatially distributed minimalist reduction of frequency domain discharge data. *Geoscientific Model Development*, *16*(1), 353–381.
20. Gawusu, S., Mensah, R.A. and Das, O. (2022). Exploring distributed energy generation for sustainable development: A data mining approach. *Journal of Energy Storage*, *48*, 104018.
21. Gillefalk, M., Tetzlaff, D., Marx, C., Smith, A., Meier, F., Hinkelmann, R. and Soulsby, C. (2022). Estimates of water partitioning in complex urban landscapes with isotope-aided ecohydrological modelling. *Hydrological Processes*, *36*(3), e14532.
22. Giudicianni, C., Herrera, M., Nardo, A.D., Adeyeye, K. and Ramos, H.M. (2020). Overview of energy management and leakage control systems for smart water grids and digital water. *Modelling*, *1*(2), 134–155.
23. Gore, S., Dutt, I., Dahake, R.P., Khodke, H., Kurkute, S.L., Dange, B. and Gore, S. (2023). Innovations in Smart City Water Supply Systems. *International Journal of Intelligent Systems and Applications in Engineering*, *11*(9s), 277–281.
24. Heck, K., Coltman, E., Schneider, J. and Helmig, R. (2020). Influence of radiation on evaporation rates: A numerical analysis. *Water Resources Research*, *56*(10), e2020WR027332.
25. Hu, J.-L. and Chuang, M.-Y. (2023). The Importance of Energy Prosumers for Affordable and Clean Energy Development: A Review of the Literature from the Viewpoints of Management and Policy. *Energies*, *16*(17), 6270.
26. Ibrahim, A.-S., Akanbang, B.A.A. and Laube, W. (2022). Sustaining decentralized collaborative governance arrangements in Africa: A case study of land management committees in the Upper West Region, Ghana. *GeoJournal*, *87*(2), 641–660.
27. Khang, A., Chowdhury, S. and Sharma, S. (2022). *The Data-Driven Blockchain Ecosystem: Fundamentals, Applications, and Emerging Technologies*. Taylor and Francis Group, p. 288.

28. Kilic, D., Rivière, A., Gallois, N., Ducharne, A., Wang, S., Peylin, P. and Flipo, N. (2023). Assessing water and energy fluxes in a regional hydrosystem: Case study of the Seine basin. Comptes Rendus. *Géoscience*, *355*(S1), 1–21.
29. Koedel, U., Schuetze, C., Fischer, P., Bussmann, I., Sauer, P.K., Nixdorf, E., Kalbacher, T., Wichert, V., Rechid, D. and Bouwer, L.M. (2022). Challenges in the evaluation of observational data trustworthiness from a data producers viewpoint (FAIR+). *Frontiers in Environmental Science*, *9*, 772666.
30. Kumar, V., Azamathulla, H.M., Sharma, K.V., Mehta, D.J. and Maharaj, K.T. (2023). The state of the art in deep learning applications, challenges, and future prospects: A comprehensive review of flood forecasting and management. *Sustainability*, *15*(13), 10543.
31. Lacity, M.C. and Lupien, S.C. (2022). *Blockchain Fundamentals for Web 3.0*. University of Arkansas Press.
32. Laledia, S.S. and Channi, H.K. (2023). An Overview of Blockchain Technology in Smart Grid. Renewable Energy Optimization, Planning, and Control In: *Proceedings of ICRTE 2022*, pp. 183–192.
33. Liu, Y. and Shang, C. (2022). Application of blockchain technology in agricultural water rights trade management. *Sustainability*, *14*(12), 7017.
34. Livieris, I.E., Pintelas, E., Stavroyiannis, S. and Pintelas, P. (2020). Ensemble deep learning models for forecasting cryptocurrency time-series. *Algorithms*, *13*(5), 121.
35. Mardiani, E. and Iswahyudi, M.S. (2023). Mapping the Landscape of Artificial Intelligence Research: A Bibliometric Approach. *West Science Interdisciplinary Studies*, *1*(08), 606–618.
36. Mejías Moreno, M. (2022). Research Group on Applied Hydrogeology and Shallow Geothermal Energy (GIHAGS): General information and research lines. In: *Proceedings of the 1st Scientific-Technical Conference of the Geological and Mining Institute of Spain-National Centre* (*IGME*), pp. 33–36. ReaearchGate.
37. Namasudra, S. and Akkaya, K. (2023). Introduction to Blockchain Technology. In: *Blockchain and Its Applications in Industry 4.0*. Springer, pp. 1–28.
38. Nithish, R., Kumar, P., and Sudha, S. (2023). Colossal Encipher Isolation Conserve Classifier. In: *2023 3rd International Conference on Advance Computing and Innovative Technologies in Engineering* (*ICACITE*), IEEE.
39. Nižetić, S., Djilali, N., Papadopoulos, A. and Rodrigues, J.J. (2019). Smart technologies for promotion of energy efficiency, utilization of sustainable resources, and waste management. *Journal of Cleaner Production*, *231*, 565–591.
40. Ollivier, C. (2019). *Characterization and Spatialization of the Recharge of Karst Hydrosystems: Application to the Fontaine de Vaucluse aquifer, France*. Université d'Avignon, France.
41. Oyinloye, D.P., Teh, J.S., Jamil, N. and Alawida, M. (2021). Blockchain consensus: An overview of alternative protocols. *Symmetry*, *13*(8), 1363.
42. Pahl-Wostl, C., Nilsson, C., Gupta, J. and Tockner, K. (2011). Societal learning needed to face the water challenge. *Ambio*, *40*(5), 549–553.
43. Pahonțu, B., Arsene, D., Predescu, A. and Mocanu, M. (2020). Application and challenges of Blockchain technology for real-time operation in a water distribution system. In: *2020 24th International Conference on System Theory, Control, and Computing* (*ICSTCC*), IEEE.
44. Palominos, P., Ortega, C., Alfaro, M., Fuertes, G., Vargas, M., Camargo, M., Parada, V. and Gatica, G. (2022). Chaotic Honeybees Optimization Algorithms Approach for Traveling Salesperson Problem. *Complexity*, *2022*. https://doi.org/10.1155/2022/8903005.
45. Pandey, Y., Iqbal, A., Singh, S.P. and Husain, M.A. (2022). Micro/Pico Hydropower Generation System Using Self-Excited Induction Generators and Applications of AI for Its Performance Improvement. *Computer Science, Technology, and Applications*, *51*.

46. Paniyil, P. (2023). *Photovoltaics, Batteries, and Silicon Carbide Power Electronics Based Infrastructure for Sustainable Power Networks*. PhD Thesis, Department of Electrical and Computer Engineering, Clemson University, USA.
47. Parhamfar, M., Sadeghkhani, I. and Adeli, A.M. (2023). Towards the application of renewable energy technologies in green ports: Technical and economic perspectives. *IET Renewable Power Generation, 17*(12), 3120–3132.
48. Schuite, J., Flipo, N., Massei, N., Rivière, A. and Baratelli, F. (2019). Improving the spectral analysis of hydrological signals to efficiently constrain watershed properties. *Water Resources Research, 55*(5), 4043–4065.
49. Shari, N.F.M. and Malip, A. (2022). State-of-the-art solutions of blockchain technology for data dissemination in smart cities: A comprehensive review. *Computer Communications, 189*, 120–147.
50. Sharma, D.K., Pant, S., Sharma, M. and Brahmachari, S. (2020). Cryptocurrency mechanisms for blockchains: Models, characteristics, challenges, and applications. In: *Handbook of Research on Blockchain Technology*. Academic Press, pp. 323–348.
51. Sharma, P.K. and Park, J.H. (2018). Blockchain based hybrid network architecture for the smart city. *Future Generation Computer Systems, 86*, 650–655.
52. Shih, Y.-C. T., Sabik, L.M., Stout, N.K., Halpern, M.T., Lipscomb, J., Ramsey, S. and Ritzwoller, D.P. (2022). Health economics research in cancer screening: Research opportunities, challenges, and future directions. *JNCI Monographs*, (59), 42–50.
53. Srinidhi, H., Shreenidhi, H. and Vishnu, G. (2020). Smart Hydroponics system integrating with IoT and Machine learning algorithm. In: *2020 International Conference on Recent Trends on Electronics, Information, Communication, and Technology*. IEEE.
54. Sunny, J., Pillai, V.M., Nath, H.V., Shah, K., Ghoradkar, P.P., Philip, M.J. and Shirswar, M. (2022). Blockchain-enabled beer game: A software tool for familiarizing the application of blockchain in supply chain management. *Industrial Management and Data Systems, 122*(4), 1025–1055.
55. Tagnon, B., Oularé, S., Kouamé, K., Kouassi, K., Kouadio, Z. and Savané, I. (2020). Fully integrated and physically-based approach for simulating water flows in a large-scale, heavily-agricultural and low-instrumented watershed. *Journal of Hydrology, 586*, 124781.
56. Taha, M.F., ElMasry, G., Gouda, M., Zhou, L., Liang, N., Abdalla, A., Rousseau, D. and Qiu, Z. (2022). Recent Advances of Smart Systems and Internet of Things (IoT) for Aquaponics Automation: A Comprehensive Overview. *Chemosensors, 10*(8), 303.
57. Thaeer Hammid, A., Awad, O.I., Sulaiman, M.H., Gunasekaran, S.S., Mostafa, S.A., Manoj, N., Kumar, B.A. Khalaf, Al-Jawhar, Y.A. and Abdulhasan, R.A. (2020). A review of optimization algorithms in solving hydro generation scheduling problems. *Energies, 13*(11), 2787.
58. Vera, E.A.S. (2022). *Alternatives to Reduce Grid Disturbances Caused by the Rapid Increase in Distributed Energy Resources*. Department of Technology Leadership and Innovation, PhD Thesis, Purdue University, USA.
59. Wu, Z., Wang, J., Zhong, H., Gao, F., Pu, T., Tan, C.-W., Chen, X., Li, G., Zhao, H. and Zhou, M. (2022). Sharing economy in local energy markets. *Journal of Modern Power Systems and Clean Energy, 11* (3), 714–726.
60. Yang, B., Sun, Y. and Wang, S. (2020). A novel two-stage approach for cryptocurrency analysis. *International Review of Financial Analysis, 72*, 101567.
61. Yin, L., Nie, J. and Han, L. (2021). Understanding cryptocurrency volatility: The role of oil market shocks. *International Review of Economics and Finance, 72*, 233–253.
62. Zakari, A., Khan, I., Tan, D., Alvarado, R. and Dagar, V. (2022). Energy efficiency and sustainable development goals (SDGs). *Energy, 239*, 122365.
63. Zhan, Y., Xiong, Y. and Xing, X. (2023). A conceptual model and case study of blockchain-enabled social media platform. *Technovation, 119*, 102610.

22 Blockchain in Intelligent Building Systems: Enhancing Security, Efficiency, and Climate Resilience

Mahendra Gooroochurn[1*] and Tulsi Pawan Fowdur[2]

One of the major consequences of climate change is extreme weather, and heightened stress on building systems to meet the indoor conditions for ensuring human wealth and comfort. Indeed a deterioration in indoor environmental conditions in the built spaces in terms of thermal discomfort, light levels, and air quality has been correlated to poor productivity, absenteeism, low performance, and negative impact on health. Therefore, it is of paramount importance to monitor indoor conditions and control building systems accordingly. However, the traditional approach of relying on active energy systems such as mechanical ventilation, air-conditioning, and artificial lighting is discouraged due to the direct coupling it creates between urbanization and carbon emissions. An emphasis is laid on bioclimatic design, where the building is designed within the context of its immediate environment, to harness natural resources as much as possible, through the application of passive solar design principles. A further step is to provide means to modulate these passive systems based on the indoor conditions and prevailing ambient conditions in a reactive and proactive manner using machine learning algorithms as knowledge base. A network of connected homes can serve as a great tool for understanding the efficacy of the control algorithms, paving the way to improving performance over time, especially as climate change leads to fundamental changes in the way the building fabric interacts with the external environment. For this to happen, the intelligent building control systems require remote data collection from sensors and transmission to one or more intermediate end-points or servers as well the transmission of control signals to the actuators. Consequently, cyberattacks can take place at several points in this interconnected network of sensors and actuators used in building controls. The risk of data tampering and more seriously tampering of control signals is therefore significant and needs to be addressed. Blockchain can be really useful in such situations as it provides edge protection and allows authentication of the sensor to server and server to actuator transmission along with stable data share, life cycle management, access control compliance of devices, and

[1] Mechanical and Production Engineering Department, Faculty of Engineering, University of Mauritius, Moka, Mauritius.
[2] Electrical and Electronic Engineering Department, University of Mauritius, Moka, Mauritius.
* Corresponding author: m.gooroochurn@uom.ac.mu

DOI: 10.1201/9781003686026-22

self-sustaining operations. Moreover, blockchain can be applied for tracking and tracing transactions through devices, which are performed during the operation, as well as to encrypt and transmit data securely. It is therefore of paramount importance to ensure trusted communications in a building control system so that its adoption does not compromise its security. This chapter reviews some examples of passive systems that can be modulated to influence the indoor environmental conditions, together with the input parameters to be monitored, and how they can be used by a controller to adapt the passive systems.

22.1 Introduction

The serious environmental, economic, and social issue caused by climate change has taken a planetary dimension, requiring multidisciplinary and transdisciplinary synergy in research for finding plausible solutions. The complexity and interconnectedness of our ecosystems is such that it will not be an understatement to say that all natural systems, including the human body, have been adversely impacted by climate change in one way or another. One area attracting high interest is the built environment, given the huge environmental implication involved with the construction, operation and demolition of buildings (Gooroochurn et al., 2021a). On a positive note, the built environment also holds the key to remedy to climate change through mitigation measures and for small island developing states like Mauritius, with a marginal carbon footprint, adaptation measures to ensure a safe and proper living environment.

As evidenced through the extensive research findings reported in literature, a passive design methodology is key to ensuring thermal comfort in interior spaces as a first priority before active systems are considered for meeting cooling or heating loads and from a carbon footprint perspective, renewable energy systems for aiming to carbon neutrality. However, the design of passive systems need to be carefully integrated with bioclimatic design because the passive design measures themselves can worsen indoor conditions under certain conditions, as reported by Gooroochurn (2023) for the tropical context of Mauritius. Indeed, Ozarisoy et al. (2019) in their study for terraced houses in the UK found that insulation and air-tight fabric were the main causes for thermal discomfort, recommending more ventilation and heat gain control through the cladding material. The follow-up study reported in Ozarisoy (2022) showed the lack of climate change consideration in the design and construction of state-of-the-art houses, badly impacted by summer extreme temperatures and heatwaves, needing retrofits through integration of passive measures. The study by Ren et al. (2022) showed the correlation between the urban development, climate change and the deterioration in the urban thermal environment.

Several studies have aimed at quantifying the increase in cooling load and decrease in heating load as well as the resultant operative and radiant temperatures under different climate scenarios. Along this line, Hosseini et al. (2022) investigated the error that can creep in simulations in using meso-scale weather data, and used a high spatio-temporal resolution building energy model, with due consideration for micro-climates, which are prevalent in the local context of Mauritius as well. The latter was found to account for a 17% rise in cooling degree day compared to the meso-scale model, which can increase by 500% in extreme weather conditions. Their projections show a staggering increase in cooling demand by four and five times, respectively, for the periods 2040–2069 and 2070–2099 with reference to 2010–2039, concluding that temperatures above the comfort zone of 26°C will become more frequent. Kaitouni et al. (2023) reported a 27% increase in Indoor Overheating Hours (IOH) in Morocco when comparing the current situation with 2050. For a study in India, Thapa et al. (2023) reported an increase of 221 and 467% in annual energy for 2050 and 2080, respectively, compared to the current context. They recommended the use of aerated bricks, decrease in infiltration rate and window-to-wall (WWR) ratio to reduce the problem of overheating.

Li et al. (2021) further studied the effect of thermal discomfort on human well-being and performance in an office environment, and reported that several sick building syndrome symptoms are intensified with rise in indoor temperature beyond the human comfort zone of 26°C. Temperatures of 25, 27, 29, 31, and 33°C were used, which are typical ambient temperatures witnessed in summer in Mauritius. The need to factor in the impact of climate change over the long-term on the indoor environmental conditions has been highlighted in several studies. For example, Jafarpur et al. (2021) in their study in Canada where changes in temperature set points was studied, clearly highlighted the importance of keeping a holistic view of impact on the indoor conditions over the long-term as opposed to seeking cost savings through short-term energy savings. The ability to adapt temperature set points with respect to the ambient conditions as a means to cut energy use is testament of the need to have adaptive systems, and the design paradigm being recommended through this chapter and demonstrated by exemplar passive systems, is the modulation of the passive measures themselves, which when combined with active system controls can achieve far higher savings (Eslamian & Eslamian, 2023).

Since intelligent building control systems require secure remote data collection from sensors and transmission to one or more intermediate end-points or servers as well the transmission of control signals to the actuators, blockchain has been applied in several such systems. For example in Bachtobji et al. (2022), a hierarchical architecture consisting of IoT, fog and cloud layers where blockchain technology was integrated at multiple layers to improve the security of the BMS (blockchain management systems) was proposed. Moreover, in Saeed et al. (2022), a blockchain-based trust strategy was designed and implemented for a BMS to collect trust evidence and provide a trust score to each device. The trust evidence and scores were securely stored and shared with other devices in the network by integrating them into blockchain exchanges. Other related applications of blockchain can be found in Fowdur et al. (2023) and Fowdur et al. (2022).

22.2 Indoor Thermal Comfort Using Passive Measures

The passive design philosophy applied in conjunction with bioclimatic design, once set aside in preference to mechanical building services solutions, namely lighting design and heating, ventilation, and air-conditioning (HVAC) systems, has regained its just place as a fundamental building design paradigm, both for new and existing constructions. It has become a key solution in tackling climate change, both for reducing the carbon footprint of the ever increasing built environment to stay within the 1.5°C target, but increasingly important to maintain reasonable indoor environmental conditions for humans to stay healthily and comfortably. Humans are known to spend at least 80% indoors as per WHO, and this figure rises to 95% for the elderly, who are even more vulnerable to the negative impacts of overheating and heat waves. It will come as a reassurance that traditional buildings, incorporating passive design measures adapted to the local context, were better in providing a thermally comfortable environment while using substantially less energy. In the Mauritian context, the traditional creole houses were known to have air vents for ongoing air exchange with the external environment and terraces around the liveable indoor spaces, which acted as buffers to regulate heat gains (Gooroochurn et al., 2021b). The challenge lies in marrying modern architecture with these traditional practices, where digitalisation, IoT (internet of things) and AI (artificial intelligence) offer valuable technologies to yield intelligent solutions.

A passive design methodology involves equipping the building envelope with elements that allow regulation of air, heat, and light transport between the indoor and outdoor environments in such a way as to achieve proper indoor conditions for the occupants with minimum use of active, carbon-emitting building energy systems, especially active cooling and heating systems which are large energy consumers. The building fabric regulates heat transfer through its thermal properties, including the thermal conductance and admittance, while glazing has the additional direct solar heat gain property calling for careful consideration of the extent to which they are added to the walls (referred to as a the window-to-wall ratio - WWR) and the orientation they are found at. As observed by Gooroochurn et al. (2021c), the presence of glazing in corner spaces leads to a compounding effect with excessive heat gains by direct solar radiation. The use of shading, both for glazed areas and for the flat roofs in Mauritius, were found to be highly effective in warding off unwanted heat gains, typically referred to as cool building strategies.

The flat cast concrete roof was found to be a major source of heat gain, being exposed throughout the day to solar radiation, and irradiated almost perpendicularly at noon during summer conditions, further worsening heat gains and radiative temperatures. The ceiling temperature was measured to be as high as 45°C in the work on ventilated wall by Gooroochurn et al. (2020), demonstrating the significance of the flat roof in causing thermal discomfort, and the justification for highly recommending the installation of solar shade as part of the circular homes concept developed for climate community action at household level by Gooroochurn (2022a). The design of external shading devices needs to be done with respect to the solar elevation with two main types of systems: (1) for blocking high elevation sun in the form of overhang for azimuth angles around north (in the southern hemisphere), and (2) for blocking low to medium elevation sun between east and north-east and west and north-west directions.

Although having fixed passive measures can help to some extent in dealing with seasonal variations in solar position and the changing need for heat gains (for example, an overhang of a given depth blocking high elevation summer sun and allowing in lower elevation winter sun), but there are cases where the passive measures when fixed and not modulable can do the opposite to what is intended (for example, a cool roof or green roof blocking beneficial heat gains through the roof in winter). When it comes to regulating heat gains through glazed areas in a flexible way, awnings, louvres, and shutters as illustrated in Figure 22.1 can be effectively used between seasons and even during a given season as per the indoor and ambient conditions, offering a powerful means to optimise the indoor thermal comfort, especially when combined with openable windows and other natural ventilation strategies. The ability to design these systems using a sensing, control and actuation paradigm, symbolic of a mechatronic system, opens the scope for optimization as discussed in the next section.

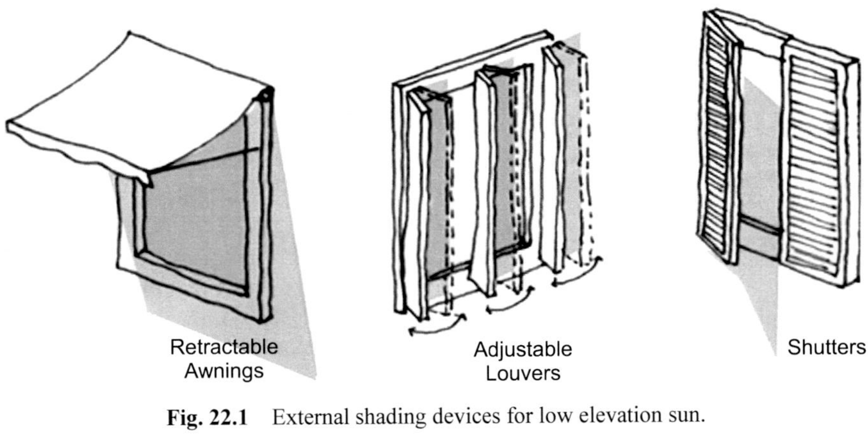

Fig. 22.1 External shading devices for low elevation sun.

22.3 Intelligent Passive Measures for Indoor Thermal Comfort Optimisation

As reviewed by Alkhatib et al. (2021), several control strategies have been applied to building systems. The framework proposed by Gooroochurn et al. (2021d) will be extended in this chapter and used as a basis to describe the automation of exemplar passive measures for the tropical context of Mauritius. Moreover, the integration of blockchain to this system so that it can provide the cybersecurity needs when communicating control and performance data over communication channels, will be explained in detail. In general, the ability to collect input factors and the desired output performance allows the development of machine learning solutions, referred to as the knowledge base in subsequent block diagrams. However, where the control routine can be set based on simple rules and logic, traditional programming and fuzzy logic can be applied to develop the software to operate the actuators. The proposed framework with blockchain integration is shown in Figure 22.2.

The residential building is equipped with sensors to monitor indoor environmental parameters such as temperature, relative humidity, air quality and air draft as well as disturbances such as the presence of occupants. Moreover, outdoor weather meters are used to capture parameters such as temperature, precipitation, relative humidity, wind speed, atmospheric pressure, luminosity, and global solar radiation. There is also a GPS module which indicates the exact geolocation of the building. All the data is then sent to a Raspberry-PI gateway which aggregates the data and uses a blockchain encoder to encrypt and mine the data before sending it to a local analytics and storage server. The use of blockchain at this point secures the data along the communication path between the sensors and gateway. The detailed explanation of the functions of the block chain encoder are given in Section 22.3.1. The local server then uses a blockchain decoder to decode the data. It then stores the data in a database and also performs predictions on the data. The real-time data and the predictions are then encoded using blockchain and sent to a centralized analytics server via the communication input interfaces. The analytics server then decodes the data using its block chain decoder and performs analytics on the data to determine the control signals that will be sent to the actuators. The different actuators used in this system for activating the different passive measures are shown in Figure 22.2. Exemplar passive measures deemed to be of high potential to control heat gains and air movement to regulate the indoor environmental conditions in the tropical context of Mauritius will be described in detail in Sections 22.3.3–22.3.8.

22.3.1 Blockchain Encoder

The system model for the blockchain encoder is shown in Figure 22.3.

The data obtained from the sensors are stored in arrays with sizes corresponding to the block size specified by the block chain used. The data are then encrypted using the private key of the client i.e., the transmitting device (which can be the Raspberry Pi gateway for communication segment X or the local server at points Y and W) and public key of the receiving device (which can be the Local Server at point X, the centralized server at point Z or the mobile devices as point W). For both the transmitter and receiver, a set of public and private keys are already generated and shared between them. The cryptographic algorithms such as the ECDH or RSA can be used for the encryption.

The mining involves the generation of a unique hash for the block using the SHA-256 algorithm. A message digest is a 256-bit hash value that SHA-256 produces. This digest can be visualized as a 64-byte hexadecimal text, which

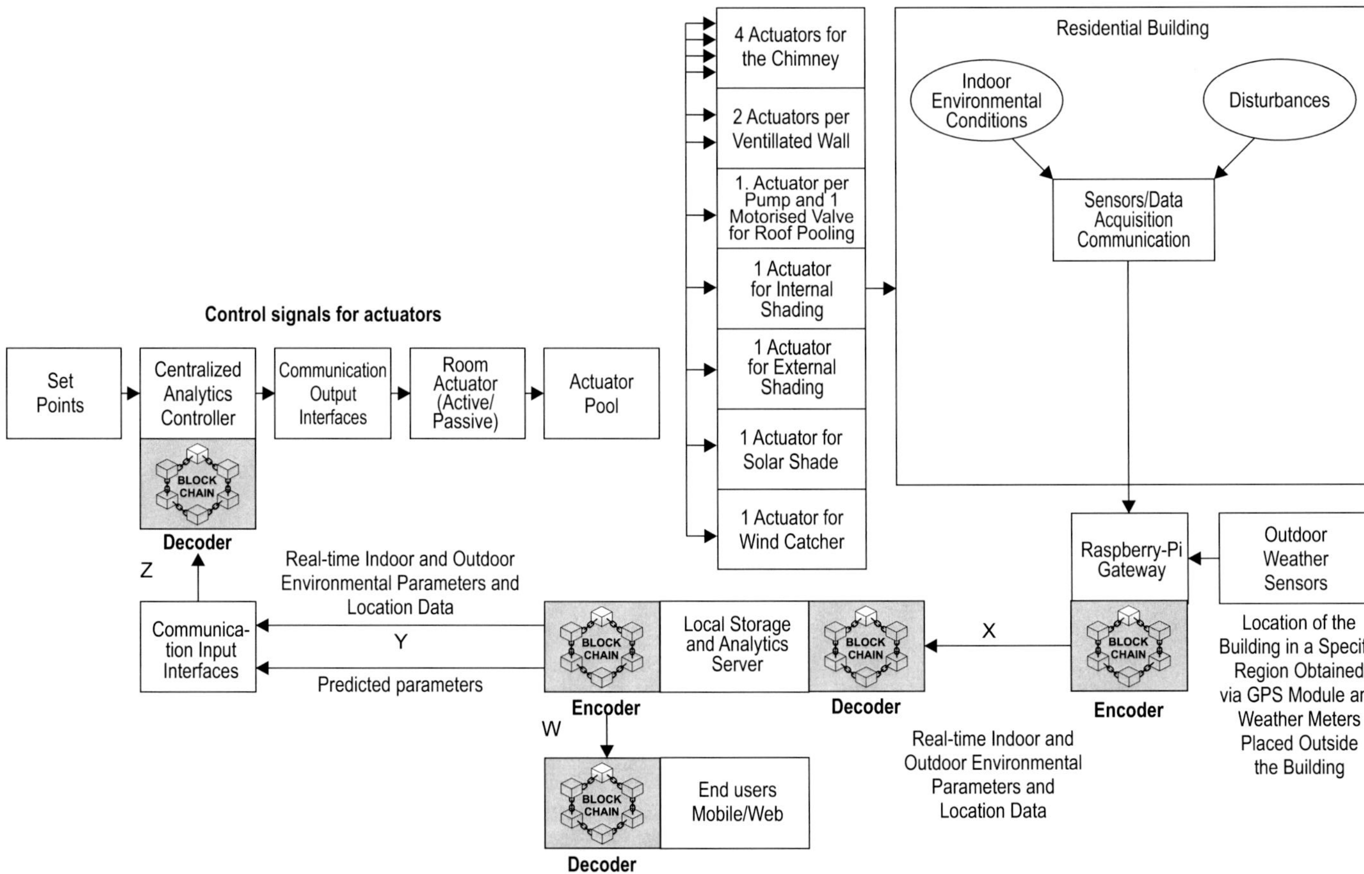

Fig. 22.2 Structure of proposed automated passive system with blockchainbased transmission.

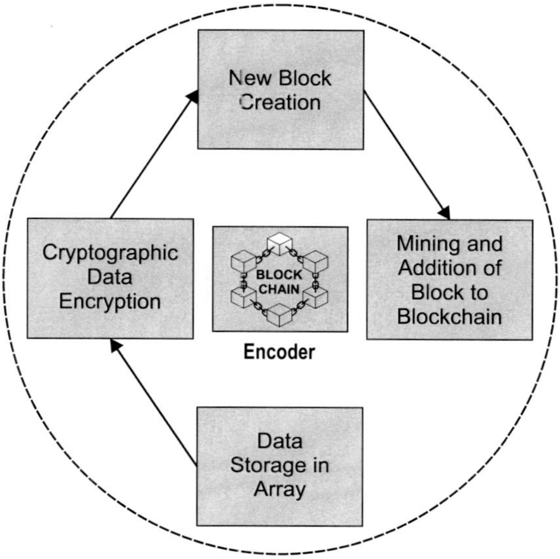

Fig. 22.3 System model for blockchain encoder.

corresponds to a 32-byte array. The data used to generate the unique hash consists of the Block's id, the hash value of the previous block, the current timestamp, the nonce value and the encrypted array of power data. For each generated block, the hash value shall start with "0000". A typical content of the block includes the id number, the current block's hash value, the previous block's hash value and the encrypted data. Each block represents the basic unit of the blockchain. The previous block's hash value, for example, can be ''0'' when the first block of Genesis is produced. After the hash is generated, the block gets appended to the blockchain. Finally, it is transmitted to the target device which can be one of the many targets shown in Figure 22.3, depending on which device is transmitting.

22.3.2 Blockchain Decoder

The system model for the blockchain decoder is shown in Figure 22.4.

At the decoder side, the received block is checked to see if it is not empty. If it contains data, a check is made to verify if its Hash matches the Hash of the previous block. This ensures that there has been no tampering of the data during transmission. If that is the case, the block is added to the chain. Eventually, the data are decrypted from the block using appropriate decryption algorithms with the cryptographic keys.

22.3.3 Ventilated Facades

The accumulation of heat inside the building fabric cavity leads to increased heat gains to the interior as well as radiative heat onto the occupants, which due to the

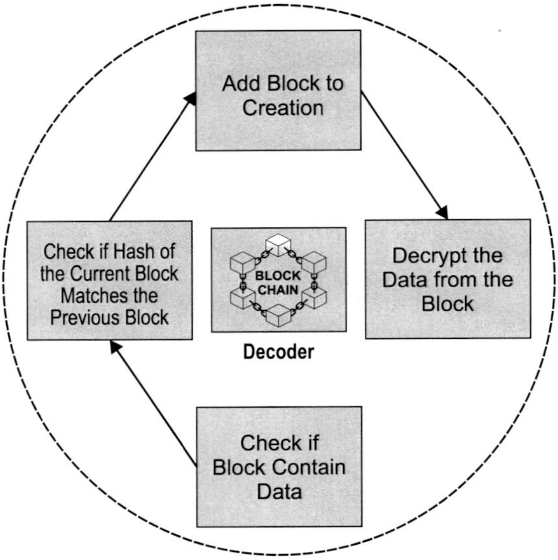

Fig. 22.4 System model for blockchain decoder.

high thermal capacity of concrete structures take time to be dissipated. The use of single-sided hollow blocks (left illustration in Figure 22.5) leads to trapped air pockets (Figure 22.5), accumulating heat when exposed to the sun during the day with a high latency to cool, and causing thermal discomfort at night to the detriment of occupants in residential buildings who are not generally at home during the day. Gooroochurn et al. (2020) developed a ventilated façade using double-sided blocks (see right illustration of Figure 22.5), with slots cut at bottom and top levels on the outside wall to connect the block cavities to the atmosphere, which is passively flushed when heated by the sun, hence not leading to heat accumulation. The direct connection of the block cavity to the external air has the added benefit of allowing a rapid cooling when temperatures drop at night. A 6°C drop in internal wall temperature in the west orientation was obtained.

Having a permanently opened louvre will lead to heat loss to the environment even in winter conditions when the heat gain needs to be channeled to the interior. With the current control strategy, the indoor temperature is cross-checked against the range for human comfort and the prevailing ambient temperature. The knowledge base can be set using machine learning approaches to develop the relationship between the interior conditions and the ambient conditions, which can also be integrated with weather prediction routines to take predictive action, given the latency in the thermal process of heat transfer through the building fabric.

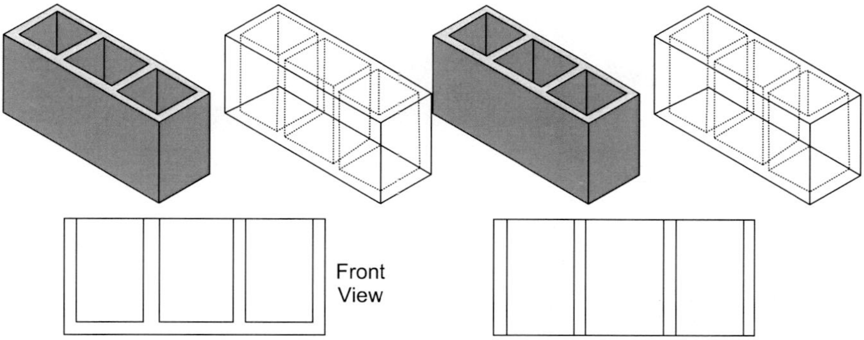

Fig. 22.5 Normal single-sided hollow block (left) and light double-sided hollow block (right).

22.3.4 Solar Chimney

The glass-based solar chimney, mountable on the roof or affixed to the wall, was proposed by Gooroochurn et al. (2021e). The wall-mounted version is illustrated in Figure 22.6a, b. As presented in Gooroochurn (2022b), the solar chimney can be operated in several modes. The solar chimney has four openings labeled A, B, C, and D (Table 22.1.). The side with openings C and D is in contact with the wall. When openings B and C are open, the system allows cross-ventilation across the interior space it is connected to, hence operates in cooling mode. When openings A and D are open, the system can transfer warm air accumulated in the solar chimney to the interior, and hence operates in the heating mode. The control flow for the system is depicted in the block diagram of Figure 22.7. The solar chimney operational rules, being easily formulated as rules and logic, therefore a fuzzy logic controller design was proposed. The system also allows operation in night flushing mode, using the low power fan in-built inside the solar chimney, providing a buffer between the indoor and outdoor, hence allowing fresh air exchange in peak summer conditions where despite the external air temperature is cool and well within the human comfort range, but this cool air does not make its way into the building due to lack to wind pressure or due to closing of windows to keep out mosquitoes. These two problems are overcome by the solar chimney. Furthermore, as described above, the system can be operated in heating and cooling modes during the day, by opening the appropriate louvres, hence exemplifying the flexibility the automation of this passive measure provides.

22.3.5 Roof Pooling

Another potential passive measure that can be used to regulate the heat gains through the roof, which has been reported to be the major cause for overheating by heat gains and radiative heat through the cast concrete, flat roofs in Mauritius, is roof pooling. In this method, a layer of liquid, typically water, is either stagnated or allowed to flow over the surface of the roof, which in itself modifies the thermal

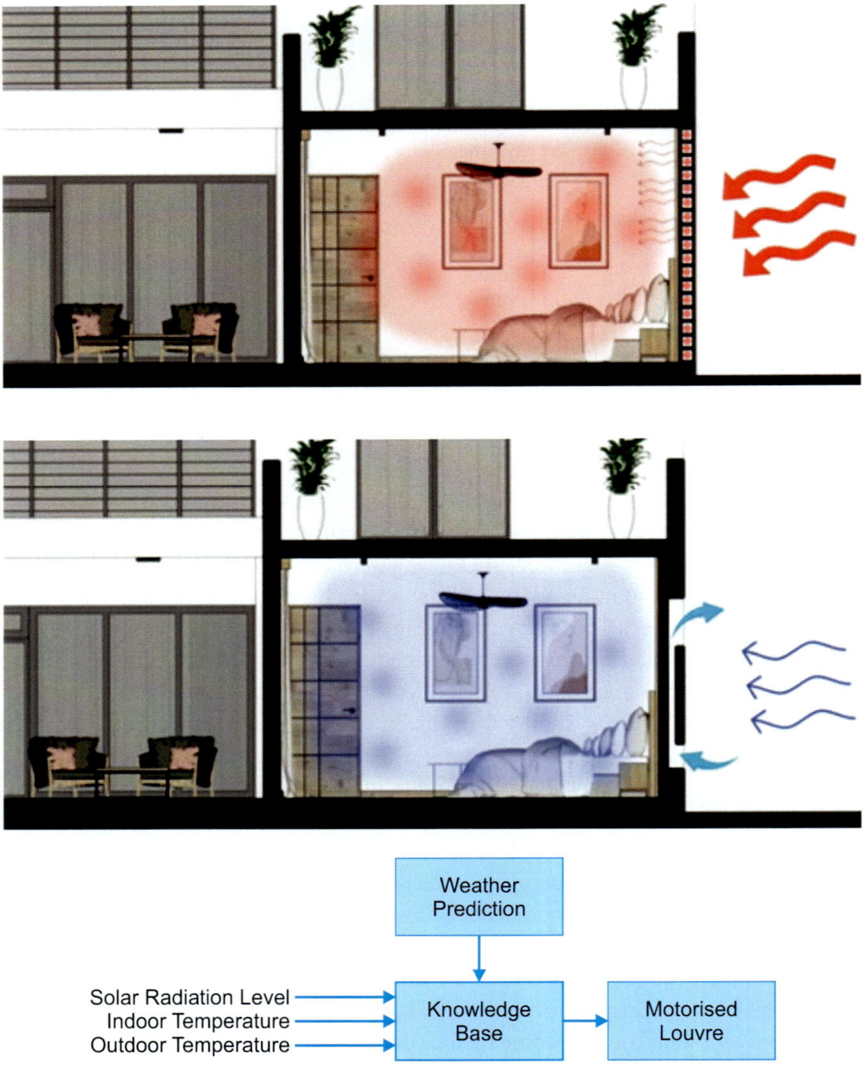

Fig. 22.6a Heat trapped in cavities with normal block and Heat control with ventilated cavity and Control for ventilated wall.

properties of the roof. The proposed control system consists of a simple pipe network to transfer water from a ground level tank to the roof by a pump, where the water is retained or allowed to flow out of the roof on the adjacent side, and therewith carry away any heat that is accumulated. The ability to stagnate the water or flow the water in itself offers a controllable factor to influence the heat flow from the outside to the interior spaces. The control of the water flow from the extreme of maximum flow to no flow (stagnation) is achieved by a motorised

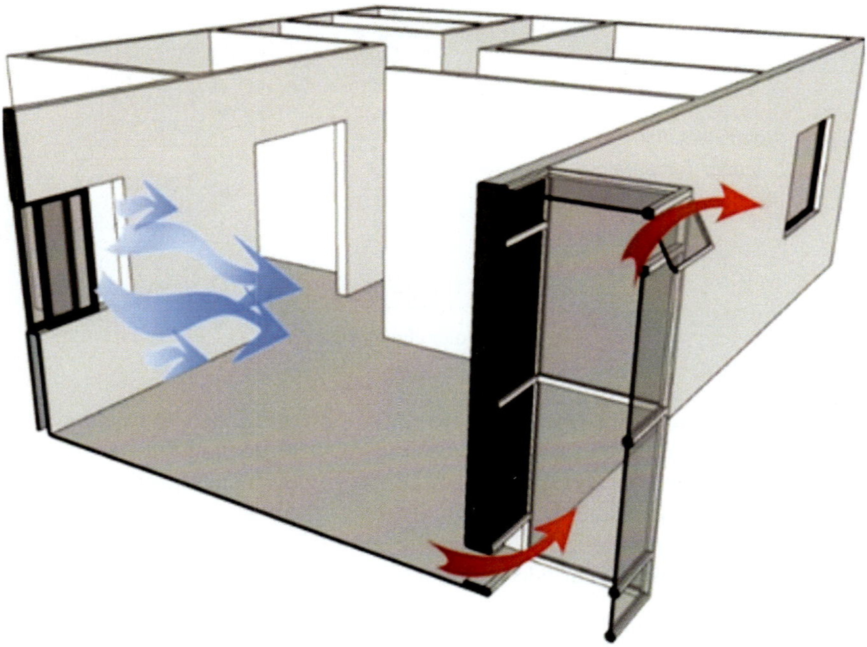

Fig. 22.6b Glass-based, wall-mounted solar chimney prototype.

Table 22.1 Solar chimney operating modes (Gooroochurn et al., 2021e).

Mode of Operation	*Openings Opened*	*Openings Closed*
Inactive	None or A and B	C and D
Cross-ventilation/cooling mode	B and C	A and D
Heating mode	A and D	B and C
Night-flushing mode	A and D	B and C

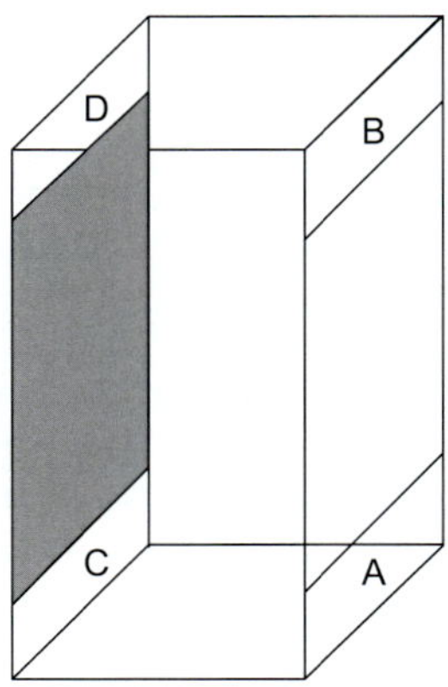

valve in the outflow pipe network, which can feed the same tank or a second tank. With a second tank, a second pump and motorised valve will be required to move water in the other direction, with a proper coordination between the two pumps and motorised valves.

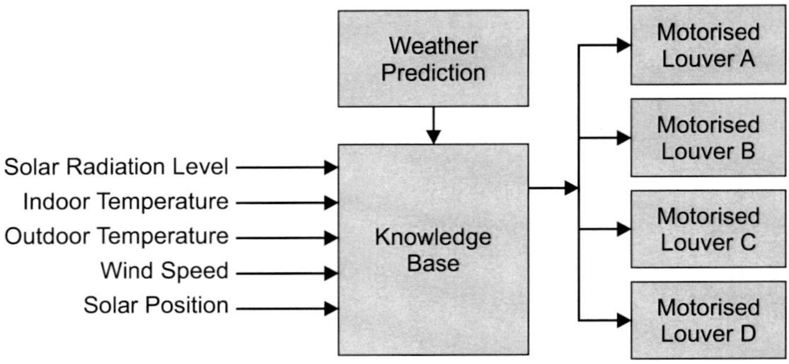

Fig. 22.7 Solar chimney control.

The control input and output are depicted in Figure 22.8. The solar radiation and indoor/outdoor temperatures are used as input to the control system as for the other systems considered above. Additionally, the wind speed is added as a control factor given the evaporative and heat transport effect it can have on the phenomenon in question. The water level in the tank(s) is also to be added in the control algorithm for adequate control of the pump. The knowledge base to be developed in this case is the relationship between the indoor and outdoor temperature for a given solar radiation and wind speed for different regimes of heat flow. With that knowledge base, it becomes possible to predict the future state of the interior spaces for a current set of indoor and outdoor conditions as well as predicted weather patterns. This system can be readily retrofitted to existing buildings, is not cost prohibitive and can benefit from harvested rainwater to avoid the use of potable mains water for the roof flooding process.

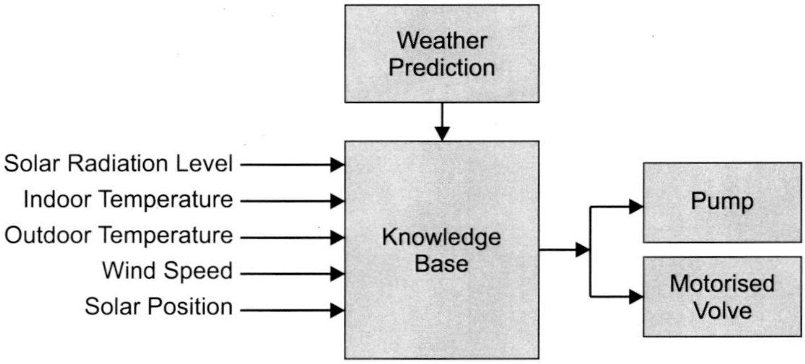

Fig. 22.8 Roof pooling control strategy.

22.3.6 Roof Shading

Shading the roof from unwanted solar radiation can be another effective passive strategy to control heat gains through the roof. Installing a permanent metal structure is a possible solution, albeit a costly one, but more importantly from a control perspective, is unfortunately an inflexible one. Installing solar shade membranes, which come with a specific solar shading factor, typically ranging from 50% to 100%, is a more cost effective option, which can be manually deployed or retracted to cope with seasonal changes or under cyclonic conditions. However, the system proposed is an automatic one, as illustrated in Figure 22.9 which can deploy and retract the solar shade as per requirement using a motor actuator and associated mechanisms. The knowledge base to be developed is the influence of the solar shade at fully retracted and fully deployed states on the indoor conditions with respect to the ambient conditions (solar radiation and temperature), with the ability to anticipate changes in weather patterns and take predictive action given the relatively large time lag of the thermal transfer characteristics of the roof. The type of solar shade is also an important factor to be considered in developing the knowledge base.

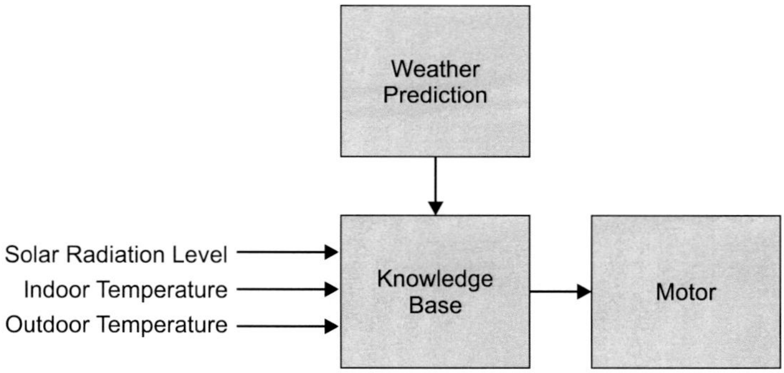

Fig. 22.9 Roof shade control strategy.

22.3.7 External/Indoor Shading Devices for Windows

Another significant source of heat gains reported in literature is through the glazed areas, characterized by the WWR. An external (e.g., fins or louvres) or internal shading device (e.g., blinds or curtains) can be designed with an actuating system to deploy or retract the shading element, and operate in a similar manner to the roof shade system described above and illustrated in Figure 22.9. Based on previous studies, external shading is more effective than internal shading, although the latter is easier to implement and less costly.

22.3.8 Wind Catcher/Vertical Shading Device

While the measures described earlier emphasized on heat gain regulation, another important factor in achieving a suitable indoor environment is adequate ventilation, commonly measured as the number of air changes per hour. Window opening and closing can itself be automated as part of the natural ventilation strategy, but

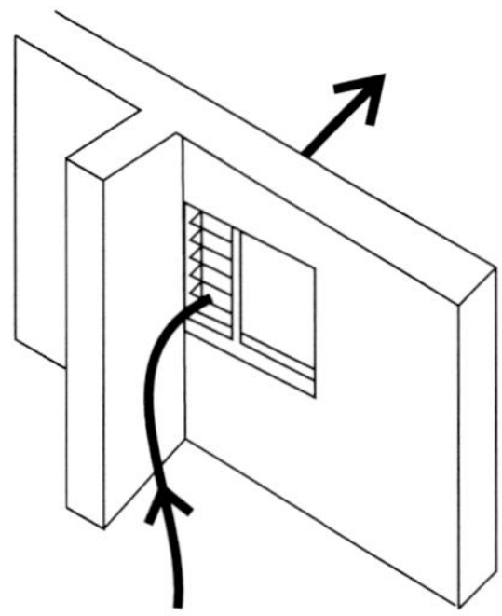

Fig. 22.10 Vertical element to serve as wind catcher and solar shade.

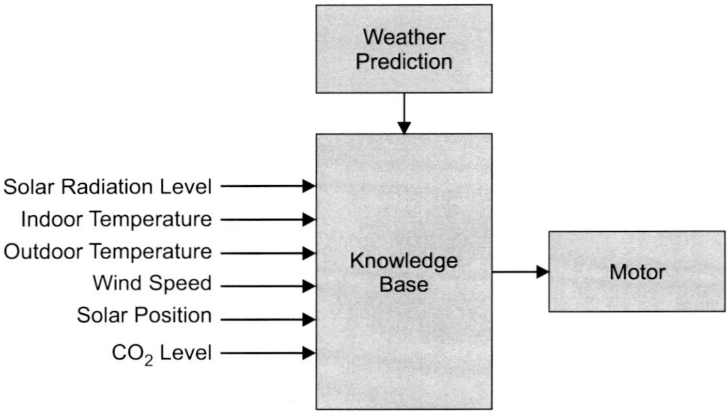

Fig. 22.11 Control strategy for wind catcher.

another passive system is presented to control air movement across the interior spaces, termed as a wind catcher as depicted in Figure 22.10. The wind catcher is a hinged vertical member that can be held in parallel or at a given angle to the wall, including at right angles. When held at an acute angle up to a right angle to the wall, it increases the amount of air that can flow into the building. It can also serve as a vertical shading element against low elevation sun, e.g., the setting or rising sun.

As part of the control strategy (illustrated in Figure 22.11), the wind speed and direction are used to determine whether a particular wind catcher on a specific orientation of the building façade needs to be hinged at an angle to the wall to promote air penetration or not. The indoor CO_2 level can also be used as an input factor to determine the need for more fresh air. The air movement across the interior spaces also creates the beneficial chill effect on the human skin to further improve thermal comfort. The outdoor air temperature is an important factor to decide on the deployment of the wind catcher. The solar position and solar radiation level are useful when the wind catcher is also used as a shading device against low elevation sun, where the solar position, namely, the solar azimuth, is used as an input factor to determine the position of the sun with respect to the glazed area. The solar position can be obtained from solar charts for the specific location of the project site for a specific time and day.

22.4 Conclusion

With the worsening of climate change as our modern society faces serious challenges to curtail its dependency on fossil fuels, leading to greenhouse gas emissions causing global warming and associated severe adverse impacts on our ecosystems and our livelihoods, ambient and indoor temperature rises to extremes are already a reality and expected to continue rising. This temperature rise will in turn affect the health, wellbeing, productivity, and performance of occupants in interior spaces, cutting across office workers, students in school and inhabitants in their homes in general. Passive design is a key cornerstone to sustainably control heat gains into building interiors before active cooling systems are prescribed. The automation of passive measures, along the lines illustrated through selected systems with their associated control strategies, for the tropical context of Mauritius, will bring further benefits in optimising the regulation of the building envelope with respect to ambient conditions. A similar methodology can be applied in other climate contexts for passive measures deemed effective and finding means to modulate their influence on the building envelope. The development of a knowledge base using machine learning methods is a vital part of the control strategies, and with such a learned engine, combined with weather prediction using real-time sensor data, the control algorithm can be rendered more effective in modulating the influence of the passive measures. The transmission of these sensor data and other information on the system performance and indoor/outdoor environmental conditions pose a threat with cybersecurity, to circumvent which blockchain technology can be integrated in the development of the communication functionalities.

As has been described in this chapter, blockchain technology can be implemented at different transmission points in the intelligent building control system to enable trust and confidentiality when data is sent from the Raspberry-Pi gateway to the local server, from local server to end users and the centralised server. It is indeed mandatory to ensure this type of trusted communications for the building control system considered since tampering of the actuator signals in particular could expose the occupants of the building to serious threats and affect the performance of the systems in achieving optimum environmental conditions control. Such technological solutions enabling low cost measures to develop resilience to climate change impacts are concrete examples of how modernization can be married with innovation, and yet support sustainable development. This has to become the design paradigm for inventions and technological breakthroughs so that we do not repeat the same mistakes and cause phenomena like climate change.

References

1. Alkhatib, H., Lemarchand, P., Norton, B. and O'Sullivan, D. (2021). Deployment and control of adaptive building facades for energy generation, thermal insulation, ventilation, and daylighting: A review. *Applied Thermal Engineering*, *185*, 116331.
2. Bachtobji, S., Kouicem, D.E. and Mabrouk, M.B. (2022). Towards Blockchain Based Architecture for Building Information Modelling (BIM), In: *2022 5th Conference on Cloud and Internet of Things* (*CIoT*), Marrakech, Morocco, pp. 53–59. doi: 10.1109/CIoT53061.2022.9766540.
3. Eslamian, S. and F., Eslamian (2025). *Blockchain Technology in Water and Environmental Systems* CRC Press, Taylor and Francis, USA, p. 282.
4. Fowdur, T.P., Indoonundon, M., Hosany, M.A., Milovanovic, D. and Bojkovic, Z. (2022). Achieving Sustainable Development Goals through Digital Infrastructure for Intelligent Connectivity. In: Boulouard, Z., Ouaissa, M., Ouaissa, M., El Himer, S. (Eds.), *AI and IoT for Sustainable Development in Emerging Countries. Lecture Notes on Data Engineering and Communications Technologies*, *105*. Springer, Cham, Germany, https://doi.org/10.1007/978-3-030-90618-4_1.
5. Fowdur, T.P., Babooram,L., Murdan, A.P., Milovanovic, D., Bojkovic, Z. and Indoonundon, I. (2023). Enabling technologies and applications of 5G/6G powered intelligent connectivity, In: D. Milovanovic, Z. Bojkovic, and T.P. Fowdur (Eds.), *Driving 5G Mobile Communications with Artificial Intelligence towards 6G*. ISBN 9781032071244, April 6, CRC Press.
6. Gooroochurn, M., Shamachurn, H., Surnam, B.Y.R. and Cadersa, A. (2020). Experimental and Simulation Analysis of the Performance of a Low Cost and Space Preserving Ventilated Façade for Hot and Humid Climates. In: *The Fourth International Conference on Efficient Building Design* (*ASHRAE*), November 2020, Beirut.
7. Gooroochurn, M., Cadersa, A. and Pattoo, P. (2021a). Circular Design in the Construction Industry in Mauritius with a Case Study for Demolition Waste Management, *Journal of the Institution of Engineers Mauritius*, *March*(15), 73–84.
8. Gooroochurn, M. and Giridharan, R. (2021b). Enhancing Energy Performance and Comfort of Built Environment in Tropical Climate, Book Chapter in *Urban Microclimate Modelling for Comfort and Energy Studies*. ISBN 978-3-030-65420-7, Springer Nature Switzerland AG 2021, Published 8 April.
9. Eslamian, S. and F. Eslamian. (2023). *Handbook of Hydroinformatics: (Volume I)*: *Classic Soft-Computing Techniques*. Elsevier, p. 461.
10. Gooroochurn, M.J.Y., Coret, J.Y., Brown, N., Venkannah, S. and Wright, A. (2021c). *Assessing the Efficacy of Passive Measures for the Tropical Context of Mauritius through Parametric Simulations and in situ Measurement*. In: *CIBSE 2021 Technical Symposium*, UK 13–14 July 2021.
11. Gooroochurn, M., Mallet, D., Jahmeerbacus, I., Shamachurn, H. and Sayed Hassen, S .Z. (2021d). A framework for AI-based building controls to adapt passive measures for optimum thermal comfort and energy efficiency in tropical climates. In: *FTC 2021 - Future Technologies Conference*, 28–29 October. Vancouver, Canada.
12. Gooroochurn, M., Lollchund, R., Rughooputh, S.D.D.V., Shamachurn, H. and Rughooputh, R. (2021e). *Assessing the Efficacy of Solar Chimneys in Mauritius through Dynamic Simulations and* in -situ *Measurements*. Mauritius Research and Innovation Council Technical Report, January.
13. Gooroochurn, M. (2022a). Circular homes: An energy-water-materials nexus for community climate engagement and action at grassroots level in the built environment. Results in Engineering, *15*, 100548. ISSN 2590-1230. https://doi.org/10.1016/j.rineng.2022.100548.
14. Gooroochurn, M. (2022b). A hybrid glass-based solar chimney to promote cross-ventilation and night flushing. In: *2022 International Conference on Electrical, Computer, Communications and*

Mechatronics Engineering (*ICECCME*), 16–18 November 2022. Maldives National University, IEEE Xplore. https://ieeexplore.ieee.org/document/9988027.

15. Gooroochurn, M. (2023). Adaptive passive measures for tropical climates: A case study for Mauritius. In: *Removing Barriers in Environmental Comfort below Tropic of Cancer*. Springer Nature Switzerland AG 2022, Published May.
16. Hosseini, M., Javanroodi, K. and Nik, V.M. (2022). High-resolution impact assessment of climate change on building energy performance considering extreme weather events and microclimate: Investigating variations in indoor thermal comfort and degree-days. *Sustainable Cities and Society*, *78*, 103634. ISSN 2210-6707, https://doi.org/10.1016/j.scs.2021.103634.
17. Jafarpur P. and Berardi, U. (2021). Effects of climate changes on building energy demand and thermal comfort in Canadian office buildings adopting different temperature setpoints. *Journal of Building Engineering*, *42*, 102725. ISSN 2352-7102, https://doi.org/10.1016/j.jobe.2021.102725.
18. Kaitouni, S.I., Chahboun, R., Bouhssine, Z., Cakan, M., Brigui, J. and Ahachad, M. (2023). Simulation-based assessment of the climate change impact on future thermal energy load and indoor comfort of a light-weight ecological building across the six climates of Morocco. *Thermal Science and Engineering Progress*, *45*,, 102137. ISSN 2451-9049. https://doi.org/10.1016/j.tsep.2023.102137.
19. Li, G., Liu, C. and He, Y. (2021). The effect of thermal discomfort on human well-being, psychological response and performance. *Science and Technology for the Built Environment*, 27(7), 960–970. doi: 10.1080/23744731.2021.1910471.
20. Ozarisoy, B. and Elsharkawy, H. (2019). Assessing overheating risk and thermal comfort in state-of-the-art prototype houses that combat exacerbated climate change in UK. *Energy and Buildings*, *187*, 201–217. ISSN 0378-7788, https://doi.org/10.1016/j.enbuild.2019.01.030.
21. Ozarisoy, B. (2022). Energy effectiveness of passive cooling design strategies to reduce the impact of long-term heatwaves on occupants' thermal comfort in Europe: Climate change and mitigation. *Journal of Cleaner Production*, *330*, 129675. ISSN 0959-6526. https://doi.org/10.1016/j.jclepro.2021.129675.
22. Ren, Z., Fu, Y., Dong, Y., Zhang, P. and He, X. (2022). Rapid urbanization and climate change significantly contribute to worsening urban human thermal comfort: A national 183-city, 26-year study in China. *Urban Climate*, *43*, 101154. ISSN 2212-0955. https://doi.org/10.1016/j.uclim.2022.101154.
23. Saeed, M., Amin, R., Aftab, M. and Ahmed, N. (2022). Trust Management Technique Using Blockchain in Smart Building. *Eng. Proc.*, *20*, 24. https://doi.org/10.3390/engproc2022020024.
24. Thapa, S., Rijal, H.B., Pasut, W., Singh, R., Indraganti, M., Bansal, A.K. and Panda G.K. (2023). Simulation of thermal comfort and energy demand in buildings of sub-Himalayan eastern India: Impact of climate change at mid (2050) and distant (2080) future. *Journal of Building Engineering*, *68*, 106068. ISSN 2352-7102. https://doi.org/10.1016/j.jobe.2023.106068.

Glossary

A

Access control – blockchain and secure data trust; security policy; identity management
Agricultural water management – blockchain applications; water rights trading
Algorithms – consensus algorithms; cryptographic algorithms; hashing; encryption algorithms
Architecture – blockchain architecture; trading system architecture; environmental conservation systems
Auditability – blockchain for secure data trust
Authentication – digital signatures; blockchain security

B

Blockchain – introduction; history; networks; protocols; architecture; theory; structure; fundamentals; adoption risks; digitization of water resources; environmental conservation applications
Blockchain applications – flood hydrology; water resources; urban water management; agricultural water; SDGs; smart grids; cyber,physical systems; climate change; IoT; hydrosystems; building systems; energy systems; environmental sustainability
Blockchain consensus – PoW; PoS; DPoS; BFT; PBFT; SBFT; DBFT; DAG; hybrid models; consensus efficiency
Blockchain generations – public blockchain taxonomy
Blockchain immutability – hashing; security
Blockchain in climate change – supply chain transparency; clean energy trade; GHG tracking; financing climate projects
Blockchain risks – adoption barriers; security challenges
Blockchain technology inside – compute, build, deploy; internal mechanisms
Blue water – virtual water types
Byzantine Fault Tolerance (BFT) – theory; PBFT; practical implementations

C

Case studies – flood forecasting; environmental conservation; water management; virtual water; hydro systems
Clean energy – trade facilitation; smart grids; renewable energy

D

E

F

G

H

Hash functions – applications; methods; quality criteria; digital signatures; mining
Hydro-informatics – information technology role
Hydro-systems – blockchain advantages; sustainability

I

Indoor thermal comfort – intelligent building systems; passive measures
Information technology – water management; flood control
Intelligent buildings – blockchain in building management systems
IoT – smart systems; blockchain for IoT; water management
Iran – virtual water; water footprint; water trade; water security
Identity management – blockchain-based; access control
Immutability – blockchain, hashing

K

Key management – symmetric and asymmetric cryptography

L

Local energy markets – blockchain-enabled trading

M

Mining – PoW; hash functions

N

Non-academic content – blockchain analysis

P

Passive climate measures – indoor comfort
Patents – blockchain frontier; patent boom

R

Resilience – hydrologic resilience; systems resilience
Risk management – flood risk; adoption risks; blockchain challenges

S

SDGs – water and blockchain contributions

T

U

V

W

Z

Index

About the Editors

Yaser Sabzevari is a highly qualified professional in the field of water engineering. He holds a graduate degree in Water Engineering from Lorestan University, where he honed his comprehensive understanding of water management principles and techniques. Driven by a passion for sustainable water resource management, Yaser pursued further academic advancement and obtained a Master's degree in Irrigation and Drainage, also from Lorestan University. Through his master's studies, he delved deeper into the intricacies of water engineering, seeking innovative solutions to optimize water usage in agricultural practices. Continuing his pursuit of knowledge and expertise, Yaser received his PhD in Irrigation and Drainage at Isfahan University of Technology. This doctoral program provides him with an opportunity to contribute to cutting-edge research in the field, exploring advanced methodologies and technologies for efficient water management for develop science and knowledge, Yaser spent a research period at institute of Hydrology, Slovak Academy of Sciences as a National Scholarship Holder in 2025. in this period, Yaser developed and modified hydrological modelling in snow dominated regions. Throughout his academic journey, Yaser has been dedicated to making meaningful contributions to the water sector. His research interests encompass the integration of emerging technologies, such as blockchain, into water management systems. By combining his expertise in water engineering with the potential of blockchain technology, he aims to address critical challenges faced by the industry and promote sustainable practices. Yaser's academic background, coupled with his commitment to environmental sustainability, positions him as a visionary in the field of water engineering. With his strong theoretical foundation and practical experience, he seeks to bridge the gap between academia and industry, fostering collaborations that drive positive change in water management practices.

Saeid Eslamian received his PhD in civil and environmental engineering from the University of New South Wales, Australia, in 1998. Saeid was a visiting professor at Princeton University and ETH Zurich in 2005 and 2008, respectively. He has contributed to more than 1,000 publications in journals, conferences, and books. Eslamian has been honored as a 2-Percent Top Researcher by Stanford University and Elsevier for total Career. Currently, he is a full professor of Resilient Climate and Water Systems and the Director of the Excellence Center in Risk Management and Natural Hazards, Isfahan University of Technology. His scientific interests are floods, droughts, water reuse, climate change adaptation, sustainability, and resilience.

Faezeh Eslamian is a PhD holder of bioresource engineering from McGill University. Her research focuses on the development of a novel lime-based product to mitigate phosphorus loss from agricultural fields. Faezeh completed her Bachelor's and Master's degrees in Civil and Environmental Engineering from Isfahan University of Technology, Iran, where she evaluated natural and low-cost absorbents for the removal of pollutants such as textile dyes and heavy metals. Furthermore, she has conducted research on worldwide water quality standards and wastewater reuse guidelines. Faezeh is an experienced multidisciplinary researcher with research interests in soil and water quality, environmental remediation, water reuse, and drought management. She is now working in GHD, Montreal, Canada.

For Product Safety Concerns and Information please contact our EU representative GPSR@taylorandfrancis.com Taylor & Francis Verlag GmbH, Kaufingerstraße 24, 80331 München, Germany